Physics and Mechanics of Ice

International Union of Theoretical and Applied Mechanics

Physics and Mechanics of Ice

Symposium Copenhagen, August 6-10, 1979
Technical University of Denmark

Editor **Per Tryde**

With 159 Figures

Springer-Verlag
Berlin Heidelberg New York 1980

Assoc. Professor PER TRYDE
Institute of Hydrodynamics and Hydraulic Engineering
Technical University of Denmark
DK-2800 Lyngby

ISBN-13 : 978-3-642-81436-5
DOI : 10.1007 / 978-3-642-81434-1
e-ISBN-13 : 978-3-642-81434-1

Offsetprinting: fotokop wilhelm weihert KG, Darmstadt · Bookbinding: Konrad Triltsch, Würzburg
2060/3020/543210

Preface

In the programme of the symposium was written:
"The International Union of Theoretical and Applied Mechanics has taken the initiative to organize the symposium. As the name of IUTAM implies, the organization brings forward achievements within the field of theoretical mechanics for application in science and engineering.

According to the rules of IUTAM that only invited persons can attend, all lecturers and participants have been appointed by the members of the scientific committee. To facilitate contact among the attending persons, it has been decided to restrict the total number to 85 persons including the lecturers. Only one session is planned, making it possible for everybody to attend all lectures.

Most scientists and engineers have realized that the knowledge attained by extensive basic research is essential in order to solve technological problems. In the process of acquiring this knowledge we often fail to understand that scientific progress is only achieved by two main principles: (1) By studying the scientific literature and applying or improving the theories in order to predict behaviour and forces correctly, or (2) by rejecting existing theories and developing new ways to cope with the problem, resulting in a more differentiated and, hopefully, more exact theory.

Ice seems to be a simple material, but it is in fact so complex and strange that it is only in the latest decades that we have come to know some of the natural laws governing its behaviour.

This symposium is one of the steps towards a better understanding of the problems of ice".

The 25 papers presented at the symposium have now been compiled in the present volume. The editor has read through the papers and corrected obvious errors both in orthography and syntax, but it has not been within my scope to rewrite any part of the papers. All theories and formulae presented by the authors are not in any way checked or evaluated by the editor. The editor has worked out a short resumé of the discussions following the presentation of the papers, partly based on notes made during the discussion and partly derived from a recording made on magnetic tape throughout the discussion part of the symposium.

As editor I wish to thank all the authors for their expert contributions both in the written and verbal presentation and their willingness to participate in the discussions.

I also wish to thank the International Union of Theoretical and Applied Mechanics and its president professor F. Niordsson for the initiative in arranging the symposium and for the assistance offered to me during the planning.

At the Institute of Hydrodynamics and Hydraulic Engineering I wish to thank the professors H. Lundgren and F.A. Engelund for their continued encouragement to do ice research at the institute.

Finally I wish to thank the committee members for their valuable contribution in proposing and selecting the authors and the participants and in assisting me to make the final selection of topics to be presented.

For the professional assistance offered me by the staff of the Institute of Hydrodynamics and Hydraulic Engineering I particularly wish to thank Ms. E.B. Jensen, Ms. L.K. Hansen and Ms. K. Djørup for executing the secretarial work with such care and accuracy.

Lyngby, October 1979 Per Tryde

General Information

Scientific Committee:

Dr. L.W. Gold, Canada
Professor S.S. Grigorian, USSR
Professor L. Lliboutry, France
Dr. M. Mellor, USA
Professor J.F. Nye, UK
Dr. J. Schwarz, Germany FR
Professor T. Tabata, Japan
Assoc. Professor P. Tryde, Denmark (Chairman)

Co-sponsors:
International Association for Hydraulic Research (IAHR)
International Commission on Snow and Ice
The Geological Survey of Greenland, Ministry for Greenland
Geophysical Isotope Laboratory, University of Copenhagen
Greenland Technical Organization, Ministry for Greenland
Danish Hydraulic Institute, Danish Academy of Technical Sciences.

Scientific Organization:
International Union of Theoretical and Applied Mechanics (IUTAM)
President: Professor F.I. Niordson, Denmark
Secretary-General: Professor Jan Hult, Sweden

Danish Organizing Committee:
Professor W. Dansgaard, University of Copenhagen
Assoc. Professor S.J. Johnsen, University of Copenhagen
Assoc. Professor N. Reeh, University of Copenhagen
Assoc. Professor P. Tryde, Technical University of Denmark.

Acknowledgements

Financial support of the symposium has been generously provided by the International Union of Theoretical and Applied Mechanics and by the following institutions:

Danish Hydraulic Institute
Danish Ship Research Laboratory
Society of Hydraulic Engineering
Greenland Technical Organization
International Business Machines A/S (IBM)
Institute of Hydrodynamics and Hydraulic Engineering
The Ministry of Education.

Secretariat:
P. Tryde
Institute of Hydrodynamics and Hydraulic Engineering
Technical University of Denmark, Building 115
DK-2800 Lyngby, Denmark.

Participants

(Authors are identified by an asterisk)

*ASSUR, A., Cold Regions Research and Engineering Laboratory, Hanover, N.H. 03755, U.S.A.

BENGTSSON, L., Div. Water Resources Engineering, University of Luleå, S-951 87 Luleå, Sweden

*BERCHA, F.C., F.G. Bercha and Associates Limited, 426-339 6th Ave. S.W., Calgary, Alberta, Canada T2P OR8

BILLFALK, L., Hydraulics Laboratory, Swedish Power Board S-810 71 Älvkarleby, Sweden

*BREPSON, R., Alsthom-Atlantique, Neyrtec, 75, Rue Général Mangin, F-38041 Grenoble Cedex, France

Brooks, L.D., Chevron Oil Field Research Company, P.O. Box 446, La Habra, California 90631, U.S.A.

*DEMPSTER, R.T., Faculty of Engrg. & Appl Science, Memorial University of Newfoundland, St. John's, Newfoundland A1B 2X5, Canada

DRUEZ, J.A., Applied Sciences Dept. University of Quebec at Chicoutimi, Chicoutimi, Quebec, G7H 2B1 Canada

*DUVAL, P., Laboratoire de Glaciologie du C.N.R.S. 2 rue Tres-Cloitres, F-38031 Grenoble-Cedex, France

EARLE, E.N., Shell Development Company, P.O. Box 481, Houston, Texas 77001, U.S.A.

*EDWARDS, R., Arctec Canada Ltd., 311 Leggett Drive, Kanata, Ontario K2K 1Z8, Canada

ELDRUP, M., Chemistry Department, Risø National Laboratory, Risø, 4000 Roskilde, Denmark

*EMERY, J., Department of Civil Engineering, McMaster University, Hamilton, Ontario, Canada L8S 4L7

FABRICIUS, J., Danish Meteorological Institute, Lyngbyvej 100, 2100 Copenhagen Ø, Denmark

*FRANKENSTEIN, G., Cold Regions Research and Engineering Laboratory, Hanover, N.H. 03755, U.S.A.

*FREDERKING, R.M., Geotechnical Section, Division of Building Research, National Research Council, Ottawa, Ontario K1A OR6, Canada

FREYSTEINSSON, S., Armuli 4, 105 Reykjavik, Iceland

*GOLD, L.W., Division of Building Research, National Research Council, Ottawa, Ontario K1A OR6, Canada

*GOODMAN, D.J., Cavendish Laboratory, Madingley Road, Cambridge CB3 OHE, UK

HASLE-NIELSEN, A. Danish Hydraulic Institute, Agern Alle 5, 2970 Hørsholm, Denmark

HERFJORD, K. Norges Skipsforskningsinstitutt, P.O.B. 4125-Valentinlyst, N-7ool Trondheim, Norway

*HUTTER, K., Laboratory of Hydraulics, Hydrology and Glaciology, Federal Institute of Technology, 8092 Zürich, Switzerland

KAPOULITSAS, G., Dodekanisou 13, Heraklion, Crete, Greece

*KERR, A.D., Department of Civil Engineering, University of Delaware, Newark, Delaware 19711, U.S.A.

KOMA, N., No. 5 Research Department, Technical Research Center, Nippon Kokan K.K., 1-Minami Watarida, Kawasaki-ku, Kawasaki, Kanaga, Japan 210

*KRY, P.R., Production Research Division, Esso Resources Canada Limited, 339-50 Avenue SE, Calgary, Alberta T2G 2B3, Canada

KUBO, Y., Japan Bridge & Structure Institute, Inc. Zenkaren Bldg. 12-4, 5-chome, Minami-Aoyama Minato-ku, Tokyo, Japan

*KURE, K., Danish Ship Research Laboratory, Hjortekærsvej 99, DK-2800 Lyngby, Denmark

LARSEN, P., Hydraulics Laboratory, Swedish State Power Board, S-810 71 Älvkarleby, Sweden

LEPPÄRANTA, M., Institute of Marine Research, PL 166 SF-00141 Helsinki 14, Finland

LINDAHL, J., Greenland Technical Organization, Hauser Plads 20, 1127 Copenhagen K, Denmark

LINDQVIST, G., Wärtsilä, Helsingforsvarvet, P.O. Box 132 SF-00151 Helsingfors 15, Finland

MANTOVANI, S., Istituto de Fisica dell'Universitá, Via Campi 203/A, 1-41100 Modena, Italy

*MELLOR, M., Cold Regions Research and Engineering Laboratory, Hanover, N.H. 03755, U.S.A.

*MICHEL, B., Department of Civil Engineering, Faculty of Sciences and Engineering, University Laval, Quebec, Quebec G1K 7P4, Canada

MIKKELSEN, A., Farvandsdirektoratet, P.O. Box 19, Ovengaden oven Vandet 62 B, 1001 Copenhagen K, Denmark

*MILLER, K.J., Dept. Mechanical Engineering, University of Sheffield, Mappin Street, Sheffield S1 3JD, England

MORLAND, L.W., School of Mathematics and Physics, University of East Anglia, Norwich NR4 7TJ, UK

MÄÄTTÄNEN, M., Dept Mechanical Engineering, University of Oulu, SF-90100 Oulu 10, Finland

*NEVEL, D.E., Cold Regions Research and Engineering Laboratory, Hanover, N.H. 03755, U.S.A.

OFFENBACHER, E.L., Department of Physics, Temple University, Philadelphia, Pennsylvania 19122, U.S.A.

PETERS, G.R., Faculty of Engrg & Appl. Science, Memorial University, St. John's, Newfoundland, Canada A1B 2X5

PILKINGTON, R., Canadian Marine Drilling Limited, P.O. Box 200, Calgary, Alberta, Canada T2P 2H8

*RALSTON, T.D., Exxon Production Research Company, P.O. Box 2189, Houston, Texas 77001, U.S.A.

RASMUSSEN, K., Danish Geographical Institute, Haraldsgade 68, 2100 Copenhagen Ø, Denmark

RASMUSSEN, S., Statens Istjeneste, Christianshavns Torv 2, 1410 Copenhagen K, Denmark

REDDY, D.V., Faculty of Engrg. & Applied Science, Memorial University of Newfoundland, St.John's, Newfoundland Canada A1B 3X5

*REEH, N., Geophysical Isotope Laboratory, University of Copenhagen, Haraldsgade 6, 2200 Copenhagen N., Denmark

*REINICKE, K.M., BEB Gewerkschaften Brigitta und Elwerath Betriebsführungsgesellschaft mbh., Riethorst 12, D-3000 Hannover 51, Germany FR

SACKINGER, W.M., Geophysical Institute, University of Alaska, Fairbanks, Alaska 99701, U.S.A.

*SAEKI, H., Faculty of Engineering, Hokkaido University, Sapporo, 060 Japan

*SCHWARZ, J., Hamburgische Schiffbau-Versuchsanstalt GmbH, Postfach 600929, D-2000 Hamburg 60, BRD-Deutschland

SEGO, D., Dept. of Civil Engineering, University of Alberta, Edmonton, Alberta, Canada T6G 2G7

SMITH, P.N., Faculty of Engineering & Applied Science, Memorial University of Newfoundland, St. John's, Newfoundland, Canada AlB 3X5

SØRENSEN, C., Rambøll & Hannemann A/S, Göteborg Alle 12, 8200 Århus N., Denmark

*TABATA, T., Institute of Low Temperature Science, Hokkaido University, Sapporo, 060 Japan

TATINCLAUX, J.C., Institute of Hydraulic Research, University of Iowa, Iowa City, Iowa 52242, U.S.A.

TINAWI, R., École Polytechnique, Box 6079 Station "A". Montreal, Quebec, Canada H3C 3A7

UDIN, I., Sveriges Meteorologiska och Hydrologiska Institut, Fack, S-601 01 Norrköping, Sweden

VALERI, S., Istituto di Fisica dell' Universita, Via Campi 203/A, 1-41100 Modena, Italy

WEIDICH, A., Geological Survey of Greenland, Ministry for Greenland, Øster Voldgade 10, 1350 København K Denmark

WOLD, K., Vassdragsdirektoratet, Hydrologisk Avdeling, Boks 5091, Majorstua, Oslo 3, Norway

ZORN, R., Danish Hydraulic Institute, Danish Academy of Technical Sciences, Agern Allé 5, 2970 Hørsholm, Denmark

Contents

Some Promising Trends in Ice Mechanics

A.Assur

U.S.Army, Cold Regions Research and Engineering Laboratory
Hanover, N.H. 03755

Summary

Ice sheets are inhomogeneous: properties vary strongly with depth. The theoretical treatment of plates with properties varying perpendicular to the plate has now been satisfactorily developed for floating ice sheets. However other problems are still waiting for solutions. The use of model ice is developing rapidly. Some suggestions how to analyze such ice are made. Breakthrough-loads on ice sheets diminish with duration of loading but no satisfactory solution is available based upon classical procedures of applied mechanics. The solution could very well come by assuming

- that deformation under creep is the failure criterion,
- that time to failure multiplied by minimum creep rate is a constant
- that the point of minimum creep rate can be assumed as a reference value for failure,
- that for modest stresses the m i n i m u m creep rate is proportional to the stress squared and
- that for the same modest time moment and modest stress it is linear with stress, allowing the assumption of linear viscosity for such cases.

It is necessary to change certain assumptions in classical theory to satisfy these conditions. Such important changes are outlined.

Introduction

Ice mechanics, or ice engineering in a broader sense, is a rapidly developing science. As such, many aspects are still controversial or not clear. Invited to speak about promising trends by the Chairman I was confronted with a dilemma but finally decided to limit myself to just a few especially important problems which are of interest to the researcher a n d of importance to the engineer a n d of significance to people working in the field of applied mechanics.

There are quite a few people present today who shaped recent trends in ice mechanics and are well known. However, I would like to acknowledge especially the presence here today of an almost unknown pioneer in ice engineering, Dr.Y.Kubo from Japan, who laid a solid foundation for the analysis of floating ice sheets, both from the theoretical and experimental points of view, supported by very fine laboratory tests. All of this was done under trying circumstances in Manchuria under war conditions [I].

We heard today about the true international spirit which prevails in the applied mechanics community as in all truly scientific endeavours. Today's gathering is another significant example.

Sea ice as an inhomogeneous plate and applications to model ice sheets

Of course, the methods of applied mechanics had a deep impact upon problem solutions in ice engineering. Let me start, however, with an example in reverse, where an ice problem had a significant impact upon subsequent developments in applied mechanics. Most principals involved are here today. I am referring to ice sheets with properties varying in depth. A good example is sea ice.

I wish I could acknowledge also the presence of another pioneer, V.L.Tsurikov, who in 1939 laid the foundation for the calculation of sea ice properties based upon structural parameters [2]. My own work eventually yielded the simple relations [3,4]:

$$\frac{\sigma}{\sigma_0} = (1-\sqrt{K_1\nu})^2 \qquad (I)$$

$$\frac{E}{E_0} = (1- K_2\nu)^4 \qquad (2)$$

$$\nu = \nu_1 S_i \qquad (3)$$

σ, σ_0 - strength of sea ice with the brine volume and zero brine volume (but same structure)

E, E_0 - Young's modulus of sea ice with the brine volume and zero brine volume.

$$K_1 = \frac{\gamma \varepsilon}{\pi / \beta_0} \qquad (4)$$

a combination of petrographic parameters based upon internal structure of ice crystals, depending upon vertical pattern (γ), ellipticity of brine pockets (ε) and relative spacing of brine pockets along a line, divided by the spacing of the brine pockets (β_0). These parameters depend upon growth conditions [5], but reasonable values lead to $K_1 \sim 1$ as a first approximation. Also approximately $K_2 \sim 1$. I am mentioning these relations since the initial linearized versions (for modest ν) are still being used, although eqs (1) and (2) give a much better representation and are currently important for the analysis of model ice sheets.

ν_1 - a function of temperature which can be calculated from phase relations [3]

S_i - ice salinity

Since $S_L = k\ S_w$ (5)

where k - partition coefficient

S_w - salinity of water

eqs (1), (2) with (5) are relevant for calculating the characteristics of model ice produced from chemically doped water.

By plotting $\sqrt{\frac{\sigma}{\sigma_o}}$ versus $\sqrt{S_w}$ or $\sqrt[4]{\frac{E}{E_0}}$ versus S_w we obtain linear relations by means of which the proper parameters can be derived. Notice also that the characteristic length of model ice sheets should vary linearly with the water salinity since it is proportional to the fourth root of E . This is important to consider in proper scaling. In most cases two straight lines result in these plots since we are dealing essentially with a two-layered system in model ice: an upper firmer crust with a random orientation of crystals and a soft lower layer with systematic orientation. I expect an explosive development in the use of model ice since a number of nations are establishing ice engineering facilities throughout the world. This again will produce a demand for refined methods in applied mechanics.

And now to the treatment of inhomogeneous ice sheets as it affected the development of applied mechanics. It became clear already in the middle fifties that certain assumptions of classical mechanics apply neither to sea ice nor to fresh ice sheets, simply because of a strong vertical orientation of relevant properties. Eq(3) for sea ice depends upon temperature and salinity, - both vary strongly in a vertical profile. Fresh ice, in many cases, consists of a snow ice layer and a black ice layer with strong differences in characteristics and a temperature profile.

I developed a method to deal with changing properties of sea ice [7,4]. I shall not repeat here the methodology. Let me just point out that Dr.D.E.Nevel, then a young soldier, quickly verified the correctness of the method. Dr.A.Kerr, then a young student, was more sceptical. Under his supervision eventually a thorough thesis was produced [8] showing the general applicability of such a treatment of inhomogeneous plates on an elastic foundation. Finally Dr.K.Hutter (all of them present) introduced the methodology as a solid contribution to the knowledge store of applied mechanics [9]. This is a fine example how a practical requirement eventually resulted in a solid contribution to applied mechanics.

An unresolved problem in the creep of ice.

Let me give you another, a tough problem, which is just beginning to resolve. I would like to invite the applied mechanics community to contribute to the final solution.

In 1961 I published the results of preliminary observations and tests [10] relating P/h^2 to t .

P [to] - breakthrough load for a fresh ice sheet
h [m] - ice thickness
t [min] - time

This showed very clearly that the breakthrough load becomes less and less as longer the load stands on the ice sheet. Obviously creep is involved. It is disturbing that up to today nobody has succeeded in explaining this phenomenon by classical methods of mechanics. Dr.Nevel, in his carefully prepared thesis, shows virtually the contrary: the stresses diminish in time, that is the load becomes safer with time. Maybe we will have to make some fundamental changes in our assumptions. An attempt to do this follows.

Just by scanning through a recent treatise on the failure of solid substances [11], in particular the time to failure under stress, it appears to me that we would do much better assuming that the deformation characterised by the product $\dot{\varepsilon}_m \times t$

$\dot{\varepsilon}_m$ - minimum creep rate , $\dot{\varepsilon} = f(t)$ -creep rate

is a useful parameter for defining failure. Of course, we have to define what we understand as beginning of failure under creep. Let us plot $\ln \dot{\varepsilon}$ versus ln t. This gives us much better insight than the customary plot of $\dot{\varepsilon}$ versus t , which gives us a poorly defined "steady state" creep rate. The logarithmic plot gives us a sharply defined picture with a well expressed $\ln \varepsilon_m$

(logarithm of the minimum creep rate). $\ln \dot{\varepsilon}$ can also be plotted versus $\ln \varepsilon$ (logarithm of strain).

The engineer, of course, is mostly interested in the transition from the regime of strain hardening (decreasing $\ln \dot{\varepsilon}$) to the regime of strain softening (increasing $\ln \dot{\varepsilon}$). The minimum $\ln \dot{\varepsilon}_m$ between these two regimes can be used as a reference value with a characteristic "time to failure" t_m or "failure strain" ε_m

Just to anticipate some criticism let me point out that the actual failure will usually occur after t_m or even before t_m (for quick loading). The engineer (and physicist) is still interested in a well defined reference value.

Let us then develop expressions for the dependence of $\dot{\varepsilon}_m$ upon stress and for the function $\ln \dot{\varepsilon}$ in the vicinity of our important reference value ε_m or t_m .
In this process we will make the simplest possible assumptions.

The behavior near time zero (quick loading) or after a long time of creep (substantially after our reference time t_m) is of no interest for our present discussion. This is simply because we are dealing with loads of long, not short duration, and because the engineer does not want to get into a state of accelerating creep, although he may tolerate decreasing creep rates.

We shall follow some classical expressions but will depart from classical assumptions at some critical points yielding to accumulated evidence.

For a long time there have been two principal directions in the study of creep:

$$\dot{\varepsilon} \propto \sigma^{n} \tag{6}$$

and

$$\dot{\varepsilon}_m \propto \sinh \frac{\sigma}{\sigma_1} \tag{7}$$

Eq (6) (Glen's law for ice) has some advantages. Its simplicity makes it tractable in mathematical analysis; also a study of dislocations leads to expressions similar to eq (6). There are two serious disadvantages. First there is plenty of experimental evidence that n itself increases with σ. In addition, for $n \neq 1$ eq (6) is not linear for small σ. The linearity would simplify the solution of differential equations by assuming linear viscoelasticity or Newtonian viscosity.

Both disadvantages disappear in eq (7). Plotting ln $\dot{\varepsilon}_m$ versus ln $\sinh \frac{\sigma}{\sigma_1}$ we would start with a slope of a power n=1 which gradually increases to n=3 and above as observed. Earlier experiments, for example[12], for small σ, indeed, seemed to indicate a linearity between $\dot{\varepsilon}_m$ and σ. However, these were done plotting ε versus t with a poorly defined $\dot{\varepsilon}_m$ mostly after a constant time. We shall below present a modified theory which shows that for a c o n s t a n t time this relation still holds for modest σ. Eq (7), based upon kinetic rate theory, seemed to be far-fetched to some people.

In addition, careful experimentation with modern methods shocked the foundation of our thinking: for small σ $\dot{\varepsilon}_m$ proved to be proportional to σ^2 rather than σ. An examination of classical theory shows that an omission was made. Let us proceed (new symbols are used).

The classical rate theory of creep, see for example [6], assumes that for a material at rest thermal oscillations occur equally in all directions, thereby producing no consequences. Thermal vibrations occur at a frequency given by $\frac{kT}{h}$

k - Boltzmann constant
h - Planck's constant
T - absolute temperature

The Boltzmann distribution gives the probability for any given unit to become activated. The product of this probability times the frequency of thermal vibration kT/h gives the frequency ν of activation

$$\nu = \frac{kT}{h} \exp \left(\frac{-\Delta F}{NkT}\right) \tag{8}$$

ΔF - activation energy, a property of the material.
N - Avogadro number.

We assume barriers to distortion which must be overcome in creep. If a directed potential, such as shear stress, is applied, then a directional bias is created, and the barrier height in the direction of motion is reduced by the work $f\frac{\lambda}{2}$ in the direction of the force and raised a like amount in the opposite direction (λ - distance between barriers or successive equilibrium positions, f - force on flow unit).

The classical theory effectively assumed that once an oscillation reaches the distance $\lambda/2$ a l l the work e v e r y t i m e is given up in heat. This is where my criticism sets in. There is a probability of transition less than one.

The reduced barrier height in the direction of force f increases the activation frequency in that direction to

$$\underset{\rightarrow}{\nu} = \frac{kT}{h} \exp\left[-\frac{(\Delta F/N - f\lambda/2)}{kT}\right] \qquad (9)$$

This is the form which must be changed to establish a new more adequate system of deformation characteristics. We multiply the work $f\frac{\lambda}{2}$ by a partition coefficient p. It shows the proportion of flow units which i r r e v e r s i b l y transfer to the next equilibrium position once the distortion reaches the distance $\lambda/2$. It is certainly zero when f = 0. The classical theory assumes p jumps to 1 (certainty) as soon as a stress is introduced; we maintain that there is only a gradual increase: the partition coefficient is a function of f . The simplest possible assumption is linearity

$$p \propto f = f/f_0 \qquad (10)$$

However we leave the option open to introduce another function for the partition coefficient if accurate tests indicate the need.

The work $f\lambda/2$ in eq(9) must be multiplied by $p(f)$. Since for the activation frequency in the opposite direction $\overleftarrow{\nu}$ we have to change the sign in front of $f\frac{\lambda}{2}p$, we obtain for the net frequency in the direction of the force

$$\overrightarrow{\nu} - \overleftarrow{\nu} = 2\frac{kT}{h} \exp\left(-\frac{\Delta F}{kT}\right) \text{Sinh}\left(\frac{f\lambda}{2kT}p\right) \qquad (11)$$

with $p = f/f_0$

The classical theory (without p) then proceeds by assuming $\dot{\varepsilon} = X(\overrightarrow{\nu} - \overleftarrow{\nu})$ (12)

with X a vague or unknown function of t i m e and structure expressing the proportion of successful barrier crossings with an adjusted λ'. We accomplished the same in a more definite way by introducing $p = f/f_0$ under the hyperbolic sine. For small arguments Sinhx=x so that p in eq (6) is equal to X in eq (12) for small arguments. The advantage of our approach is to show that the deformation rate is proportional to $\text{Sinh}\left(\frac{\sigma}{\sigma_0}\right)^2$ rather than $\text{Sinh}\left(\frac{\sigma}{\sigma}\right)$ as previously assumed. This is supported by modern observations.

Let me skip now a few steps and write eq(11) in terms of minimum creep rate with D - the deviator stress under triaxial stress conditions

$$\dot{\varepsilon}_m = \frac{kT}{h} \exp\left(-\frac{\Delta F}{RT}\right) \text{Sinh}\,\frac{D^2}{4SD_0 kT} \qquad (13)$$

R = Nk - the gas constant $\qquad f_0 = D_0/2S$

S - number of "flow units" per unit area.

Such an equation alone should be used only for the time when minimum creep rate occurs (steady state creep in previous understanding). The structure of the material itself changes during strain hardening (adjustments between crystals and grains) and strain softening (largely rearrangement of slip lines within crystals). Only at the moment of equilibrium between these two processes (no structural change) can eq(11) be used, strictly speaking. For the time before and after we have to make additional assumptions.

Plotting ln $\dot{\varepsilon}$ versus ln t, for example Fig 14.67 for soil [6] we find at first a linear decrease, or well defined minimum $\dot{\varepsilon}_m$ and then a rapid increase. Ice behaves not much differently, at least in the area of our interest. The positions of the minima proceed with a slope of -1 versus ln t. This simply means that the product $\dot{\varepsilon}_m \times t_m = \varepsilon_m$ a constant nominal strain. This is a very <u>important</u> conclusion and probably the key to the solution of our dilemna mentioned above. It is therefore reasonable to express the process in terms of nondimensional time t/t_m or in fractions of the nominal failure strain $\dot{\varepsilon}_m$.

Let us discuss a differential equation describing the process. The relative changes $\ddot{\varepsilon}/\dot{\varepsilon}$ in the creep rate in terms of nondimensional time characteristics vice versa changes in the structure of the material. There is an equilibrium (no change) at the point of minimum creep. If we set $\ddot{\varepsilon}/\dot{\varepsilon} = \alpha$ a constant then we have the simplest possible expression for strain softening which, of course, starts already at the beginning of the process (when loading starts) and eventually overwhelms the initial strain hardening. The strain hardening starts initially very fast and eventually disappears in time. A suitable simple expression is $\ddot{\varepsilon}/\dot{\varepsilon} = -\frac{\beta}{x}$ where $x = t/t_m$ is our nondimensional time, or time expressed in terms of the "reference time to failure". Combining both expressions we have the simple differential equation

$$\frac{\ddot{\varepsilon}}{\dot{\varepsilon}} = \alpha - \frac{\beta}{x} \tag{14}$$

as a first attempt to describe the process. Setting $\ddot{\varepsilon}/\dot{\varepsilon}$ to zero for x=1 by definition we see that $\alpha = \beta$. Integrating (14) we have

$$\ln \dot{\varepsilon} = \beta \frac{t}{t_m} - \beta \ln \frac{t}{t_m} + \ln C \tag{15}$$

or

$$\dot{\varepsilon} = C \left(\frac{t_m}{t}\right)^{\beta} e^{\beta t/t_m} \tag{16}$$

The integration constant C, in general, must be

$$\frac{\dot{\varepsilon}_m}{e^{\beta}} = \frac{\varepsilon_m}{t_m e^{\beta}} \tag{17}$$

and in our case

$$\dot{\varepsilon} = A \sinh \left(\frac{\sigma}{\sigma_0}\right)^2 \left(\frac{t_m}{t}\right)^{\beta} e^{\beta t/t} \tag{18}$$

since for the reference value $t_m/t = 1$ eq(16) must give the minimum creep rate derived above. If our assumption $\dot{\varepsilon}_m t_m = \varepsilon_m$ as a constant is not entirely correct we will have a slight change in the powers used. Another alternative is to set $\int_0^{t_m} \dot{\varepsilon}\, dt$ as a constant.

The engineer, of course, watching a creeping ice sheet under long term loading, is interested in the state of the plate at a definite time t

$$\text{Since} \quad t_m / t = \varepsilon_m / \dot{\varepsilon}_m t = \frac{\varepsilon_m}{t\ A\ \text{Sinh}(\sigma/\sigma_o)^2} \tag{19}$$

and according to preliminary results the creep rate for small stresses is linear to stress, provided it is observed at a definite but modest time after beginning of loading, $\beta = 1/2$ would give us a condition for linear viscosity in such a case. This preliminary conclusion still requires experimental verification.

We could also plot $\ln \dot{\varepsilon}$ versus $\ln \varepsilon$ rather than ln t and be confronted with more complexity than eq [15].

Finally let me say a few words about our partition coefficient p as a function of stress. By plotting $\ln \dot{\varepsilon}_m$ versus ln p based upon careful experiments under otherwise equal conditions (same temperature, same composition and structure) we find a slope of 2 for small σ. By defining the shape of this relationship and its departure from eq [13], or a similar expression, we will eventually be able to achieve an expression for the partition function, more precise than eq [10]. A possible candidate to explore is

$$\dot{\varepsilon}_m = B\ \text{Sinh}^n \frac{\sigma}{\sigma_o} = B'\ \text{Sinh}\ (p \frac{\sigma}{\sigma_o}) \tag{20}$$

with $p = f(\sigma)$. One can also use the concept of a restoring force, rather than the probability argument in eq 11 and achieve the same result. σ_0 is very likely proportional to Young's modulus.

I have presented the substance of the changes in creep equations informally before three different audiences since the fall of 1978, and have compared the developing conclusions with experimental evidence. I would like especially to acknowledge useful discussions with the following gentlemen: R.T.Martin, A.M.Fish, M.Mellor and K.Hutter.

References

1. [Kubo Y. a.o.] (1941): Study on river ice. South Manchurian Railway Co. 4 volumes, English translation (1955): 778 p.

2. Tsurikov V.L. (1939): Problema pročnosti morskogo l'da [The problem of sea ice strength]. Severnyj Morskoj Put' 16:45-74.

3. Assur A. (1958): Composition of sea ice and its tensile strength. Nat. Ac. Sci. - NRC Publ. 598:106-138.

4. Weeks W., Assur A. (1967): The mechanical properties of sea ice USA CRREL Monogr. II-C3 XIV + 80 p.

5. Assur A., Weeks W.F. (1963): Growth, structure, and strength of sea ice. IASH Publ. 61:95-108

6. Mitchell F. K. (1976): Fundamentals of soil behaviour F. Wiley & Sons, NY XVI + 422 p.

7. Assur A. (1967): Flexural and other properties of sea ice sheets. Physics of snow and ice. Intern. Conf. Hokkaido: 557-567.

8. Palmer W.T. (1971): On the analysis of floating ice plates. PHD thesis, New York Univ. VIII + 94 p.

9. Hutter K. (1973) On the fundamental equations of floating ice plates. Mitt. Versuchsanstalt Wasserbau, Hydrologie and Glaziologie No8 XVIII + 150 p.

10. Assur A. (1961): Traffic over frozen or crusted surfaces. Mechanics of soil-vehicle systems. Intern.Conf. Torino: 913-923

11. Regel'V.R., Slucker A.T., Tomaševskij E.E. (1974): Kinetičeskaja priroda pročnosti tverdych tel. [Kinetical nature of the strength of solid materials] Moscow.

12. Butkovich T.R., Landauer T.K. (1959): The flow law for ice. USA CRREL RR 56. 7 p.

The Development and Application of Multimodal Ice Failure Theory

F.G. BERCHA
F.G. Bercha and Associates Limited
Calgary, Alberta, Canada

Summary

Multimodal ice failure behavior is important for accurate characterization of numerous ice mechanics phenomena. Ridge formation, pile-ups, and the associated macroscale ice cover properties as well as interactions of ice formations with structures under conditions potentially admitting more than one failure mode require multimodal characterization. In this paper, theoretical approaches with associated results for quantification of multimodal behavior are presented. Simplified approaches applicable directly by designers are outlined and corroborated with a more rigorous probabilistic approach.

Introduction

Pile-up and ridge formation, dynamic icebreaking, quasistatic action of ice on steep cones, or nonconservative moored ship interactions with ice are some of the system types which appear to manifest ice failure phenomena not characterizable by a single pure mode. Naturally, many systems are quite adequately characterizable by a unique mode, but this is a subject covered in other papers [3, 7]. The problem of characterizing a failure mode mix was generally referred to by the author in [6, 7] and was first explicitly noted by him in 1976 in a paper [8] describing observation and measurement of ice interactions with a steep sloped, instrumented, conical light pier located at Yamachiche Bend in the St. Lawrence River in Eastern Canada. Here, it was visually observed that the ice sheet giving peak load appeared to be failing simultaneously, but at different perimeter locations, in both flexural and crushing modes. That is, multimodal behavior was observed. Indeed, the cone was steep enough--75° from the horizontal--that flexural loads could be large, but deterministic calculations still clearly favored occurrence of a flexural failure. How can the observed mode mix be analyzed?

Probabilistic analysis can be used to demonstrate that, within existing interaction system information quality constraints, a unique valued load prediction for each failure mode is quite unrealistic; rather, a probability load density distribution, about some preferred value range for each pure mode, but capable of overlapping with other mode loads, is a more accurate description. Thus, for a given set of ice-structure interaction system characteristics, within the bounds of empirical data inputs, more than one interaction mode is simultaneously admissible. It also appears to be possible to predict, approximately or rigorously, proportions and load levels of each admissible load likely to be generated in the interaction.

Implications of multimodal behavior predictability include a better ability to analyze and quantify interaction parameters for phenomena such as those mentioned above, as well as rubble fields, ridges, pileups, grounding and beaching, jamming, and other phenomena constituting links in both localized interaction and macroscale ice system models.

In this paper, following this brief introduction, two approaches to and associated results of multimodal analyses will be presented; namely, a simple quasiprobabilistic approach, and a more rigorous probabilistic one. It should, however, be emphasized at the outset, that the present paper is primarily a heuristic one--it seeks to mobilize activity in an area requiring further attention rather than purporting to present the final solution. Accordingly, although some results are presented, much of the paper is dedicated to presenting concepts of approaches to the problems so that other investigators might pursue them further.

Multimodal Theory and Applications

1. General Concept

Consider an ice-structure interaction process in which the total horizontal load, F_i, corresponding to the ice failure mode, may be characterized by a functional relationship among n system parameters, p_j.

$$F_i = F_i(p_j) \tag{1}$$

in a manner such as that given in any of references [1, 2, 3, 4, 5, 6, 7]. The p_j are variables such as structure width, ice thickness, flexural strength, coefficient of friction, and so on. Define the line load, f_i, as the load per unit projected frontal width of structure. Then, if N failure modes are admissible within the interaction, the total line load, f, may be expressed as

$$f = \sum_{i=1}^{N} f_i \Phi_i \tag{2}$$

where Φ_i is defined as the mix coefficient for the ith failure mode, and all Φ_i must meet the normality condition

$$\sum_{j=1}^{N} \Phi_j = 1 \; . \tag{3}$$

The total horizontal load, F_x, is then expressible as

$$F_x = fb \tag{4}$$

where b is the frontal planar projected structure width.

It follows then, that the total interaction load for multimodal interactions is expressible as a mix of pure mode loads proportioned by means of a set of appropriately defined mix coefficients, Φ_i. The quantification of these mix coefficients is the subject of most of the balance of this section. Two methods are outlined; an approximate one and one based on probabilistic theory.

2. Approximate Method and Applications

The variation in total load, F_x, for a sheet-cone interaction, as a function of cone angle, for a system admitting flexural and crushing loads only is shown in Figure 1. Clearly, the flexural load tends to dominate the lower angle region, because the associated load is significantly smaller than a crushing load so that the sheet is likely to fail in flexure before the higher crushing load can be generated. However, as some critical angle, α_c, is approached, the pure crushing and pure flexure loads approach each other in magnitude, suggesting more than one failure scenario alternative. Indeed,

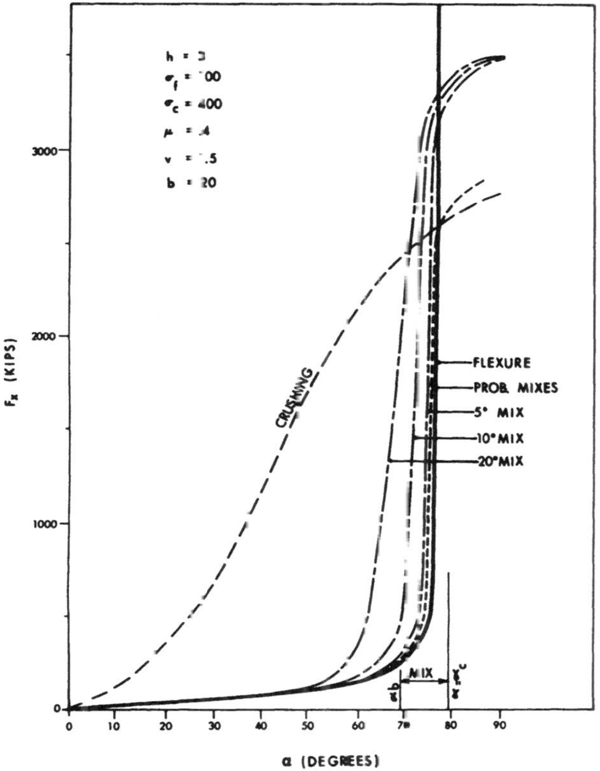

FIGURE 1 DETERMINISTIC PURE AND DIFFERENT MIXED MODES

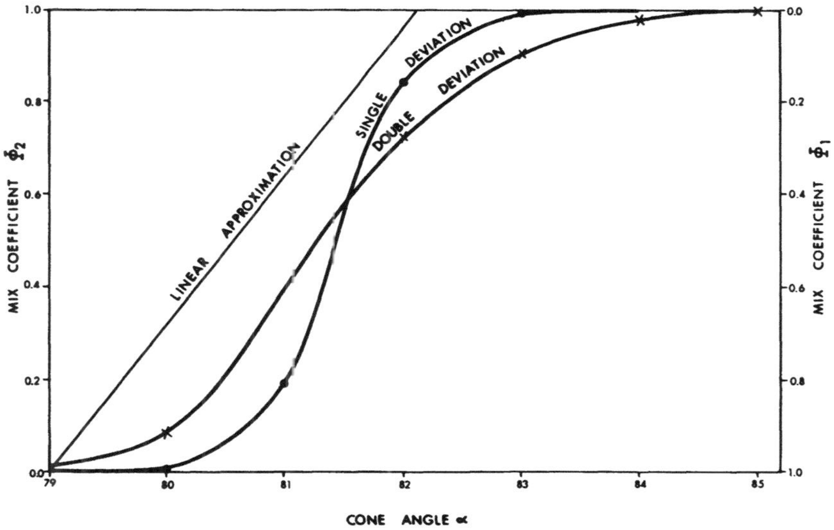

FIGURE 2 VARIATION OF MIX COEFFICIENTS WITH SLOPE

for any angle α in the mix region, the total load may be expressed as

$$F_x = b[f_F \Phi_F + f_C \Phi_C] \tag{5}$$

where now the subscripts F and C correspond to flexure and crushing, respectively, and the mix coefficients Φ_i take on some value between 0 and 1, expressible as a function of cone angle α between α_c and α_b, where α_b is the angle at which mixing begins. The most obvious approach to evaluating the mix coefficients is to vary them linearly over the mixing range as shown in Figure 2. Results based on such a variation are shown in Figure 1 for a 5, 10, and 20° mix. As will be noted later, the linear mix coefficient, although not rigorously correct, gives adequate answers for engineering purposes--and it certainly resolves the problem of load singularity at the critical angle [7].

Simple mode mixing theory can also be applied to evaluation of inplane sheet loads sustained by pile-ups or first-year ridges. Consider the case where a sheet has fragmented against an obstacle yielding a consequent pile of fragments such as that shown in Figure 3a. Define the angle between the fragment faces and the horizontal by α, and assume a uniform frequency distribution for this angle, as shown in Figure 3b. That is, assume that the ice face orientation is equiprobable for all values of α, for $0 \leq \alpha \leq \pi/2$. In reality, certain preferred directions become evident, and the out-of-plane rotation of fragments may have an effect, but at a first approximation level, the above assumption can be justified. Now, consider the variation in horizontal interaction load F_x with slope angle α, for pure crushing and pure flexure, as shown in Figure 3c. Recall the resolution factor ξ, described in [7], relating the vertical and horizontal interaction force components as

$$f_x = \xi f_z , \tag{6}$$

where

$$\xi = \frac{\mu + \tan\alpha}{1 - \mu\tan\alpha} \tag{7}$$

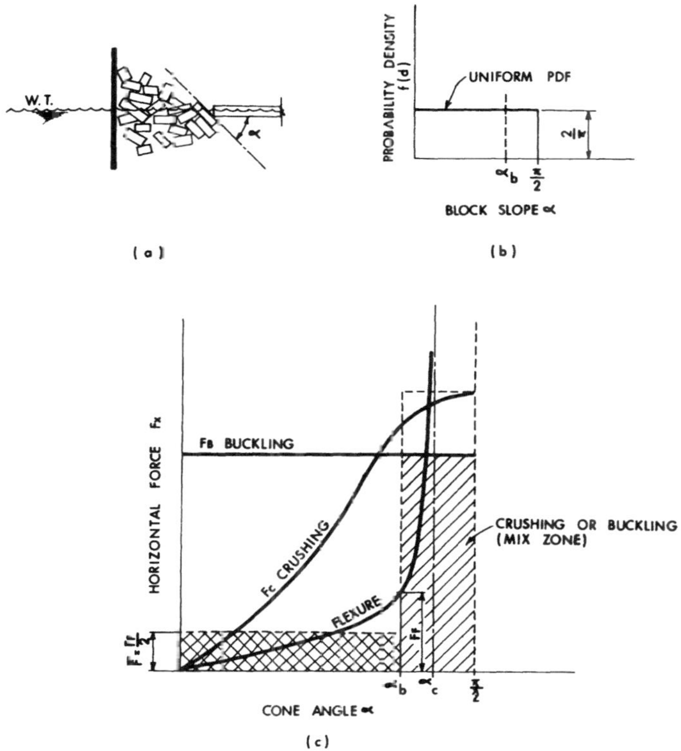

FIGURE 3 PILE-UP

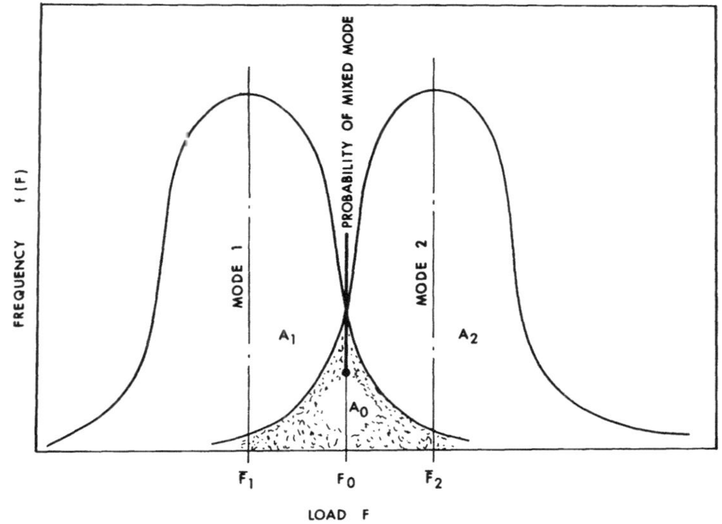

FIGURE 4 MULTIMODAL WARNER DIAGRAM

and μ is the coefficient of friction.

The critical angle α_c is defined as the angle beyond which no sliding can occur and failure occurs principally by crushing. This angle is given by $\tan^{-1}\frac{1}{\mu}$. In reality, however, mixing of failure modes occurs before the critical angle is reached, and for current interaction conditions, linear mix proportioning 10° below the critical angle is chosen. Now judge that for $\alpha_c - 10° \le \alpha \le \alpha_c$ crushing is increasingly more likely and for $0 \le \alpha \le \alpha_c - 10°$ flexure is predominant. Then the following mix coefficients Φ_F and Φ_C for flexural and crushing modes, respectively, may be defined,

$$\Phi_F = \frac{\alpha_c - 10}{90} \tag{8a}$$

and

$$\Phi_C = 1 - \Phi_F \, . \tag{8b}$$

The final pile-up horizontal force may now be determined as the mix of the mean flexural and crushing forces, f_F and f_c, as follows:

$$f_P = \Phi_F \cdot f_F + \Phi_c \cdot f_c \tag{9}$$

where f_F and f_c are given by:

$$f_F = \frac{1}{2}\,\xi_{\alpha_c - 10} \cdot P_{max} \tag{9a}$$

$$f_c = \sigma_c \cdot h \cdot \sin^2\alpha \tag{9b}$$

where P_{max} is the critical vertical flexural load given by [7],

$$P_{max} = \frac{2\pi\,\sigma_f\, I}{L_0 h} \tag{10}$$

and the characteristic length L_0 is $\sqrt[4]{\frac{4EI}{\rho_w}}$, where I is moment of inertia per unit width, h is ice sheet thickness, σ_f and σ_c are flexural and crushing strengths of sheet ice, E is the elastic modulus, and ρ_w is density of water.

Should it be found that the ice sheet slenderness ratio is small enough to mobilize a buckling mode before crushing, then the unit crushing load, given below, should be used instead of Equation (9b),

$$f_B = \frac{\sqrt{3}}{4} h\sqrt{E \rho_w h} \ . \tag{11}$$

That is, if f_B is smaller than f_c, buckling will occur before crushing and therefore f_B should be used in Equation (9) instead of f_c, with the appropriate mix coefficient Φ_B.

Numerical results based on the above pile-up theory were compatible with full scale ice pile-up loads measured from drillships, while those based on other theories [10] were found to be too low.

3. Refined Method

Recall the theory of normally distributed random variables [9], with which physical quantities are characterized by a spectrum of values rather than as unique point values. Such spectra of values, when plotted in frequency distributions, generally resemble a known distribution type which can be specified as a functional dependence on two or three variables. A normally distributed variable, F, can be specified by a pair of values, $\bar{F}$ and S_F, representing the mean value and standard deviation, respectively.

Figure 4 shows the juxtaposition of two load spectra, constituting what is termed a Warner Diagram [9]. In such diagrams, for correctly normalized abscissae, it is demonstrable that the shaded overlap area represents the probability of co-occurrence of the two normally distributed random variables. If each of the distributions represents an admissible failure mode, the overlap area signifies the probability of co-occurrence of the two modes, or the probability of multimodal behavior. Define the overlap area as A_0, and define the balance of the distribution areas corresponding to each of the two modes as A_1 and A_2. Assume loads in the mix region, A_0, are equiprobable for either mode, while in the pure regions, A_1 and A_2, loads are caused entirely by the relevant mode. Then, the mode 1 mix coefficient Φ_1 can be expressed as

$$\Phi_1 = A_1 + \frac{A_0}{2} = 1 - \frac{A_0}{2} \tag{12a}$$

and from the normality condition, the mode 2 coefficient is

$$\Phi_2 = 1 - \Phi_1 = \frac{A_0}{2} \, . \tag{12b}$$

In order to evaluate the mode mix explicitly, then, one needs an expression for the overlap area A_0.

Assume the random variable F to be randomly distributed, so that its frequency distribution may be expressed as

$$f(F) = \frac{1}{\sqrt{2\pi}\, S_F} \, e^{-\frac{1}{2}\left[\frac{F - \bar{F}}{S_F}\right]^2} \quad \text{for } -\infty < F < \infty \, . \tag{13}$$

First, the condition for intersection of the two distributions can be obtained from equality of ordinates as follows,

$$F_0^2 \left(\frac{1}{S_{F_1}^2} - \frac{1}{S_{F_2}^2}\right) + 2F_0 \left(\frac{\bar{F}_1}{S_{F_1}} - \frac{\bar{F}_2}{S_{F_2}}\right) + \left(\frac{\bar{F}_1^2}{S_{F_1}^2} - \frac{\bar{F}_2^2}{S_{F_2}^2} - 2L_n \frac{S_{F_1}}{S_{F_2}}\right) = 0 \tag{14}$$

where F_0 is the value of the random variable F at which the two normal distributions intersect, $\bar{F}_i$ and S_{F_i} are the mean value and standard deviation for the ith distributions, respectively. Then, the area A_0 is given by

$$A_0 = A_0^1 + A_0^2 \tag{15}$$

where A_0^1 is the area under the first normal distribution for $F > F_0$ and A_0^2 is the area under the second distribution for $F < F_0$. These values may be calculated using three point Gaussian quadrature to within 99.7% of closed form accuracy, using the following formula:

$$A_0^i = 0.5 - \frac{Z_i}{2\sqrt{\pi}} \left\{ \frac{5}{9} \left[e^{-Z_i^2 (1 - \sqrt{3/5})^2/8} + e^{-Z_i^2 (1 + \sqrt{3/5})^2/8} \right] + \frac{8}{9} \, e^{-Z_i/8} \right\} \tag{16}$$

where Z_i is the value of the standardized normally distributed random variable at the point of intersection, F_0, given by

$$Z_i = \frac{F_0 - \bar{F}_i}{S_{F_i}} . \tag{17}$$

Table 1 summarizes salient results of calculations of mix coefficients based on a set of crushing and flexure loads calculated from random variable inputs obtained on the basis of field data [1, 2]. Loads, their standard deviations, and calculated values of Φ_i based on both one and two standard deviations are given. For all cases, significant mixing does not begin until approximately 5° before the critical angle, confirming choice of 5° mixes suggested earlier. Mix coefficients do appear to vary nonlinearly over the mix interval, as shown in Figure 2, but the general trend represented by the linear approximations is certainly borne out. A few aspects of the results, however, are not entirely clear at this time. Where mean values of loads coincide, but have different standard deviations, mix coefficients are not 50/50, but rather tend to favor the higher standard deviation load. Information dispersion theory suggests similar results. Also, for certain combinations of random variables, overlap areas can extend from left to right distribution boundaries. The significance of such areas is not clearly apparent. Results for the 5° mix based on the probabilistic mix coefficients are also shown in Figure 1, and may be considered compatible with the ones based on linear approximations.

Conclusions

Approaches to the prediction of proportions of different ice failure modes constituting a multimodal failure interaction were presented. First, a simplified theory developed on the basis of observation and judgment was presented and discussed. An extension of the theory to include pile-up interactions with ice sheets was also outlined. Next, a more rigorous theory of multimodal behavior based on methods of probabilistic analysis was derived. The two theories generally agree. The latter may be extended to incorporate pile-up in a manner analogous to that of the simplified theory.

The multimodal theories do eliminate certain unnatural load discon-

Table 1. Mix Coefficient Probabilistic Calculations

N.S.*	α	$\bar{F}_F$	S_{F_F}	$\bar{F}_c$	S_{F_c}	F_0	Z_1	Z_2	A_0	Φ_1	Φ_2
1	70	165.49	68.38	2396.80	565.29	440.70	4.020	-3.460	0.0064	0.9968	0.0032
	75	252.28	104.22	2532.50	597.29	636.22	3.684	-3.175	1.0074	0.9963	0.0037
	80	508.94	210.53	2632.50	620.87	1110.72	2.858	-2.451	0.0120	0.9940	0.0060
	81	1995.00	236.80	2644.70	623.70	2343.80	1.473	-0.482	0.3850	0.8070	0.1930
	82	3481.00	263.10	2656.90	626.60	3090.00	-1.487	0.691	0.3130	0.1570	0.8430
	83	4967.00	289.50	2669.20	629.50	4182.90	-2.710	2.405	0.0137	0.0068	0.9932
	84	6453.00	315.80	2681.40	632.40	5160.40	-4.093	3.920	0.0053	0.0027	0.9973
	85	7939.40	342.17	2693.70	635.31	6077.56	-5.441	5.326	0.0000	0.0000	1.0000
2	75	252.28	208.00	2532.50	1200.00	749.50	2.390	-1.486	0.0780	0.9610	0.0390
	80	508.94	421.00	2632.50	1242.00	1280.00	1.830	-1.090	0.1710	0.9140	0.0860
	81	1995.00	474.00	2645.00	1247.00				1.0000	0.5000	0.5000
	81.5	2738.00	250.00	2651.00	625.00	3126.00	1.552	0.760	0.2840	0.1420	0.8580
	82	3481.00	526.00	2657.00	1253.00	2786.00	-1.322	0.103	0.5520	0.2760	0.7230
	83	4967.00	579.00	2669.00	1259.00	4016.00	-1.642	1.070	0.1900	0.0963	0.9037
	84	6453.00	632.00	2681.00	1265.00	5053.00	-2.215	1.875	0.0440	0.0220	0.9780
	85	7939.40	685.00	2693.70	1271.00	6001.00	-2.830	2.602	0.0100	0.0050	0.9950

*Number of standard deviations used.

tinuities in interaction load curves, in consonance with physical measurement and observation. They also admit observed concurrent diverse mode failures and provide a method for quantifying associated load levels. Finally, pile-up or ridge load levels quantified utilizing the theories appear to conform to measurements better than those based on previously available predictions.

The preliminary nature of the investigations leading to the postulation of the multimodal theories presented herein suggests that caution should be exercised in their application pending further corroboration. Such corroboration should include a study of physical observations, laboratory modelling, application of modified theoretical approaches, and development of a better understanding of the significance of certain apparent anomalies contained in the present theories.

References

1 Bercha, F.G.: Applications of probabilistic methods in ice mechanics. ASCE Meeting, Chicago, 1978.

2 Bercha, F.G., Danys, J.V., and Rokne, J.G.: Probabilistic ice-structure interaction theory. IAHR Symposium, Lulea, Sweden, 1978.

3 Bercha, F.G. and Danys, J.V.: On forces generated by ice acting against bridge piers. Third National Hydrotechnical Conference, Laval, Canada, 1977.

4 Bercha, F.G.: On the scale effect in ice mechanics. Sixth Canadian Congress of Applied Mechanics, Vancouver, Canada, 1977.

5 Bercha, F.G. and Danys, J.V.: Friction in ice structure interactions. NRC Ice Problems Workshop, Calgary, Canada, 1977.

6 Bercha, F.G.: Mathematical simulation of ice-structure interactions. Fifth Canadian Congress of Applied Mechanics, Fredericton, Canada, 1975.

7 Bercha, F.G. and Danys, J.V.: Prediction of ice forces on conical offshore structures. Marine Science Communications, 1 (5), 1975.

8 Danys, J.V. and Bercha, F.G.: Investigations of ice forces on a conical offshore structure. Ocean Engineering, 3 (5), 1976.

9 Haugen, E.B.: Probabilistic approaches to design. John Wiley & Sons Inc., 1968.

10 Parmerter, R.R. and Coon, M.D.: Model of pressure ridge formation in sea ice. J. Geophysical Research, 77 (33), 1972.

Numerical Model of Penelope Ice Viscometer Flows

R. BREPSON

NEYRTEC - ALSTHOM-ATLANTIQUE - GRENOBLE

and LABORATOIRE de GLACIOLOGIE - C.N.R.S. - GRENOBLE - FRANCE.

SUMMARY

The plane flow of ice over a bump as observed with the so-called Penelope ice viscometer has been calculated.
The cavity which forms in the lee of the bump is included as an initial condition. The deformation of the ice is assumed to be the sum of the linear elasticity an the non-linear viscosity resulting from the Glen creep law.
A finite difference method based on orthogonal curvilinear co-ordinates is used to determine the stresses and deformation from unit ε_0 and σ_0, which are computed at the end of the procedure from the experimentally measured strains and torque.
The model yields a very low Young's modulus (E = 1250 bar) and a fluidity parameter $B = \tau^{-3}\, y^{-1} = 0.079$ bar^{-3} year^{-1}.
However, the fact that the outer ice remains bubbly makes the computed radial velocities unrealistic.

SETTING UP THE PROBLEM

The sliding of a temperate glacier over an obstacle has recently been simulated within the so-called PENELOPE ice viscometer [1], which is in effect a very large Couette-type viscometer. The simulation consisted in dragging a ring of ice over two diagonal bumps and showed that a cavity forms in the lee of each bump. The flow is two-dimensional and will be modelled here using orthogonal curvilinear axes (α, β), the inner and outer walls of the viscometer being represented by α contours.
In the steady state, ice creep is assumed to follow the Glen law.

Calling the strain rates $\dot{V}_{ij}$, the stresses σ_{ij} , the stresses deviation σ'_{ij} , and setting $3I'_2 = \sigma_*^2$, in which I'_2 is the second invariant of σ'_{ij} , then the Glen law may be written as :

$$\dot{v}_{ij} = \frac{B}{2} (I'_2)^{\frac{n-1}{2}} \sigma'_{ij} = 3^{\frac{-n+1}{2}} B \sigma_*^{n-1} \left(\frac{3}{2} \sigma'_{ij} \right) \quad (1)$$

During transient processes, Andrade creep is observed which obey the same law with $B = B(t)$. Andrade creep will not however be taken explicitly into account. Rapid transient reversible creep is also observed and this produces a total deformation proportional to stress with an apparent shear modulus of 1500-2000 bar [2] . Since the pressure rises to about 100 bar on the bumps and vanishes in the cavities, this pseudo-elasticity cannot be neglected. The ice is thus considered to be viscoelastic in the Maxwellian sense. Letting total deformation be λ_{ij} , we can write :

$$\dot{\lambda}_{\alpha\alpha} = \frac{\dot{\sigma}_{\alpha\alpha}}{E} - \frac{\nu}{E} (\dot{\sigma}_{\beta\beta} + \dot{\sigma}_{\gamma\gamma}) + \dot{v}_{\alpha\alpha} \quad (2)$$

....

We shall use reduced variables [3] with ε_0 as unit deformation, $\sigma_0 = E\, \varepsilon_0$ as unit stress ans $t_0 = 3^{\frac{n+1}{2}} (E\, B \sigma_0^{n-1})^{-1}$ as unit of time. Letting :

$$\Lambda_{ij} = \lambda_{ij} / \varepsilon_0 \ , \ \Sigma_{ij} = \sigma_{ij} / \sigma_0 \ , \ dT = \frac{dt}{t_0} \ ,$$

Eqs (2) becomes :

$$\frac{d\Lambda_{\alpha\alpha}}{dT} = \frac{d\Sigma_{\alpha\alpha}}{dT} - \nu \frac{d\Sigma_{\beta\beta}}{dT} - \nu \frac{d\Sigma_{\gamma\gamma}}{dT} + \Sigma_*^{n-1} \left[\Sigma_{\alpha\alpha} - \frac{1}{2} \left(\Sigma_{\beta\beta} + \Sigma_{\gamma\gamma} \right) \right]$$

....

$$\frac{d\Lambda_{\alpha\beta}}{dT} = (1 + \nu) \frac{d\Sigma_{\alpha\beta}}{dT} + \Sigma_*^{n-1} \cdot 3 \Sigma_{\alpha\beta} \quad (3)$$

Assuming Poisson's ratio ν and exponent n to be known, values of Σ_{ij} and of reduced displacements $U_i = u_i / \varepsilon_0$ during an arbitrary time interval $\Delta T = \Delta t / t_0$ are sought at all mesh points. Then σ_0 is determined from the measured torque, and the measured angle of rotation of the plastic markers above the cavity [1], that is $2\lambda_{\alpha\beta} = 2\varepsilon_0 . \Lambda_{\alpha\beta}$ provides ε_0 and hence $E = \sigma_0 / \varepsilon_0$. (Halving ΔT doubles both σ_0 and ε_0 - and the true displacements and stresses are unchanged). Finally, the measured peripheral displacement $u_\beta = \varepsilon_0 \; U_\beta$ yields t and hence fluidity parameter B.

BOUNDARY CONDITIONS

At the outer geometrical boundary U_β = a given constant, and $U_\alpha = 0$. At the inner geometrical boundary, the conditions are as follows :

- Bumps : $U_\alpha = 0$ and $\Sigma_{\alpha\beta} = f \Sigma_{\alpha\alpha}$ Results are given here for $f = 0$; further results have been obtained for $f = 0.1$;

- Cavities : The conditions are $\Sigma_{\alpha\alpha} = 0$ and $\Sigma_{\alpha\beta} = 0$.

The procedure starts by computing the stresses ignoring viscous deformation. This is a pure elastic problem for which U_β is given at the outer boundary, $U_\alpha = 0$ and $\Sigma_{\alpha\alpha} = \Sigma_{\alpha\beta} = 0$ in the cavities. Viscous deformation is then calculated under the following boundary conditions :

outer boundary : $\dot{U}_\alpha = \dot{U}_\beta = 0$

inner boundary : $\dot{\Sigma}_{\alpha\alpha} = \dot{\Sigma}_{\alpha\beta} = 0$ (cavity wall) ;

$\dot{U}_\alpha = 0 \quad \dot{\Sigma}_{\alpha\beta} - f \dot{\Sigma}_{\alpha\alpha} = 0$ (contact area)

EQUATIONS IN CURVILINEAR COORDINATES

The inner and outer ice boundaries as plotted at the end of an experiment have been expressed as a Fourier series with even terms up to Cos 18φ and Sin 18φ, say $S_i(\varphi)$ and $S_o(\varphi)$. Curves of constant α are calculated by interpolation, and the orthogonal curves of constant β are readily derived from these by an iterative procedure using the Corfou program.

From lengths x_α, x_β along α and β- lines, the metric tensor is calculated as :

$$g_{\alpha\alpha} = \left(\frac{\partial X_\alpha}{\partial \alpha}\right)^2 + \left(\frac{\partial X_\beta}{\partial \alpha}\right)^2 = e_\alpha^2 \quad (4)$$

....

The physical components of the strain tensor are then [4]

$$\Lambda_{\alpha\alpha} = \frac{1}{e_\alpha}\frac{\partial U_\alpha}{\partial \alpha} + \frac{1}{e_\alpha e_\beta}\frac{\partial e_\alpha}{\partial \beta} U_\beta$$

.... (5)

$$\Lambda_{\alpha\beta} = \frac{1}{2}\left[\frac{1}{e_\alpha}\frac{\partial U_\alpha}{\partial \beta} + \frac{1}{e_\alpha}\frac{\partial U_\beta}{\partial \alpha} - \frac{1}{e_\alpha e_\beta}\left(\frac{\partial e_\alpha}{\partial \beta} U_\alpha + \frac{\partial e_\beta}{\partial \alpha} U_\beta\right)\right]$$

Next, the stresses must be calculated from Eqs (3).
Setting $\sum_*^{n-1}\left(\frac{3}{2}\Sigma'_{ij}\right)\Delta T = V_{ij}$. The stresses are computed using a loop procedure.
Since $\Lambda_{\gamma\gamma} = 0$ (plane strain) and $V_{\alpha\alpha} + V_{\beta\beta} + V_{\gamma\gamma} = 0$, the three Eqs (3) are replaced by :

$$\Lambda_{\alpha\alpha} = (1-\nu^2)\Sigma_{\alpha\alpha} - \nu(1+\nu)\Sigma_{\beta\beta} + (1-\nu)V_{\alpha\alpha} - \nu V_{\beta\beta} \quad (6)$$

and a similar equation formed by exchanging α and β.

It follows that :

$$\Sigma_{\alpha\alpha} = \frac{1-\nu}{(1-2\nu)(1+\nu)}\Lambda_{\alpha\alpha} + \frac{\nu}{(1-2\nu)(1+2\nu)}\Lambda_{\beta\beta} - \frac{1}{1+\nu}V_{\alpha\alpha}$$

$$\Sigma_{\beta\beta} = \;.....\qquad (7)$$

$$\Sigma_{\gamma\gamma} = \frac{\nu}{(1-2\nu)(1+\nu)}\left[\Lambda_{\alpha\alpha} + \Lambda_{\beta\beta}\right] + \frac{1}{1+\nu}\left(V_{\alpha\alpha} + V_{\beta\beta}\right)$$

$$\Sigma_{\alpha\beta} = \frac{1}{1+\nu}\left(\Lambda_{\alpha\beta} - V_{\alpha\beta}\right)$$

Lastly, the Σ_{ij} must be introduced into the equilibrium conditions without external forces :

$$e_\beta \frac{\partial \Sigma_{\alpha\alpha}}{\partial \alpha} + \frac{\partial e_\beta}{\partial \alpha}\left(\Sigma_{\alpha\alpha} - \Sigma_{\beta\beta}\right) + e_\alpha \frac{\partial \Sigma_{\alpha\beta}}{\partial \beta} - \frac{2\,\partial e_\alpha}{\partial \beta}\Sigma_{\alpha\beta} = \qquad (8)$$

and another equation with α and β exchanged.

NUMERICAL COMPUTATION

The mesh shown in Fig. 1 was obtained from seven values of α and 29 values of β , including the virtual points, owing to the diametral symmetry. Numbering is done from the ends, on the left and on the right in turn, to reduce the band width in the stiffness matrix. The number of points in the model is rather low, but on the other hand, partial derivatives in the above equations have been replaced by accurate finite difference equivalents of the 5th order [5] near the boundaries.

In the two equilibrium equations, the procedure given in [6] was used. The finite difference equivalents are split into a first order approximation (centred finite differences) and corrective terms which are transferred to the right-hand side of the equilibrium equations.
The boundary conditions are dealt with in similar fashion.

To compute the corrective terms, the preceding approximate values found at the mesh points are used ; at the start, they are set equal to zero and convergence is obtained after five or six iterations.
Initially, the V_{ij} are made to vanish and a pure elastic solution is obtained. The stresses so computed allow one to find a first approximation to the $\frac{d\,V_{ij}}{dT}$.

All terms including creep are placed on the right-hand side of the equilibrium equations and, after convergence, the first value of $d\Sigma_{ij} / dT$, say $(d\Sigma_{ij} / dT)^{(1)}$, is obtained.
The value of ΔT is then determined by computing, at a previously selected mesh point $\Delta T^{(1)} = \frac{(\Sigma_{\alpha\alpha}\ \text{elastic})}{C\dot{\Sigma}_{\alpha\alpha}^{(1)}}$ (9)
The viscoelastic solution may now be computed by successive approximations. This method is similar to the one given in [7] for plastic situations. The procedure converges for any value of C, at least in the 10-100 range investigated.

RESULTS

Computer runs were made for $n = 3$, 3.5 and 4, and for $\nu = 0{,}30$. (Since $\nu = \frac{1}{2}$ for pure pseudo-elasticity, this value would have been more suitable but results for this case are not yet available). The February 1973 experiment with bubbly ice made from tap-water [1] was used to calculate the values of ε_0 and σ_0. A drawback is however that,at the end of the experiment, a layer of bubble-free ice approximately 3 cm thick appeared near the inner boundary.
The layer was soft and therefore the ice cannot be considered uniform throughout its thickness.

From the measured torque, σ_0 is found to be 3075 bar, so that the true stress is approximately 270 bar at the point of initial contact and 90 bar at the point of separation. Above the cavities, the stresses are weak as pointed out in the earlier paper [1]. If $\lambda_{\alpha\beta}$ is measured in the blue ice layer between mesh points 93 and 94, then tan $(2\lambda_{\alpha\beta})$ is 0.70 and tan $(2\lambda_{\alpha\beta})$ for $n = 3$ is 1.71, whence ε_0
$\varepsilon_0 = 2{,}45$ and $E = \frac{\sigma_0}{\varepsilon_0} = 1250$ bar. This agrees with DUVAL's value for the pseudo-elastic modulus.

Finally, B is found to be 0.079 bar^{-3} $year^{-1}$, which is plausible. On the other hand, the values of U_α above the cavities at the inner boundary turn out to be incredibly high, leading to U_α = 1.05 cm at mesh point 198 instead of zero.

Fig. 1 shows the model with its diametrial symmetry.
Plotted on Figures 2 to 6 are :

(a) the values of U_α and U_β on five local α - co-ordinates : points 93-98 at the point of initial contact, intermediate points 65-70, points 37-42 at the point of escape, and points 44-49 and 58-63 above the cavity ;

(b) $\Sigma_0 = (\Sigma_{\alpha\alpha} + \Sigma_{\beta\beta} + \Sigma_{\gamma\gamma}) / 3$ and Σ_* .

Fig. 7 shows the stress variation $\Sigma_{\alpha\alpha}$ along the contact area.

Considering the high radial velocities, the existence of bubbly ice over most of the ice ring cannot be neglected.
If $\lambda_{\alpha\beta}$ is measured in the bubbly ice, tan $(2\lambda_{\alpha\beta})$ is approximately 10 , ε_0 = 0.171 and E = 18 000 bar (which is still less than the sonic vibration value).
Displacement U_α at the inner boundary is then very much lower and the initial assumption that the cavity is steady becomes approximately true. Nevertheless, a large part of the elasticity of bubbly ice comes from the bubbles themselves, the result being a low Poisson's ratio and delayed elasticity of the Kelvin type. Computations for a Kelvin body are in progress.

o 0 o

Acknowledgments

Thanks are due to Professor L. LLIBOUTRY for advice during the preparation of this paper, to the Neyrtec research centre for providing computer time, and to Miss M.L. Clément who helped with the programming.

REFERENCES

1. BREPSON R. *Simulated glacier sliding over an obstacle* , Journal of Glaciology, 23 - 1979 -

2. DUVAL P. *Lois du Fluage transitoire ou permanent de la glace polycristalline pour divers états de contrainte* - Annales de Géophysique - 32 - 1976 - 335/350 -

3. SIM R.G. *Reference stress concepts in the analysis of structures during creep* - International Journal of Mechanical Science - Vol 12 - N° 6 - 1970 - 561 .....

4. CEMAL ERINGEN A. - Continuum Physics - Vol. 1 - Academic Press - 1971 -

5. FOX L. *The numerical solution of elliptic differential equations.*- Proceedings of Royal Soc. A - 242 - 1950 -

6. BICKLEY W.G., and MILLER J.C.P. - The London, Edinburgh & Dublin Philosophical Magazine - Seven Series - Vol. 33 -

7. ILIUSHIN *Traité de Plasticité* -(Traduit du Russe) - Dunod - 1956 -

o 0 o

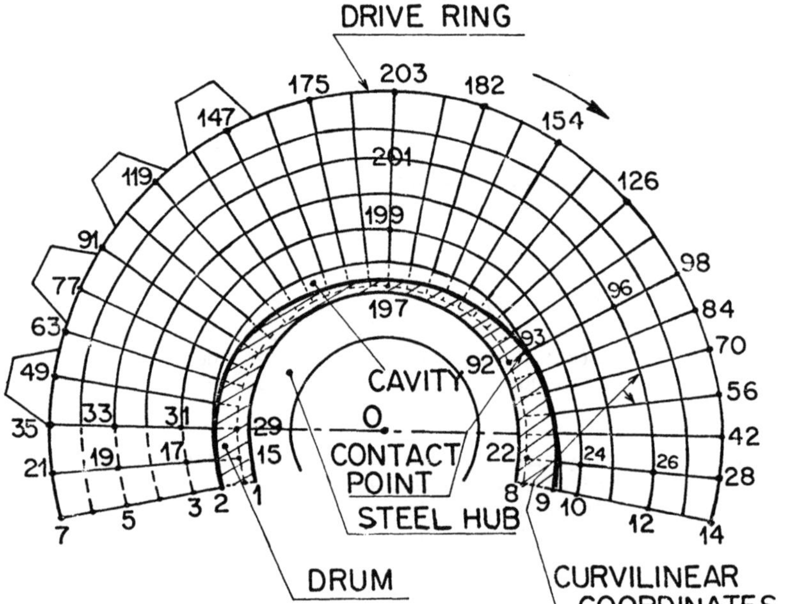

Fig. 1 NUMERICAL REPRESENTATION OF ICE FIELD

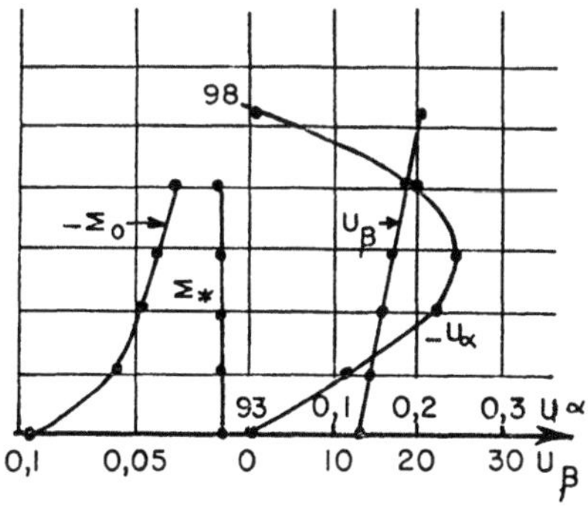

Fig 2 Displacem. and stresses

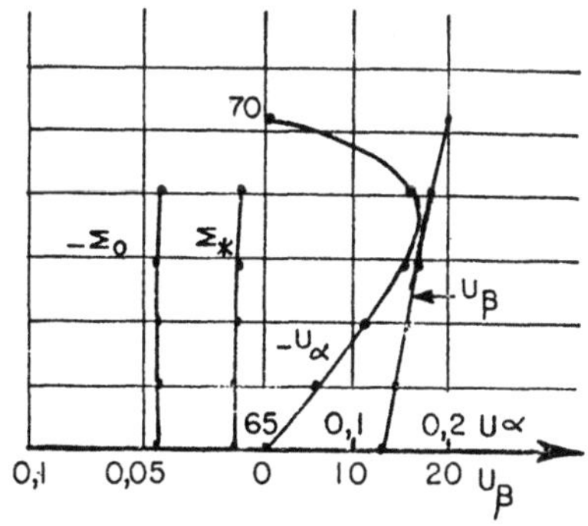

Fig 3 Displacem. and stresses

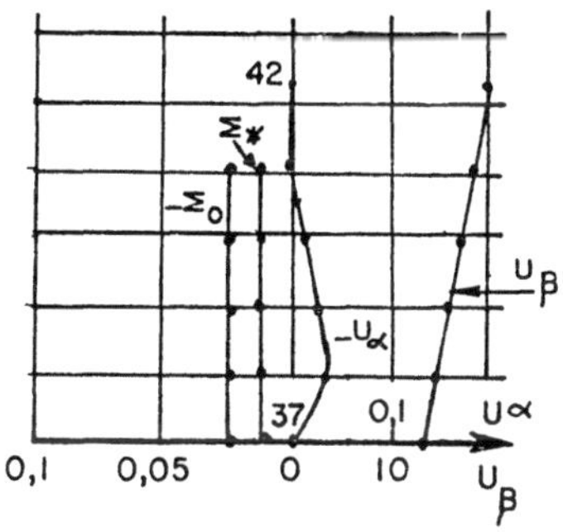

Fig 4 Displacem. and stresses

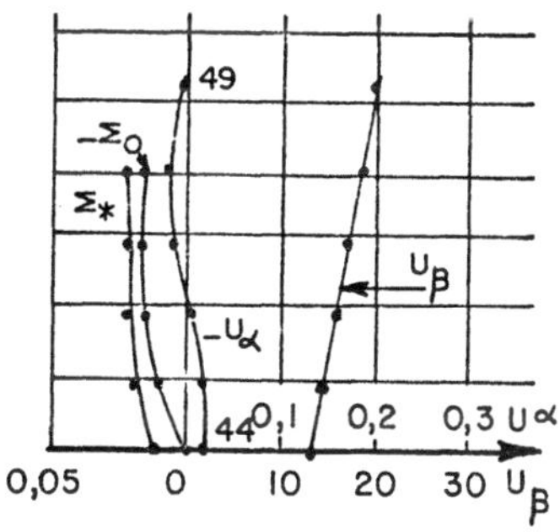

Fig 5 Displacem. and stresses

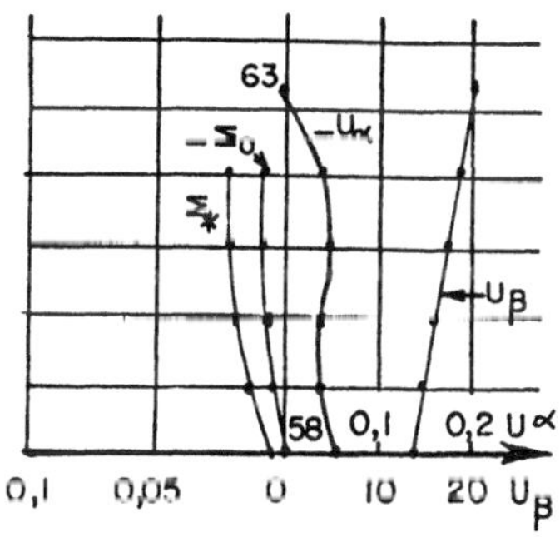

Fig 6 Displacem. and stresses

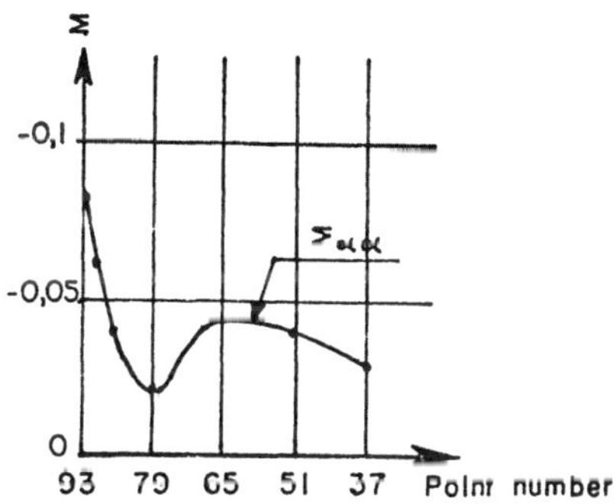

Fig 7 Normal stress along contact area

Characteristics of Iceberg Mechanics

R.T. DEMPSTER,
Dean,
Faculty of Engineering & Applied Science,
Memorial University of Newfoundland,
St. John's, Newfoundland, Canada.

SUMMARY

Data on iceberg size characteristics, iceberg towing and iceberg drift from experiments conducted offshore Newfoundland are presented and discussed. The general outline of operational procedures which can be used by oil rig operators to avoid iceberg collision is described.

Introduction

In 1970, a group of four petroleum companies, Amoco, Imperial Oil, Mobile, and Canadian Superior, contributed $229,000 to fund an iceberg dynamics study conducted by Memorial University. The primary objective of the study was to develop operational techniques for towing icebergs and to gain information on iceberg drift which would lead to a detailed drift prediction model. In three subsequent years, Memorial University continued the drift studies on its own initiative.

A rig operator working in iceberg infested waters requires advance warning of possible collision with icebergs, so that appropriate action can be taken, either to tow the iceberg until it no longer constitutes a threat, or to remove the rig from the drilling location.

The danger zone around a rig is specified by the amount of time required by the operator to take action. In most situations, between 10 and 24 hours advance warning of a possible collision is required. An iceberg moving at a typical speed of 0.5 knots would, therefore, be considered a danger at distances of between 20 and 50 miles upstream of the rig. These length and time scales are important because they define in general terms the nature and accuracy of the model or models required for drift and towing prediction.

The International Ice Patrol[1] has a functioning model for drift prediction which is based on dynamic, as opposed to kinematic principles. The accuracy of their prediction is limited by the accuracy of the input variables. At the present time, the precise effects of currents and wind on an iceberg cannot be computed because currents in the region of the iceberg are usually not known, and neither is the iceberg's shape and drag coefficient.

The author uses a hybrid dynamic/kinematic predictor which is similar in some respects to the Ice Patrol model. Actual iceberg drift data, which can be obtained using radar mounted on an offshore rig or platform, is used as input for the combined effect of currents, wind, and Coriolis effect on the iceberg, thus precluding the need for computation of these effects. This approach is adequate for prediction in the vicinity of an oil rig, but is not suitable for prediction of drift over periods of several weeks.

Clearly, there is a need for the large-scale prediction model of the Ice Patrol, which can give a general indication of the number of icebergs that can be expected in the vicinity of an oil rig, and the author's model which can be finely tuned to give accurate detailed predictions for the danger zone around the rig.

Iceberg Size and Shape Characteristics

Most studies of icebergs, whether related to drift, towing, or scour, require a knowledge of iceberg sizes and shapes. Although essential, these data are sparse in even rudimentary form, and almost non-existent in detailed form.

The International Ice Patrol of the U.S. Coast Guard has developed five descriptive categories for icebergs, namely, Blocky, Drydock, Dome, Pinnacled, and Tabular, and a complementary scaling system based on the length and height above water of the icebergs. Measurements of the underwater draft of icebergs by the Ice Patrol have produced a mean value of 1 to 2.6 for the ratio height above water to draft.

More recently, Benedict[2] has profiled the underwater part of a number of North Atlantic and Arctic icebergs using a modified side-scan sonar. He was able to profile only one or two sections of each iceberg studied because statistical data on iceberg drafts were the primary objective of the study, and shape a secondary consideration. Blenkarn and Knapp[3] developed a simple technique for determining the size of icebergs. Using measurements of iceberg length (L), width (W), and height above water (H), they computed volume above water from L x W x H multiplied by a shape factor S which they generated from examination of numerous photographs of icebergs. The shape factor generally lies between 0.07 and 0.35, but can range higher. A specific gravity of 0.9 for ice was chosen from the general literature and used in conjunction with the above-water volume to compute the mass of the iceberg.

During the Memorial University towing experiment, airborne stereo photographs were taken of selected icebergs to provide a means of estimating iceberg size and shape. Contour maps of the icebergs, such as are shown in Figure 1, were produced, and from these the iceberg characteristics were computed. While the method proved effective, the results were not immediately available during the field experiment. A second method using shipboard photography also proved effective and produced results relatively quickly. The length, width and maximum height of the iceberg were measured, using a sextant and radar ranging while the ship was at some distance from the iceberg. Several more sextant readings of height and width at closing ranges were made as the ship sailed toward the iceberg to check the initial long distance measurements. The iceberg was circumnavigated and polaroid photographs taken of several views of the iceberg, using a fixed camera mounted 41 feet above sea-level on the bridge of the ship. As each photograph was taken, the direction from the camera to a prominent point or feature on the iceberg was determined, using the ship's siting compass, and recorded on the photograph.

It is possible to produce an accurate plan view of an iceberg and an estimate of the above-water volume from these field measurements by a projection technique which uses the compass directions to prominent

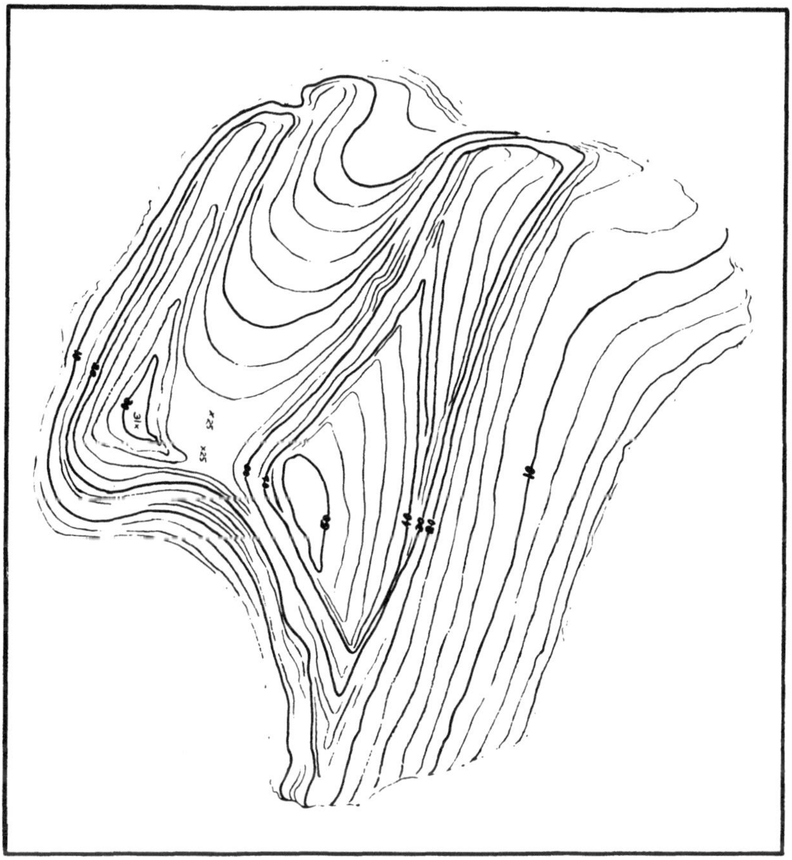

FIGURE 1. Iceberg Contours.

features of the iceberg, as projection lines, and a knowledge of the focal length of the camera. With practice, the method produces relatively good results. Comparisons of results from airborne stereo photography and the shipborne technique were remarkably good.

The dimensional data for the icebergs studied are shown in Figure 2 as the waterline area plotted against the volume above water to the power 2/3. Relationships for geometric shapes which are considered appropriate descriptors of the icebergs are superimposed on the field data. Points A, B, C, and D correspond respectively to the field data for four icebergs towed during the study and described in this paper. No definitive statement can be made on the basis of Figure 2 alone regarding which geometric shape best fits the experiental data. The equivalent cube or sphere which best fits the measured parameters of the iceberg gives approximately the same value for the projected frontal area, which is the unknown quantity required. In the case of Iceberg B, the difference in the area values for a sphere and a cube is 13%, which is not significant in view of the accuracy associated with the field measuring technique.

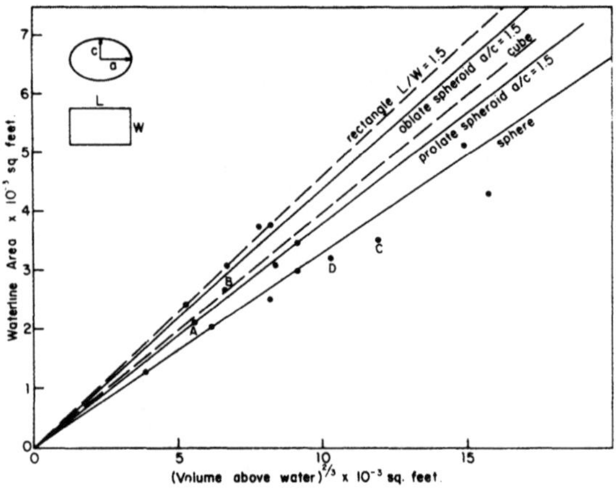

FIGURE 2 Dimensional characteristics

For the towing experiment, a sphere, which best fitted the values of waterline area and above-water volume, was used to compute the projected frontal area of the icebergs for drag calculations.

There exists a unique relationship between the above-water and below-water dimensions of a spherical iceberg having the same specific gravity as glacier ice. Values for the specific gravity of ice, in the general literature, vary from a low of 0.85 to a high of 0.9. Samples of ice taken from different icebergs in the towing experiment produced a value of 0.88, or a specific weight of 56 lbs/ft^3. A simple investigation of these samples also revealed that the tiny air bubbles in the ice were separated by approximately 2-1/2 to 5 bubble diameters, which confirms the experimental value of the specific gravity. Icebergs, therefore, float with one eight of their total volume above water. A spherical iceberg would have a projected frontal area equal to 1.22 times the waterline area and a maximum height-to-draft ratio of 1 to 3.5 in comparison with the measured mean value of 1 to 2.6 quoted by the International Ice Patrol.

Studies of calving mechanisms in glaciers can offer useful information on the initial size and shapes of icebergs, and also indicate the most probable upper bound on size. Reddy et al[4] are examining this problem, using a viscoelastic finite element analysis of thick ice plates.

Iceberg Towing

The techniques and operational procedures used during the Memorial University towing experiment have been described in detail by Bruneau and Dempster[5].

Iceberg motion under tow and natural drift conditions were recorded on film and on graphic plots, using a Kelvin-Hughes Photoplot radar system. The display presented North-up, regardless of the ship's heading, and a digital computer in the system automatically added together echoes from moving targets to form tracks from which course and speed could be measured. The display was projected onto a 24-inch viewing screen for plotting purposes.

A continuous record of the towing force was produced from the output of a calibrated load cell in the single towing line, and the tow line bearing was measured periodically using a sighting compass.

Measurements of comparable iceberg drift in the vicinity of the tow, and of the iceberg after release from the tow, provided the data for determination of the combined effect of current, wind, and related Coriolis on the towed candidate. Wind effects were small during the experiments because winds were never more than 20 knots, nor did they persist for any significant length of time. Current profiles were also measured from the ship while at anchor, but were several hours before or after towing experiments. Attempts to measure relative motion while the ship was towing proved to be impractical. Bottom anchored current meter arrays were not employed for a number of practical reasons.

The field data were analysed, using a mix of dynamic and kinematic considerations. For instance, iceberg motion in a stationary fluid field, while under the action of the towing force, was computed, using an iterative technique to solve the equation

$$M \frac{dv}{dt} = F_{tow} - (\tfrac{1}{2} C_d \rho A V^2)_{drag} \qquad (1)$$

The Coriolis motion perpendicular to the motion in the direction of towing was computed in a similar manner, and added vectorially to produce a resultant motion. Finally, the combined effect of currents, wind and implicit Coriolis was added in kinematic form, using the measured iceberg drift data.

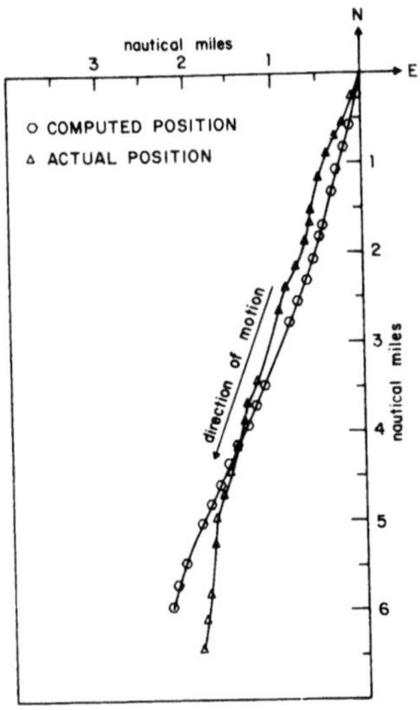

FIGURE 3 Tow path Iceberg A.

All of the parameters in the analysis, with the exception of the drag coefficient C_d, are based upon field measurements. C_d, on the other hand, has to be assumed, and is normally taken as 0.5, 0.6 or 0.7. While these values produced the correct trends of motion, a higher value of 2.0 achieved the best fit between the computed and actual data. The results with C_d=2 are shown in Figures 3, 4, 5, and 6.

The implications of the high C_d value should be considered with caution. It is probable that the coefficient in this analysis is a composite of a drag coefficient and a correction factor to compensate for errors in the estimation of the system parameters. Obviously, there is a need for data on iceberg drag coefficients and underwater shapes.

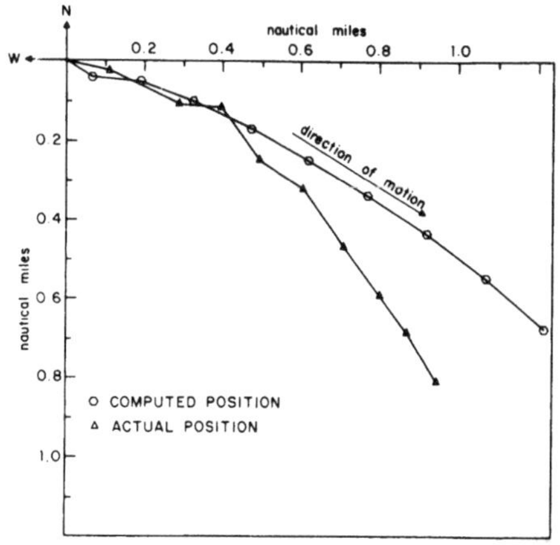

FIGURE 4 Tow path Iceberg B

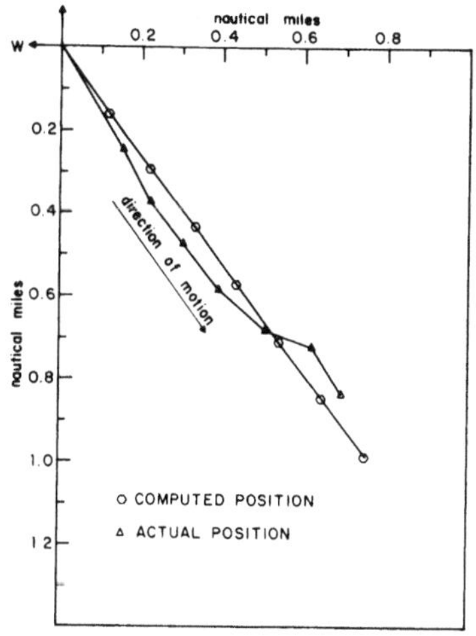

FIGURE 5 Tow path Iceberg C

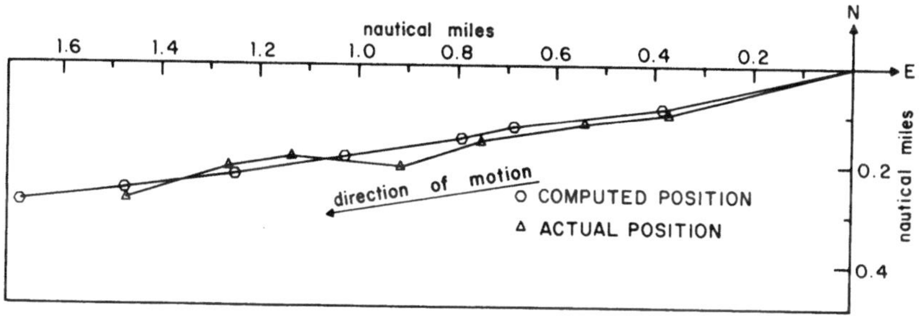

FIGURE 6. Tow path Iceberg D.

ICEBERG CODE	MASS (TONS)	PROJECTED FRONTAL AREA (SQ. FT.)	TOWING FORCE (LBS.)	TOWING DIRECTION* DEG. NORTH	TOWING PERIOD (HRS.)	CURRENT AND WIND EFFECT (KNOTS/DEGREES)	COEFF.
A	99,700	23,700	26,500	200°N	~7	0.7/187°N	2.0
B	132,300	28,400	25,000	60°N	~2	0.4/177°N	2.0
C	277,000	46,500	26,500	330°N	~2	0.65/150°N	2.0
D	226,000	41,100	28,500	320°N	~2.5	0.65/230°N	2.0

* Gross indication of tow direction because of significant variations

Field measurements on the scale of the iceberg towing experiment are difficult to achieve with great accuracy. The potential for error is large, owing to the present state of the art, the large masses of ice involved, and the small changes in velocity which are significantly large percentages of small absolute velocities. It is remarkable, in view of these difficulties, that such good agreement between the computed and the measured positions of towed icebergs can be obtained by adjustment of a single coefficient.

In general, decreasing the coefficient C_d caused a lengthening of computed iceberg path and a rotation of the path, either clockwise or anticlockwise, depending upon the relationship of the towing force direction and the current/wind/Coriolis vector. Equivalent percentage changes in the projected frontal area produced similar effects on the computed path. Significant changes in the iceberg mass, of the order of 20 percent, hardly affected the computations. Finally, neglecting the Coriolos effect from the computations caused an anticlockwise rotation of the path by a few degrees. The Coriolis effect is not significant for the sizes of the towing forces and towing periods involved. However, it should be considered in any analysis for scientific correctness.

The difference between the computed and actual positions of the towed icebergs in Figures 3, 4, 5, and 6 increases with time in some instances more than others. This is understandable because the predicted positions rely upon a step-by-step process in which errors have a tendency to accumulate, rather than cancel. Where input quantities are relatively constant, as in the case of Iceberg 4, the predicted motion and the actual motion agree extremely well. However, in the case of Iceberg B, the towing force direction varied appreciably, at an approximate rate of 24° per hour, and the iceberg motion relative to the average current direction also changed appreciably during the tow. Under such conditions, the potential for error is great, because angular measurements are particularly difficult to make accurately. A better fit of the data in Figure 4 could have been achieved with a C_d of 2.5 instead of 2.0.

On the basis of towing experiments, the following procedure is suggested for rig operators:

1) As soon as an iceberg is detected, its drift path should be plotted at one-half-hourly intervals to determine speed and direction of drift.

2) If the iceberg appears to be a hazard, it should be surveyed to determine its size. Measurements of waterline area and above-water volume, in conjunction with a geometric model chosen because it best describes the shape of the iceberg, can be used to obtain a value for the projected frontal area of the iceberg.

3) A series of predicted paths for the iceberg can be computed, using as inputs:

   i) the drift data from Step 1 to account for the combined effect of current, wind, and Coriolis;

   ii) various towing forces and directions within the capability of the towing vessel;

   iii) an assumed value for the drag coefficient of the iceberg in the range of 0.5 to 2.0.

   The towing force and direction which give the greatest degree of safety should obviously be chosen.

4) Within one hour after commencement of the tow, depending upon field conditions, it should be possible to obtain a corrected value for the coefficient of C_d by matching the computed motion of the iceberg with the actual motion.

5) Adjustments can be made throughout the tow to the tow force and direction to achieve the desired results.

Towing can be, and has been, done without prior estimation of the iceberg's path, but the technique suggested here will increase the confidence of the operator, reduce towing time, and increase safety.

Iceberg Drift

The lore of iceberg drift abounds with myths and conflicting commentary because existing field data is so incomplete. While significant data on iceberg drift have been generated in recent times for particular geographical locations, essential complementary data on currents and meteorological conditions, iceberg shape and size, are either too sketchy to be of any use, or are non-existent.

The author has measured detailed drift of icebergs during the towing experiments in 1970, and for three successive years, 1972, 1973, and 1974, using a shore-based radar at Saglek, Labrador. In 1972, data on currents and meteorological conditions were recorded, but, unfortunately, no data on the underwater shape of the icebergs. Figure 7 illustrates the general drift pattern over a period of several days for the same area, and points A, B, C & D indicate the locations of four strings of current meters.

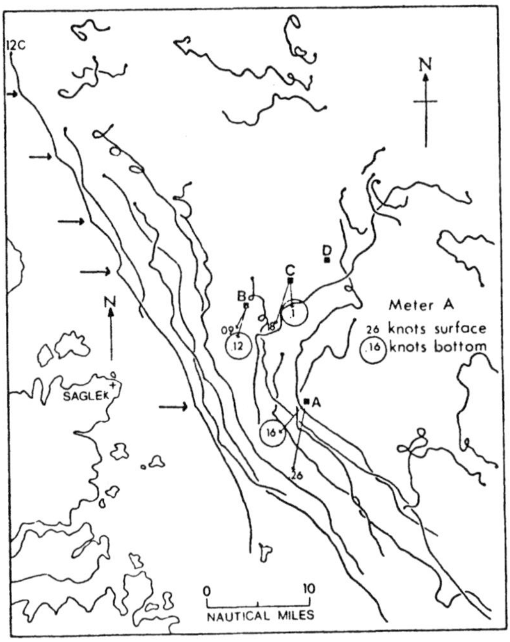

FIGURE 7. Iceberg drift pattern.

In general, iceberg drift correlates well with the current structure of an area. Wind does not appreciably affect iceberg drift unless the currents are weak, or the wind is strong for long durations. The terminology here is descriptive because the various effects are relative. Winds of 10 to 15 knots have negligible effect on medium to large icebergs in a current of 0.5 knots, but significant effect on the same icebergs in a current of 0.1 knots, or less. Also, the relative effect of translatory currents and rotary currents is illustrated in Figure 8. When the ratio, R, of rotary current to translatory current is less than unity, the iceberg path tends to be sinusoidal, or even straight, and when R is greater than unity, the drift path tends to a series of loops.

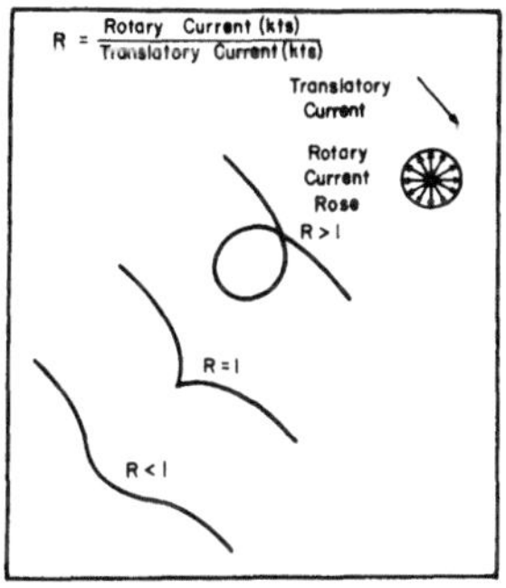

FIGURE 8 Current effects on drift

The drift pattern in Figure 7 exhibits these different paths. Iceberg 12C moved along a sinusoidal path because the inshore current was much stronger than the tidal currents. The time intervals between the points indicated by arrows on Path 12C correspond directly with the semi-diurnal tide, hence the variations in distance between the arrows show that the translatory current also varied along the coast line. Icebergs at distances greater than 20 miles from the coast were in a weak current field, and so drift paths are looped and irregular. In fact, the icebergs drifted northwards in a weak eddy structure. The magnitude of tidal currents in the upper layers of the ocean, as measured by the current meter arrays, was on average 0.3 knots. This would cause a particle of fluid to move under ideal conditions around the perimeter of a circle of approximately one nautical mile in diameter. It is interesting to note that the loops in the iceberg drift paths have length scales of this order.

The response time of icebergs to variations in environmental conditions is an important consideration in drift prediction. While it is difficult to quantify this parameter from existing iceberg data, it would appear that they follow very closely the variations of current in time and space. Calculations for hypothetical spherical icebergs with drag coefficients of 0.3 under the action of a 50,000 lbs. towing force result in response times of 6 minutes for an iceberg of 120,000 tons, 9 minutes for 400,000 tons, and 18 to 20 minutes for 1 million tons. All of the icebergs studied in the Memorial towing experiment returned to their free drift condition within an estimated 20 to 30 minutes. The nature of the tracking method did not lend itself to a more accurate estimation of response times. Figure 9 illustrates, for example, the towed path and the drift path after release from towing.

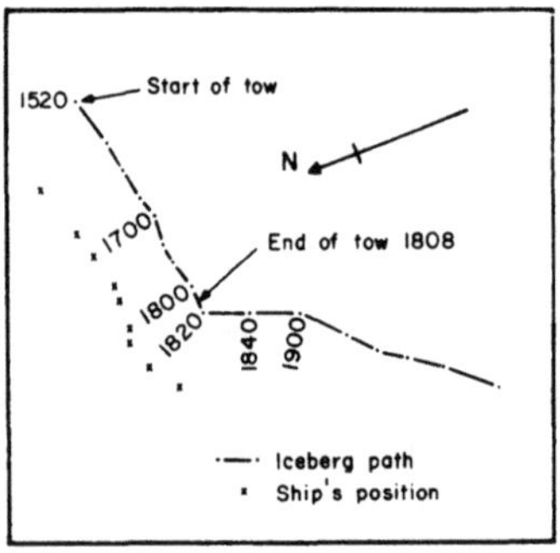

FIGURE 9. After-tow drift Iceberg D

It is the author's belief that offshore rig operators can develop satisfactory predictions of iceberg drift in their specific region from a knowledge of the local current structure, combined with data from current meters at the rig, detailed iceberg tracking data using radar, and data from any towing operations undertaken. The method involves a step-by-step approach which successively improves the accuracy of the model. While this may offend scientific principles, it makes good sense from an operational point of view when the state of the art is in its early stages.

REFERENCES

1. Mountain, D.G.: On Prediction of Iceberg Drift. Iceberg Dynamics Symposium, June 1979, St. John's, Newfoundland.
2. Benedict, C.P.: Icebergs & Canadian East Coast Exploration. ASME publication 76 - Pet - 103.
3. Blenkarn, K.A.; Knapp, A.E.: Ice Conditions on the Grand Banks. CIMM Ice Seminar, May 1968, Calgary, Alberta.
4. Reddy, D.V.; Bobby, W.; Arockiasamy, M.: Viscoelastic Analysis of Calving Glaciers. Paper in preparation for the 2nd Conference on Use of Icebergs, Cambridge, England.
5. Bruneau, A.A.; Dempster, R.T.: Iceberg Dynamics Project, E.P.O.A. Report, Project No.2, Esso Resources, Ltd., Calgary, Alberta.

Constitutive Relations for the Non Elastic Deformation of Polycrystalline Ice

H. LE GAC and P. DUVAL

Laboratoire de Glaciologie - CNRS

2, rue Très-Cloîtres

38031 Grenoble Cedex France

Summary

Constitutive relations for the non-elastic deformation of polycrystalline ice are proposed. Two types of hardening are introduced (isotropic and kinematic hardening). These case variables evolve by defined laws and are affected by two counteracting processes : strainhardening and softening. All the transient behaviour is described by this model. Quantitative predictions are compared with the corresponding experimental data for creep recovery tests.

Introduction

Constitutive equations for the non-elastic behaviour of polycrystalline ice are of much interest for glacier studies and ice engineering. A large amount of work has been done on the mechanical properties of polycrystalline ice. Creep tests were the most frequently performed [1] [2]. For permanent creep, there was established a relationship between strain rates $\dot{\epsilon}_{ij}$ and deviatoric stresses τ'_{ij} given by :

$$\dot{\epsilon}_{ij} = \frac{\tau'_{ij}}{2\eta} = \frac{B}{2} \tau^{n-1} \tau'_{ij} \qquad (1)$$

where η is the viscosity , B a parameter which depends principally on temperature and τ the effective shear stress defined by $\tau^2 = 1/2 \sum_{ij} (\tau'_{ij})^2$.

The equation (1) has been verified by Duval [3] with creep tests performed in torsion, compression and torsion-compression. In the range of stresses studied (0.5 to 5 bars), the exponant n was about three. For temperatures near melting point, a steady state is obtained after the development of non-random fabrics for strains greater than 2-3 %. The increase of strain rate after the so-called secondary creep results principally from the formation of fabrics and from recrystallization processes [4] .

The equation (1) is generally not verified for the time dependent deformation. Thus the tensors $\dot{\epsilon}_{ij}$ and τ'_{ij} are not proportional after an increase of load [3]. More, when unloaded, ice exhibits creep recovery. The anelastic strain (time dependent component of recoverable deformation after release of the stress) can be 10 times the elastic strain [5] .

A realistic deformation model should predict the material behaviour for non steady state. An approach to this problem was done by different authors [6] [7] for metals. Relationships were derived in terms of specific state variables which have a physical interpretation.

The purpose of this paper is the presentation of a deformation model for polycrystalline ice under stresses for which the dislocation creep mechanism is dominant, in order to arrive at stress strain relations in a manageable form.

Activation energy for the creep of polycrystalline ice

From Barnes and others [8] , the apparent activation energy of secondary creep was about 78 kcal/mole below -10°C, but was much higher above -10°C. These authors suggested that pressure melting between grains was responsible for this rise in apparent activation energy . For monocrystals, the activation energy was found be constant through the temperature range -20 to -0.2°C by Jones and Brunet [9]. Thus, it was concluded in favour of grainboundary effects above -10°C.

It appears that activation energies for creep in monocrystals and polycrystals are equal below -10°C [10]. From many authors [8] [10] [11], there is thus a strong indication that the rate limiting mechanism is the glide of dislocations across the basal plane. Duval [3] proposed a deformation model based upon the ideas suggested by Bailey [12] and Orowan [13]. In this model, steady state creep is a direct consequence of the competing effects of strainhardening and recovery. The strainrate limiting mechanism is the recovery processes. By studying the effect of temperature upon transient and steady state creep, Duval [5] showed that the recovery controlled creep ought to apply to the deformation of polycrystalline ice. From Mc Lean [14], strain rate $\dot{\varepsilon}$ can be written as :

$$\dot{\varepsilon} = \dot{\varepsilon}_0 \exp - v \frac{(h\varepsilon - rt)}{kT} \qquad (2)$$

where $\dot{\varepsilon}_0$ is strain rate at time $t = 0$, h the workhardening coefficient, r the recovery rate, v the activation volume, k the Boltzman constant and T the Kelvin temperature. For steady state flow, we have :

$$\dot{\varepsilon} = \frac{r}{h} \qquad (3)$$

The coefficient h is very little temperature dependent [5]. So the activation energy for steady state creep corresponds to the variation of recovery rate with temperature.

The deformation model

The constitutive equations that we shall describe are relations only for the non-elastic strains. They should describe all the transient strain behaviour. The deformation model is developed for a multiaxial stress state. The equations are formulated for the case of isotropy and small deformation. The type of anisotropy which depends directly on the deformation history of specimen (ex : Bauschinger effect) is described in the model.

1 - The constitutive equations

We want to model two types of hardening. In one type (isotropic hardening), the strength of the material remains equal in all directions. Isotropic hardening is represented by the Scalar state variable σ^*. This variable is affected by two counteracting processes : strainhardening and softening. Physcial mechanisms which might contribute to this type of hardening include principally dislocation tangles and subgrains.

The second type of hardening is directional and is represented by a tensor state variable S_{ij}. This variable is substracted from the applied stress deviator σ'_{ij} and is also affected by strainhardening and softening.

The constitutive equations are relationships among the current values of the applied stress deviator σ'_{ij}, the non-elastic strain rate $\dot{\epsilon}_{ij}$, the variables σ^* and S_{ij} and the time rates of change of σ^* and S_{ij}.

The flow equations are given by :

$$\dot{\epsilon}_{ij} = \frac{\dot{\gamma}}{2\tau^*} (\sigma'_{ij} - S_{ij}) \tag{4}$$

where $\dot{\gamma}$ is the effective shear strain rate defined by :

$\left(\frac{\dot{\gamma}}{2}\right)^2 = \frac{1}{2} \sum_{ij} (\dot{\epsilon}_{ij})^2$, τ^* is an effective shear stress defined by : $\tau^{*2} = 1/2 \sum_{ij} (\sigma'_{ij} - S_{ij})^2$.

In the range of stresses for which the equation (1) is verified, $\dot{\gamma}$ may be written in the following form :

$$\dot{\gamma} = B'(\tau^* - \sigma^*)^n$$

So, according to (4), we have :

$$\begin{cases} \dot{\epsilon}_{ij} = \dfrac{B'}{2} \dfrac{(\tau^* - \sigma^*)^n}{\tau^*} (\sigma'_{ij} - S_{ij}) \\ \dot{\epsilon}_{ij} = 0 \text{ if } \tau^* < \sigma^* \end{cases} \tag{5}$$

The rates of change of S_{ij} and σ^* are given by :

$$\frac{dS_{ij}}{dt} = 2\,h\,\dot{\epsilon}_{ij} - r_{ij} \qquad (6)$$

and $$\frac{d\sigma^*}{dt} = h'\,\dot{\gamma} - r \qquad (6\,\text{bis})$$

where the recovery rates can be described by :

$$r_{ij} = \alpha\, S_{ij}\,(S)^{m-1} \quad \text{with } S^2 = \frac{1}{2}\sum_{ij}(S_{ij})^2$$

and $$r = \alpha^*\,\sigma^{*\,m'}$$

By assuming that the two hardening coefficients h and h' are independent on stresses, we have m = m' = n. For steady state, it can be assumed that :

$$(S_{ij})_s = k\sigma'_{ij} \quad \text{with } k < 1$$

and $$(\sigma^*)_s = k'\,\tau^* \text{ with } k' < 1$$

The steady state strain rates are :

$$\dot{\epsilon}_{ij} = \frac{B'}{2}\,\tau^{n-1}\,[1-k]^n\,[1-k']^n\,\sigma'_{ij} \qquad (7)$$

By comparing the equations (1) and (7) we have

$$B = B'\,[1-k]^n\,[1-k']^n \qquad (8)$$

The equations show that a negative strain rate can be obtained when S'_{ij} is higher than σ'_{ij}. Thus, the recovery creep after load removal is treated by the model.

2 Application of the model : creep curves after partial and total load removal

The model was applied to creep tests in torsion for polycrystalline ice.

In torsion and by taking n = 3, the equation become :

$$\begin{cases} \dot{\epsilon}_{\theta z} = \dfrac{B'}{2} \dfrac{[|\tau - S_{\theta z}| - \sigma^*]^3 [\tau - S_{\theta z}]}{|\tau - S_{\theta z}|} \\ \dot{\epsilon}_{\theta z} = 0 \text{ if } |\tau - S_{\theta z}| < \sigma^* \end{cases} \tag{9}$$

$$\frac{dS_{\theta z}}{dt} = 2\,h\,\dot{\epsilon}_{\theta z} - \alpha\,(S_{\theta z})^3 \tag{10}$$

$$\frac{d\sigma^*}{dt} = 2\,h'\,|\dot{\epsilon}_{\theta z}| - \alpha^*\,(\sigma^*)^3 \tag{11}$$

Creep curves after load removal were calculated by supposing that for times t = 0, the steady state creep was attained. The internal stress $S_{\theta z}$ (t = 0) can be obtained experimentally by determining the smaller stress decrease in order to observe negative creep. The workhardening coefficient h was defined by Duval [5] as an anelastic modulus and had values about 3,000 bars. Figure 1 shows the experimental and computed creep curves after load removal. Figure 2 shows the corresponding variations with time of $S_{\theta z}$ and σ^*. The agreement is good with the following values of the parameters ; $S_{\theta z}$ (t = 0) = 1.55 bar, σ^*(t = 0) = 0.2 bar $B = 0.1\ \text{bar}^{-3}\ \text{an}^{-1}$, h = 3,000 bars and h' = 30,000 bars.

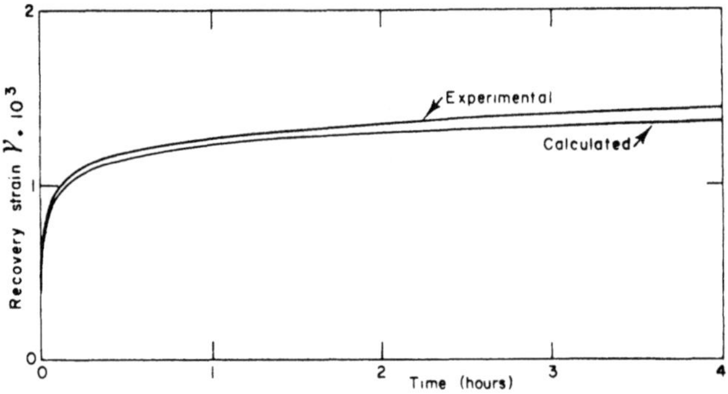

Fig.1 -Anelastic recovery strain as a function of time. Shear stress before unloading : 2 bar. Temperature T = -1.0°C

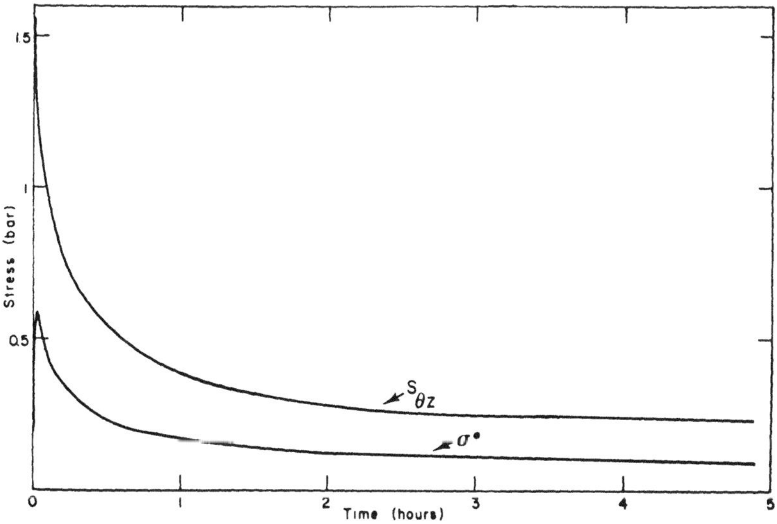

Fig. 2 - Variations with time of internal stresses σ^* and $S_{\theta z}$ after unloading. Shear stress before unloading : 2 bar.

Figure 3 illustrates the behaviour of creep curves following various stress drops with the values of parameters given above. For a small stress decrease, a zero creep rate period is obtained. This result is in accordance with the experimental results obtained by Duval [3].

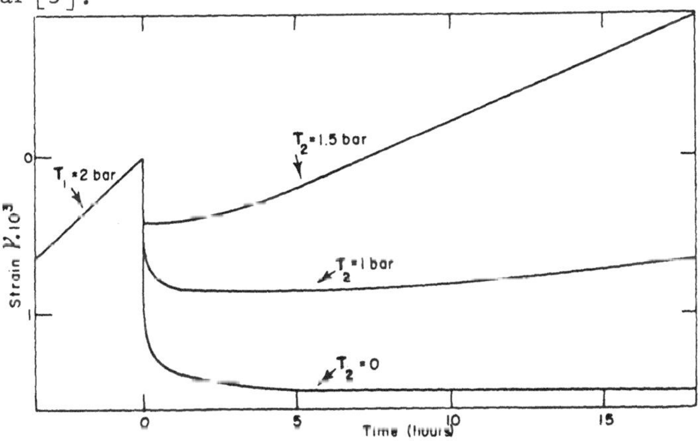

Fig. 3 - Calculated creep curves after various stress drops. Temperature T = -1.0°C.

Discussion

The preliminary tests described above do not permit to conclude in favour of the entire validity of the model. It must be tested in various tests and specially for multiaxial stress state.
A mechanical model of creep of polycrystalline ice has been previously developed by Michel [15]. It is based on the principle of dislocation multiplication and immobilization. It differs from our theory in several ways. For example, the activation energy for creep is the one for the velocity of glide of dislocations, whereas, in our model, the activation energy for steady state creep corresponds to the one of recovery rate.

References

1 Glen, J.W. : The creep of polycrystalline ice. Proc. Royal Society, Ser.A, 228 (1955) 519-538.

2 Steinemann, S. : Experimentelle Untersuchungen zür Plastizität von Eis. Beiträge zür Geologie der Schweiz, Hydrologie, 10 (1958).

3 Duval, P. : Lois du fluage transitoire ou permanent de la glace polycristalline pour divers états de contrainte. Annales de Géophysique, 32 (1976), 335-350.

4 Duval, P. : Creep and recrystallization of polycrystalline ice. Bull. Mineral. 102 (1979) 80-85.

5 Duval, P. : Anelastic behaviour of polycrystalline ice. Journal of Glaciology, 21 (1978) 621-628.

6 Miller, A. : An inelastic constitutive model for monotonic, cyclic, and creep deformation : Part I - Equations development and analytical procedures. Journal Engineering Materials and Technology, 96 (1976) 97-104.

7 Hart, E.W. : Constitutive relations for the non elastic deformation of metals. Journal Engineering Materials and Technology, 96 (1976) 193-202.

8 Barnes, P. ; Tabor, D.; Walker, J.C.F. : The friction and creep of polycrystalline ice. Proc. Royal Society, Ser.A, 324 (1971) 127-155.

9 Jones, S.J. ; Brunet, J.G. : Deformation of ice single crystals close to the melting point. Journal of glaciology. 21 (1978), 445-455.

10 Homer, D.R. ; Glen, J.W. The creep activation energies of ice. Journal of Glaciology. 21 (1978) 429-444.

11 Goodman, D.J. : Creep and fracture of ice and surface strain measurements on glaciers and sea ice. Ph.D. Thesis, University of Cambridge 1977.

12 Bailey, R.N. : Note on the softening of strain hardening metals and its relation to creep. Journal of Institute of Metals. 35, (1926) 27-40.

13 Orowan, E. : The creep of metals. Journal of West Scot. Iron and Steel Institute. 54 (1946) 45-53.

14 Mc Lean, D. : The physics of high temperature creep in metals. Reports on Progr. in Physcis. 29 (1966) 1-33.

15 Michel, B. : A mechanical model of creep of polycrystalline ice. Canadian Geotechnical Journal, 15 (1978) 155-170.

Modelling the Interaction between Ice and Ships

Roderick Y. Edwards, Jr.
President
Arctec Canada Limited
Ottawa, Canada

ABSTRACT

Physical modelling of the steady motion of a ship through level, unbroken ice has developed to the point where experiments can be carried out routinely. Nonetheless, there are in the literature very few illustrations of the degree to which estimates of performance derived from model tests in simulated ice correlate with full scale performance for the corresponding prototype.

This paper presents results of extensive level ice resistance model experiments at 1/48th and 1/36th scale for CCGS LOUIS S. ST. LAURENT in conjunction with full scale transitting performance of that ship in Goose Bay. Model test results at 1/36th scale for CCGS NORMAN McLEOD ROGERS are presented and compared with limited full scale data in the St. Lawrence River. Model test results for USCGC MACKINAW are presented with corresponding full scale results.

In the cases presented, model/full scale correlation is very good despite difficulties in correctly scaling down troublesome ice properties. Continued acquisition of full scale data and research into improving model ice properties is urged.

INTRODUCTION

Scale modelling is a cost effective method of developing estimates of performance of candidate systems under conditions too complex for mathematical models to provide a complete analysis.

The interaction between ships and ice features is a "messy" process and cannot, at present, be completely and accurately modelled mathematically. Certainly, as our understanding of principal mechanisms of ice failure and processing, etc., becomes better and better, math models will become more and more accurate. In the meantime, small scale physical modelling of icebreaking phenomena is a fruitful approach, provided model/full scale correlation can be established.

For the present, physical modelling techniques have been developed for the more elementary ship-ice interactions. Modelling steady motion of a ship through level ice has received the most attention and is most highly developed. Modelling of one year ridge penetration [8], maneuverability in level ice [2], and propulsion system performance are under intense development by various ice testing facilities.

SCALING LAWS

The laws of similitude for modelling the ship's interaction with ice have been covered at length in the literature. The requirements are tabulated below:

$\frac{h_p}{h_m}$	=	λ	Thickness
$\frac{v_p}{v_m}$	=	$\lambda^{1/2}$	Speed
$\frac{R_p}{R_m}$	=	λ^3	Force
$\frac{\sigma_{fp}}{\sigma_{fm}}$	=	λ	Bending Strength
$\frac{E_{fp}}{E_{fm}}$	=	λ	Modulus of Elasticity
$\frac{\sigma_{cp}}{\sigma_{cm}}$	=	λ	Crushing Strength
$\frac{f_p}{f_m}$	=	1.0	Coefficient of Friction
$\frac{\rho_p}{\rho_m}$	=	1.0	Ice and Fluid Density

where h is ice thickness
R is resistance or force
σ_f is flexural or bending strength
E_f is flexural modulus of elasticity
σ_c is compressive strength
f is friction factor
λ is geometric scale factor

Subscript p = prototype
Subscript m = model

It should be noted that equality of Reynolds Number is required in addition to the above laws. It is essentially impossible to meet this requirement using "conventional" ice test facilities. As a consequence, any force components which arise from viscous flow will be exaggerated when extrapolated to full scale using the above laws. Most facilities today use ice frozen on the surface of an aqueous solution of some salt, usually NaCl. Synthetic ices are also being developed. All of these materials are imperfect to some extent, i.e. while their thickness and bending strength can be produced at levels 50 to 75 times smaller than full scale values, their compressive strengths, bending strengths, modulus of elasticity and coefficient of friction between them and ship model materials cannot be simultaneously reduced by the same factor except at small scale ratios (very large models).

Two parallel approaches to this problem have been taken: acquisition of full scale data to permit an evaluation of the degree to which scaling imperfections affect predictive capabilities; and continuous devel opment of methods for achieving simultaneous scaling of the "problem parameters" (E, σ_c, f).

FULL SCALE DATA

Mutually good sets of model and full scale data are hard to find. Table 1 lists non-proprietary sources which were readily available.

USCGC MACKINAW

Model and full scale trials were reported in detail by Edwards et al [1] in 1972. However, model full scale correlation was not clear-cut as presented. Going back to basics as it were, an equation was fitted to the raw dimensional model data simply to smooth the data. Only parameters which were actually varied were used, and those in combinations and at powers which were not necessarily physically meaningful but which provided the best fit. The following equation fit the data best:

$$R = 26134.6h^2 + 7.26\sigma h + 684.5hv^{1.5} \quad (1)$$

	FULL SCALE					MODEL SCALE				
Ship Name	No. of Useful Data Points	Variation in Ice Thickness (inches)	Ice Type	Source	Problems	Scale	No. of Useful Data Points	Thickness Variation	Planned Strength Variation	Good Friction ?
USCGC MACKINAW 1977	52	6-24	Fresh	Edwards et al [1]	No friction data acquired on ship	1/48	38	Yes	Yes	One friction estimate
CCGS LOUIS S. ST. LAURENT 1971 & 1972	?	6-70	Sea	Edwards et al [2] Bradford [3]	Ramming - no friction [2] Non-uniform ice [3]	1/36 1/48	165	Yes	Yes	Yes
CCGS PIERRE RADISSON (in process)	100	18-40	Fresh & Sea	Not yet available	Friction measured, no problems, data excellent	1/36	36	Yes	No - model strength too high	No
USCGC STATEN ISLAND	28	10-30	Sea	Lewis & Edwards [13]	No friction data - snow on ice	1/36	35	Yes	No	No

TABLE 1 - NON-PROPRIETARY FULL AND MODEL SCALE DATA SOURCES

where R = model level ice resistance in (N)

h = model thickness (metres)

σ = model bending strength (kPa)

v = model speed (m/sec)

The standard error was 19% of the mean. Equation (1) was converted to full scale by scaling the coefficients appropriately to the dimensions of the full scale equation.

$$R = 1.25h^2 + 0.00035\sigma_f h + 0.0865hv^{1.5} \qquad (2)$$

where R in MN

h in metres

v in m/sec

σ in kPa

Full scale data was extracted from [1] and corrected. In [1] total propeller shaft thrust was reported. Thrust has been reduced to account for thrust deduction data acquired in model tests of the ship in open water.

In Figure 1 the full scale data has been plotted in dimensional form and equation (2) superimposed for average values of full scale data shown in Figure 1. Except at low ship speeds, the model overpredicts the full scale results by 10%, a fairly encouraging result.

CCGS NORMAN McLEOD ROGERS

A 1/36th scale model of this ship was tested in level ice in 1972. The results are reported by Edwards et al [4] 1973. During the winter of 1975 ice impact tests were carried out on the ship in the St. Lawrence Seaway. Continuous icebreaking tests were not possible because the ship was not quite able to proceed at full power in the minimum ice thickness available - 0.45 metres. Penetration runs were long (4-5 ship lengths) but eventually the ship stopped. At the time the propulsion plant was developing 13,200 SHP and producing a total of 0.988 MN of thrust. Model predictions and the full scale observation are plotted in Figure 2.

FIGURE 1
RESISTANCE VERSUS SHIP SPEED

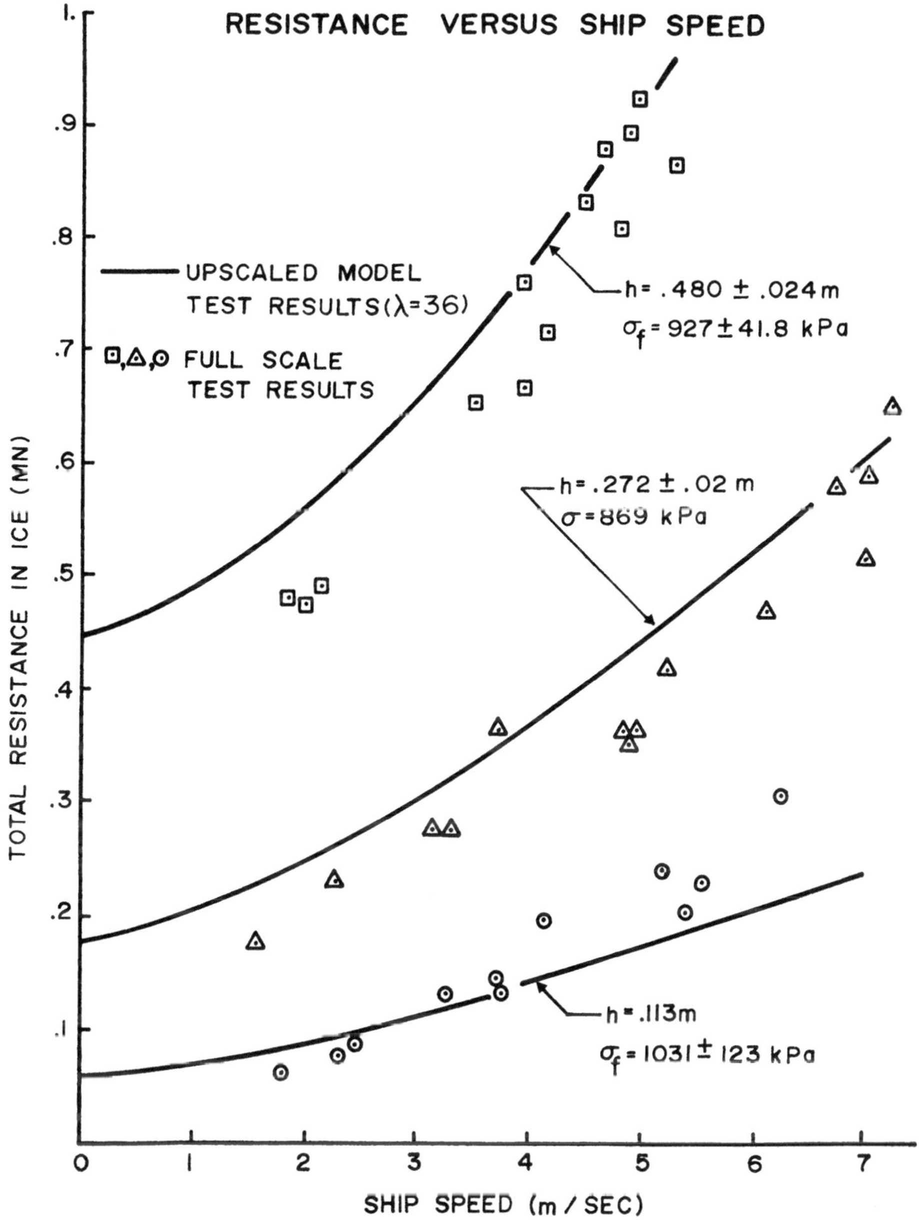

U.S.C.G.C. MACKINAW

FIGURE 2

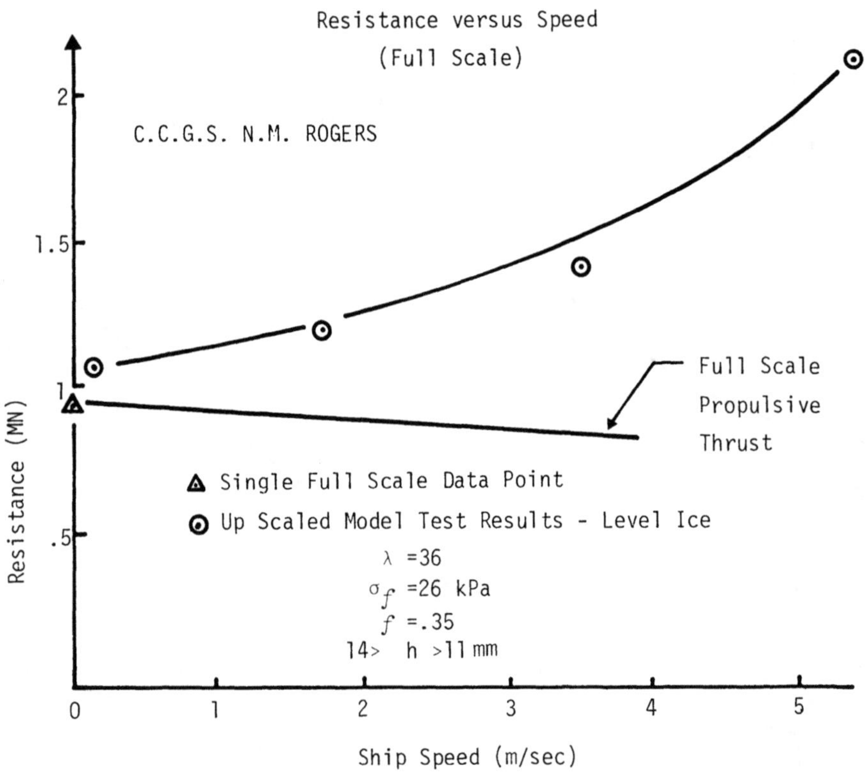
Resistance versus Speed
(Full Scale)
C.C.G.S. N.M. ROGERS
Full Scale Propulsive Thrust
Single Full Scale Data Point
Up Scaled Model Test Results - Level Ice
λ =36
σf =26 kPa
f =.35
14> h >11 mm
Resistance (MN)
Ship Speed (m/sec)
2
1.5
1
.5
0
1
2
3
4
5

CCGS LOUIS S. ST. LAURENT

An extensive ice model test program was carried out on CCGS LOUIS S. ST. LAURENT. The program is described in detail by Edwards et al [1]. Successful continuous mode level ice tests have yet to be conducted on CCGS LOUIS S. ST. LAURENT. In an effort to check model data, we have extracted data from operational tests in close pack ice off Labrador. The observations were reported by Bradford [3]. The observations are plotted in Figure 3. They are predominantly at 100% power (24,000 HP nominally - actually 22,000). Ice conditions are far from uniform. The ice index is the product of ice depth in inches and the local surface coverage in tenths. A one metre thick shorefast ice would have an index of 393.6. Superimposed on this figure are the scaled up results of the aforementioned level ice model tests.

The model level ice tests produce a rough prediction of the performance of the ship.

OVERVIEW

For the tests previously discussed, we see that extrapolation of measurements of model resistance to full scale by the scaling laws presented earlier produces remarkable agreement with limited full scale data. Remarkable because:

- Modulus of elasticity for all of these old model tests were 5 to 20 times lower than the scaled down value for fresh or sea ice ($E/\sigma << 3000$)
- Model ice compressive strength is suspected to be higher than it should be in relation to bending strength. (No serious measurements of the value were made.)
- While sliding friction was measured on all models, friction factors for the full size ships involved were known only approximately.
- The full scale propulsion plant is assumed to produce propelling force as it would in open water without consideration of ice/propeller interaction. (The ships considered were tested in ice which was thin compared to the draft, so this effect is probably small.)

FIGURE 3 - Ship Speed versus Ice Severity

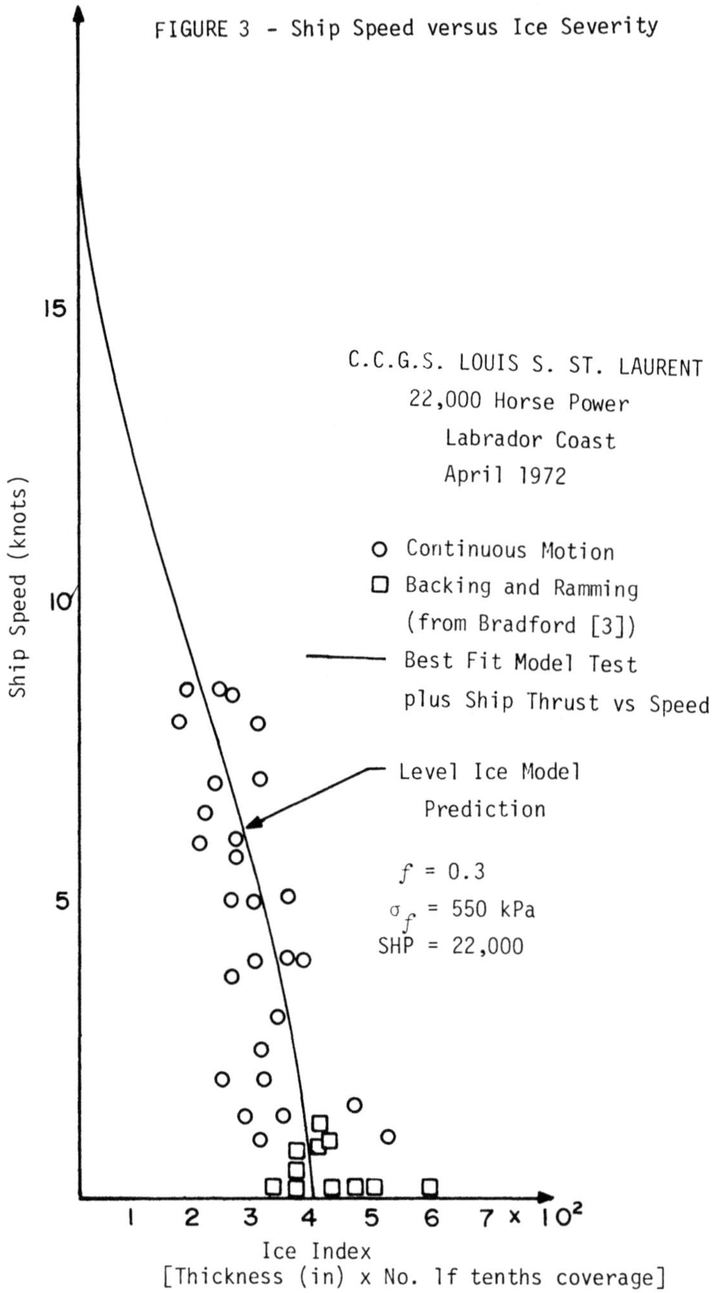

It is possible for all of the above shortcomings to combine fortuitously to produce reasonable model-full scale agreement. On the other hand, the results could be interpreted to indicate that proper scaling of ice thickness, bending strength and maintaining model hull/ice sliding friction coefficient near the values suspected for full scale ships, is sufficient to account for a majority of the resistance phenomena.

Do model tests predict the correct influence of modest changes in bow shape?

This fundamental question is still not proven. Since one of the supposed functions of model tests is to identify optimum ship designs, this problem needs to be addressed.

To conclude this paper, the state-of-the-art in achieving and measuring ice mechanical and frictional properties will be discussed.

FLEXURAL STRENGTH

Flexural strength is a quasi-property of a material. It is estimated by creating a bending failure in the material and measuring failure force and material geometry. In ice, it has been demonstrated that different methods of creating a failure moment result in different values of "bending strength". In the absence of any suitably defined standard, my firm employs the cantilever beam test. Numerous full scale measurements have been made with this technique. The results appear consistent between investigators (Weeks and Assur [6] and Dykins [7]).

In full scale, ice bending strength varies mainly with ice temperature and salinity. It ranges from 200 kPa for young sea ice to 1000 kPa for old freshwater ice. Tests have been conducted in model ice sheets with values of this property as low as 5-10 kPa. However, at very low levels of strength, two problems arise -- the ratio of modulus of elasticity falls well below values which are expected in full scale conditions, i.e. 1,000 to 5,000.

To employ simple cantilever equations for computing strength, the ratio of cantilever length to ice thickness must be chosen judiciously. Figure 4 shows that if the length of the fluid supported cantilever is kept below one half of the characteristic length of the ice cover, the simple equation:

$$\sigma_f = \frac{6PL}{Bh^2} \tag{3}$$

may be used,

where σ_f = bending strength

P = failure load

L = length of cantilever

B = cantilever width

h = thickness of ice.

It is advisable to determine the characteristic length of the ice sheet first (first step in determining modulus of elasticity), then measure strength using suitably sized beams. Beam width should be chosen on the basis of ASTM suggestions for strength tests of brittle materials.

BENDING MODULUS OF ELASTICITY

The modulus of elasticity of model ice must be reduced below the full scale value by the geometric scale ratio in order for the ratios of gravity and dynamic forces to elastic forces to be correct. Furthermore, if for a given value of bending strength, modulus of elasticity is decreased, the strain at failure increases, and the energy required to create failure is increased.

The modulus of elasticity of model ice is derived from the relationship between vertical load and deformation of the ice cover. Bending modulus of elasticity is a quasi-property and appears to be strongly dependent upon the method of measurement. Techniques which have been compared are:

End-loaded Cantilever Tests - Simultaneous measurement of the load and deflection of a floating cantilever.

FIGURE 4 - FLEXURAL STRENGTH OBTAINED USING ELLIPTICALLY SUPPORTED CANTILEVER

$$\frac{\sigma_f}{\left(\frac{6PL}{bh^2}\right)} = \frac{1}{X} = \left(\frac{\sinh X \cos X + \cosh X \sin X}{\cosh^2 X + \cos^2 X}\right)$$

$$X = \frac{L}{\sqrt{2}(1-\mu_\rho^{\;2})^{1/4}\ell_c}$$

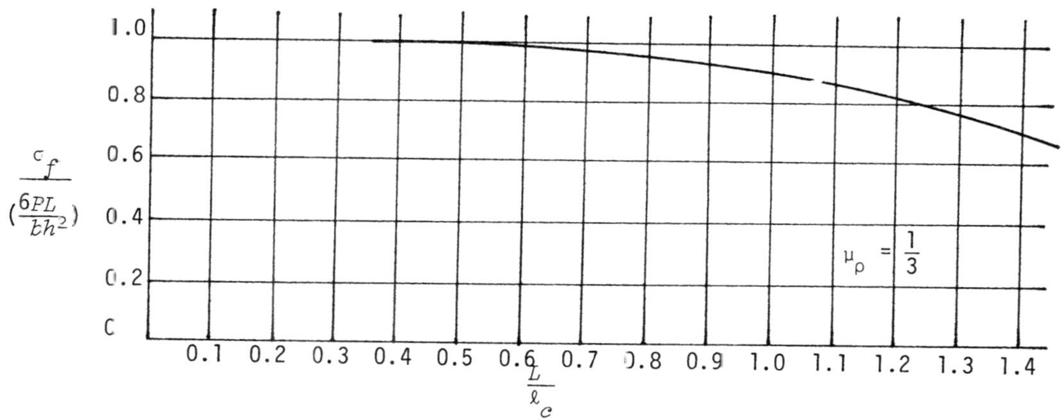

Central Distributed Load on an Infinite Plate - Simultaneous measurement of the load and deflection at the centre of an ice plate of infinite extent. (E_o)

Remote Deflection in the Neighbourhood of a Distributed Load on an Infinite Plate. (E_r)

Full scale measurements of elastic modulus in bending are scarce. Dykins has obtained modulus of elasticity ranging from 1,212 MPa to 3,720 MPa using isothermal laboratory grown sea ice beams [7]. These moduli correspond with ice strengths of 337 and 909 kPa, resulting in values of E/σ_f varying from 3,600 to 4,000.

Early studies were conducted at Arctec by Edwards and Kotras [9] to determine the influence of model ice properties on strength and modulus of elasticity.

During these tests, several methods of estimating modulus of elasticity were tried. The ratio of bending modulus of elasticity to strength was found to depend on strength and ice thickness. No method was found in the experiments to achieve E/σ above 1000 for strengths lower than 50 kPa. However, the following was concluded:

° Estimation of model ice flexural properties from brine volume models was impossible. Deviation in such relationships were so great as to warrent measurement of strength and bending modulus at the time of each test.

° Modulus of elasticity was most conveniently and consistantly estimated from the deflection at a point remote from a distributed load applied to the centre of an infinite plate. (E_r)

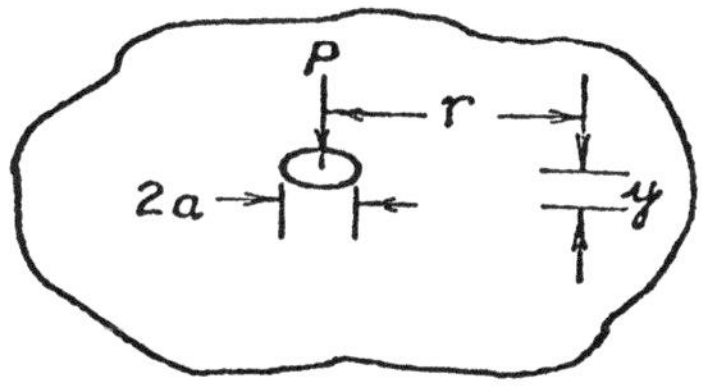

Characteristic length (ℓ_c) is estimated by solution of equation (4). Modulus of elasticity is derived from ℓ_c using well known equations.

$$\pi a^2 \rho_w g \frac{y}{P} = \frac{a}{\ell_c} \left[ber' \left(\frac{a}{\ell_c}\right) ker \left(\frac{r}{\ell_c}\right) - bei' \left(\frac{a}{\ell_c}\right) kei \left(\frac{r}{\ell_c}\right) \right] \qquad (4)$$

Various other techniques were tried. Table 2 illustrates the difference between estimates obtained from the end deflection of end loaded cantilevers and central deflection of a central loaded ice sheet.

As a result of continuing need to improve ice properties, studies were conducted at Arctec to settle properties measurement conflicts. Plastic sheets with well-known bending strength and bending modulus were floated on water and subjected to a wide range of properties tests. In the meantime, the materials were tested using ASTM standard tests. For bending modulus, the deflection at a point a short distance from a centrally located distributed load with subsequent application of elastic supported plate theory, produced results identical to the ASTM tests. For bending strength, the short floating cantilever technique produced results consistant with ASTM tests.

The flexural properties of model ice at Arctec Incorporated and Arctec Canada Limited are now assessed prior to each test with these techniques.

HIGH E/σ MODEL ICE

The measurement tests were only part of a larger program which sought to increase the model of E/σ. Figure 5 shows the results of elastic modulus and strength measurements using two ice preparation processes ("new" and "old"). The results apply only to fine crystalline ice grown at -50°C using liquid nitrogen spray. The open symbols in Figure 5 are for ice preparation procedures used prior to September 1978. The solid symbols refer to procedures now in effect at Arctec.

TABLE 2

Comparison of Modulus of Elasticity
Estimates from Two Types of Tests

Deflection of End Loaded Cantilever							Deflection of Center of Infinite Ice Sheet Exposed to a Central Load			
Bending Strength Kpa	Load/Deflection N/M	L	B (mm)	h	E MPa	$\frac{E}{\sigma}$	Load/Deflection N/M	Characteristic Length mm	E_f MPa	$\frac{E_f}{\sigma_f}$
11.8	137.6	160	80	38.7	.49	41	6624	291	12.9	1093.
15.6	194.	170	75	37.8	.94	60.3	6624	291	12.9	827.
13.0	194.	145	75	37.1	.62	47.6	6624	291	12.9	992.

FIGURE 5

BENDING MODULUS OF ELASTICITY
BENDING STRENGTH VERSUS MODEL ICE THICKNESS

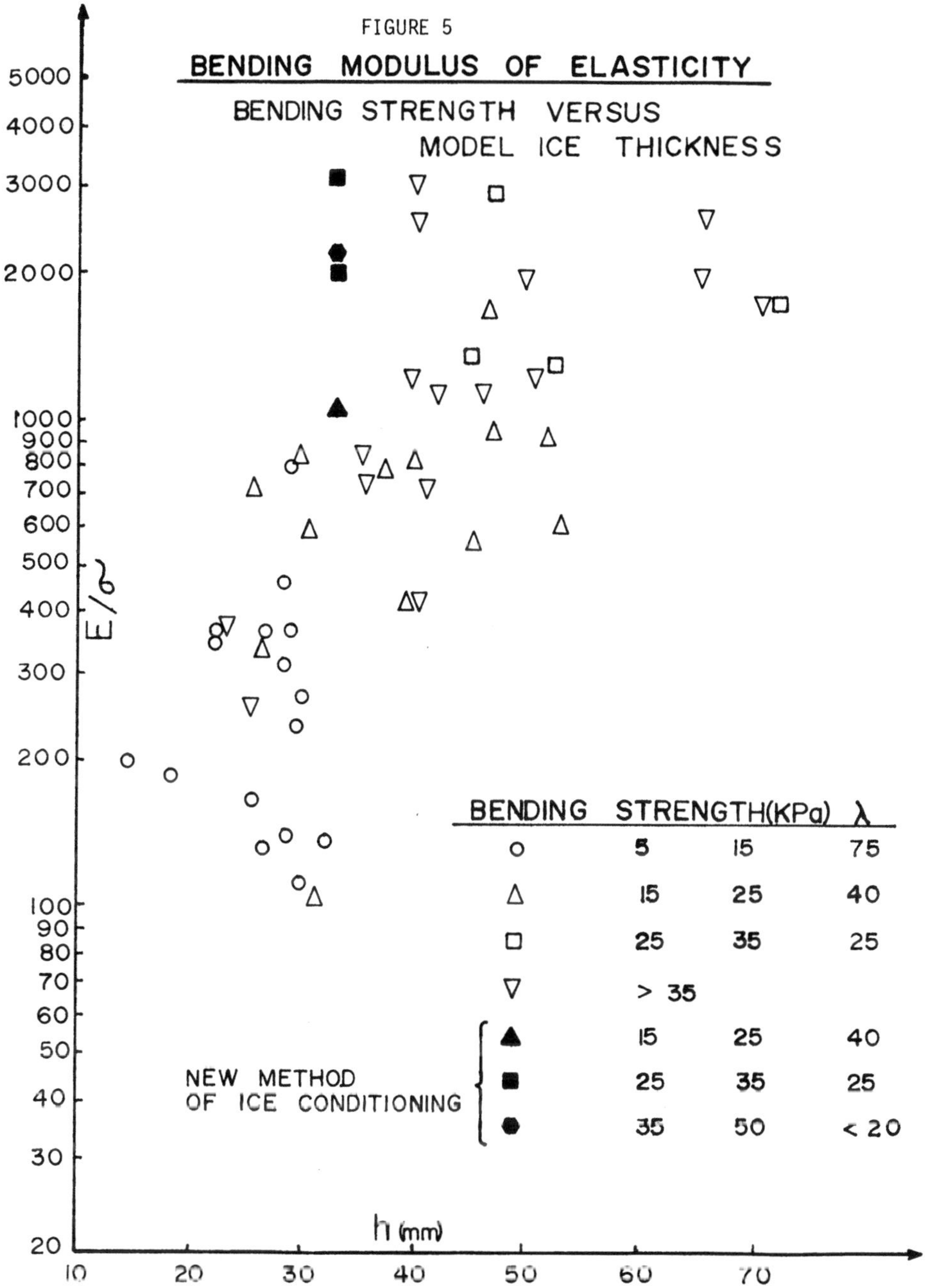

At scales of 25, it appears feasible to model E and σ simultaneously. At larger scales (smaller models) either an incorrect (low) value of E must be accommodated, or an incorrect (high) value of σ must be tolerated and corrected for.

While these results are encouraging for large ships, testing costs force us to use scales of 50 to 75. Our efforts to increase E/σ will continue until it can be maintained at 2000-3000 at ice strengths of 10 kPa.

FRICTION FACTOR

The strong influence of the coefficient of friction between the hull of an icebreaker and ice on resistance in level ice has been proven conclusively - at full scale by Makinen [10] and at model scale by Lewis [11], Edwards [1] and Lecourt et al [12]. According to [12], at model scale, the effect of increasing the coefficient of sliding friction between a model hull surface and ice from 0.05 to 0.2 doubles the resistance at low speed.

Makinen [10] described adequately the types of ice contact possible around the hull. The contact pressures range from extremely low - caused by buoyant forces (100 N/m^2) per metre of ice thickness) to the compressive failure strength of ice (4-5 MPa) in the crushing zone around the ship's forebody.

Friction tests have been carried out in both categories of contact pressure. Figure 6 shows results of cold room tests on uncorroded mild steel plate. In most cases the ice was dry and was crushed against the plate as the plate was translated. The velocity dependence of friction factor is clear.

A device has been developed for testing the friction factor on ship's hulls while in ice. It mobilizes a crushing failure while sliding. These tests are conducted in conjunction with hull roughness measurements.

FIGURE 6

EFFECT OF SPEED ON FRICTION FACTOR
FOR BARE PLATE AT -10°C

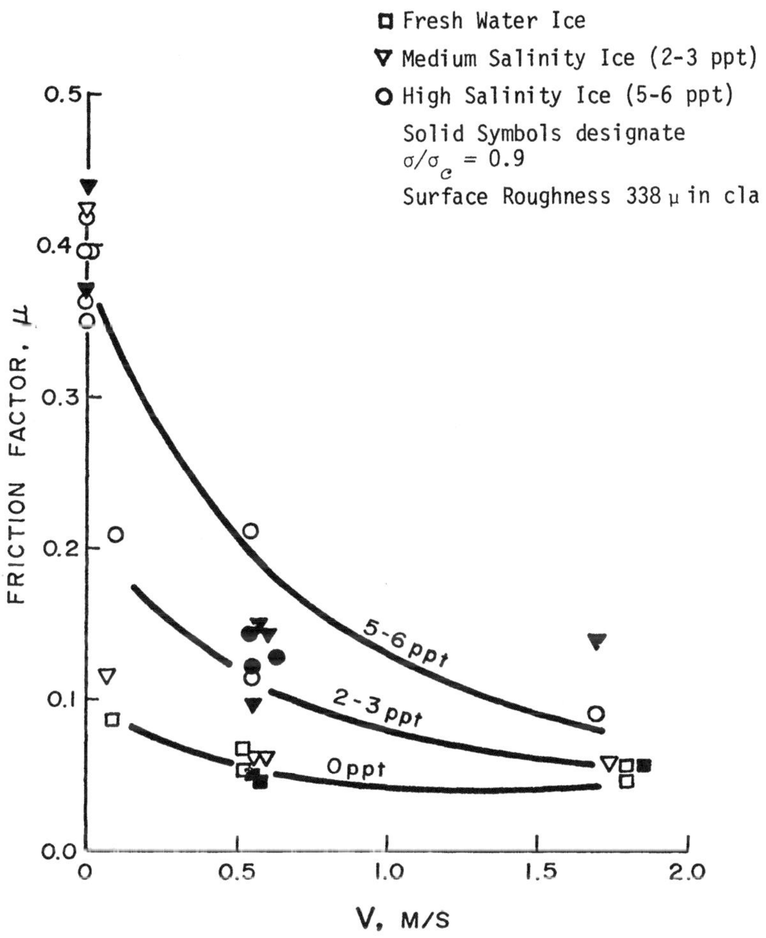

Model tests should be conducted at several friction factors. This permits later comparison with actual full scale results. Furthermore, that approach permits the assessment of the sensitivity of a hull's resistance in ice to increasing friction factor. This sensitivity will be indicative of the degree to which subsequent corrosion will degrade ship performance, and may provide insight into the influence of snow on resistance.

At Arctec, we have conducted extensive test programs to determine the best model surface coatings to achieve desired hull ice friction coefficients. Durable coatings have been found which will retain their surface preparation during repeated use in towed tests in ice.

COMPRESSIVE STRENGTH

As a ship travels through ice, the hull encounters cusp shaped projections of the ice cover. Depending upon the slope of the side shell at the contact point, the speed of the ship, the ice sheet bending stiffness, and associated mass and added mass of the deforming ice edge, a certain amount of crushing of the salient edges will take place before the ice sheet deflects significantly.

The magnitude of initial peak force on the hull is influenced by the *compressive strength of the ice*. The magnitude of the peak increases with increasing compressive strength. On the other hand, the total energy under the initial pulse increases with *decreasing* compressive strength. Intuitively, ice compressive strength should be expected to influence level ice resistance.

The compressive strength of saline model ice was investigated in the course of cylinder penetration tests. That data, presented in Table 3, demonstrates that:

° The penetration pressure of model saline ice is very sensitive to the aspect ratio of the indenter.

TABLE 3

Compressive Strength and Penetration Strength of Saline Model Ice

Indenter (Cylinder) diamet mm	Average Force Newtons	Ice Thickness mm	Unconfined Compressive Strength KPa	Flexural Modulus of Elasticity MPa	Flexural Strength KPa	Penetration Pressure KPa	$\frac{\sigma_u}{\sigma_f}$	$\frac{\sigma_p}{\sigma_{uc}}$	$\frac{E}{\sigma_f}$
19.05	12.94	18.5	17.1	1.2	-	36.6	-	2.2	-
38.10	19.6	20.0	17.9	2.4	-	25.7	-	1.41	-
152.40	162.7	33.7	42.3	11.9	20.0	31.7	2.11	.75	595
38.10	787.4	37.5	120.2	10.0	12.9	551.1	9.3	4.6	775
19.05	253.7	37.0	34.8	14.4	19.3	366.8	1.83	10.6	746
19.05	380.4	41.5	34.8	14.4	19.3	480.5	1.83	13.8	746
19.05	2157.5	61.6	104.9	59.7	29.63	1838.8	3.54	17.5	2015
19.05	2681.7	59.0	104.9	59.7	29.63	2383.	3.54	22.7	2015
38.1	2374.7	65.0	91.9	24.3	28.3	959.	3.24	10.4	859
38.1	1391.6	60.9	91.9	24.3	28.3	599.7	3.24	6.5	859
76.2	867.9	56.2	91.9	14.1	28.3	202.7	3.24	2.2	498
		h	σ_{uc}	E_f	σ_f	σ_p			

FULL SCALE RANGES

$\frac{E}{\sigma_f}$ 1000 - 5000 σ_f 200 - 900 KPa

σ_{uc} 1.2 - 3 MPa σ_p 2.0 - 20 MPa

$\frac{\sigma_{uc}}{\sigma_f} \simeq 3.-6.$

° The ratio of unconfined compressive strength (full scale) to the corresponding model scale properties is similar to the ratio of full scale flexural strength to model scale bending strength.

While superficially it appears that unconfined compressive strength and bending strength are scaled simultaneously in model ice, the severe confinement effects noted in model tests suggest that the compressive failure mechanisms are not similar in full and model scale.

Therefore, it is desirable to conduct tests in model ice to assess the penetration pressure for model ice for indentation geometries similar to those mobilized by the impingement of the hull surface on the ice edge.

CONCLUSIONS

Where model and full scale level ice resistance data are available for the same hull, model full scale correlation has been demonstrated to be good. Physically modelling the interaction between ships and ice at small scale is a fruitful approach to assessing the relative performance of distinctly different ships in a complex medium.

A very important use for modelling is to aid in optimizing new hull designs. Due to lack of full scale proof that small changes in ship shape can be detected correctly in models, it is vital that:

° Model ice properties be carefully measured, understood and improved to approach rigorous conformance to scaling laws.

° The acquisition of full scale data be continuously pursued.

REFERENCES

[1] Edwards, R.Y.Jr. et al, "Full Scale and Model Tests of a Great Lakes Icebreaker", Transaction of the Society of Naval Architects and Marine Engineers, Volume 80, 1972.

[2] Edwards, R.Y.Jr. et al, "Influence of Major Characteristics of Icebreaker Hulls on Their Powering Requirements and Maneuverability in Ice".

[3] Bradford, J.D., "Icebreaker Probes to Lake Melville, Labrador in 1972", Manuscript Report Series No. 54, 1979, Marine Sciences Directorate, Dept. of Fisheries & Oceans, Ottawa.

[4] Edwards, R.Y. Jr., German, J. Gordon, and Lawrence, R.G.A., "Comparative Model Test of the Performance of Two Canadian Coast Guard Icebreakers", Second Int'l. Conf. on Port & Ocean Engineering Under Arctic Conditions, Iceland, 1973.

[5] Enkvist, E., "On Ice Resistance Encountered by Ships in the Continuous Mode of Icebreaking", The Swedish Academy of Engineering Sciences in Finland, Report No. 24, 1972.

[6] Weeks, W.R. and Assur, A., "Mechanical Properties of Sea Ice", U.S.A. CRREL, September 1967.

[7] Dykins, J.E., "Elastic Behavior of Saline Ice", Naval Civil Engineering Laboratory, 1967.

[8] Keinonen, Arno, "An Analytical Method for Calculating the Pure Ridge Resistance Encountered by Ships in First Year Ice Ridges", Helsinki U. of Tech., Ship Hydrodynamics Lab., Finland, Report, No. 17, 1979.

[9] Edwards, R.Y. and Kotras, T., "Results of Model Ice Properties Research", 12 July 1972, ARCTEC Report 0057.

[10] Makinen, F. Ero, Lahti, Antti and Rimppi, Martin, "Influence of Friction on Resistance, Search for Low Friction Surfaces", Ice Tech 75, Symposium on Icebreaking Technologies, April 1975.

[11] Lewis, J.W. and Benze, D.L.,"Technical Progress Report of Model Ice Resistance Tests of USCGC MACKINAW", ARCTEC Report 00771-3, 1971

[12] Lecourt, E.J., Major, R.A., Thomas, H.L., Nargle, J.N., "Recent United States Coast Guard Efforts to Improve Icebreaking Efficiency" SNAME STAR Symposium, April 1978.

[13] Lewis, J.W. and Edwards, R.Y.Jr., "Methods for Predicting the Resistance of Icebreakers and Icebreaking Ships", Transactions Society of Naval Architects, Volume 78, 1970.

Finite Element Method Simulation of Large Ice Mass Flow Behaviour

J.J. EMERY and F.A. MIRZA
Department of Civil Engineering and Engineering Mechanics
McMaster University, Hamilton, Ontario, L8S4L7, Canada

Summary

The use of explicit and implicit finite element methods to simulate the flow behaviour of ice masses is described. Basal sliding and discontinuities are represented by soft elements and joint elements, respectively. Steady-state flow studies of the Barnes Ice Cap have resulted in a number of refinements to the simulation approach such as large deformations, transient thermal conditions, and functional flow relationships. Other work is in progress to simulate realistically glacier flow features such as surging.

Introduction

Extension of finite element method geomechanics creep studies to include ice flow behaviour reduces many constitutive and boundary condition limitations associated with analytical formulations [1]. It is not intended to review the finite element method or ice dynamics here, but rather to report progress in developing a two-dimensional, transient thermal regime, large deformation, incremental finite element approach for large ice mass flow simulation. The ultimate aim is to consider realistically ice mass response to climatic changes, particularly the important instability feature of surging associated with the Barnes Ice Cap [2]. This will eventually involve additional features such as accumulation/ablation, three-dimensional influences, and determination of insitu stresses from a steady-state flow model utilizing the known surface flow rates as velocity boundary conditions.

Preliminary Studies - Ice Slopes

An incremental, initial strain, finite element approach, where the stress-time curve is approximated by a series of steps in a forward marching procedure (Figure 1a) [3], was adopted to simulate the flow of idealized ice slopes [1,4]. The following assumptions were made: plane strain; homogeneous, isotropic, elastic material; properties of temperate glacier; generalization of flow relationships to multiaxial case in terms of effective stress and strain rate; Prandtl-Reuss condition applicable for flow; strain-hardening cumulative strain rate followed for changing stress (Figure 1b); and self-weight flow. Considerable simplification was introduced, but as indicated later, the finite element method has the flexibility to incorporate a number of simulation improvements.

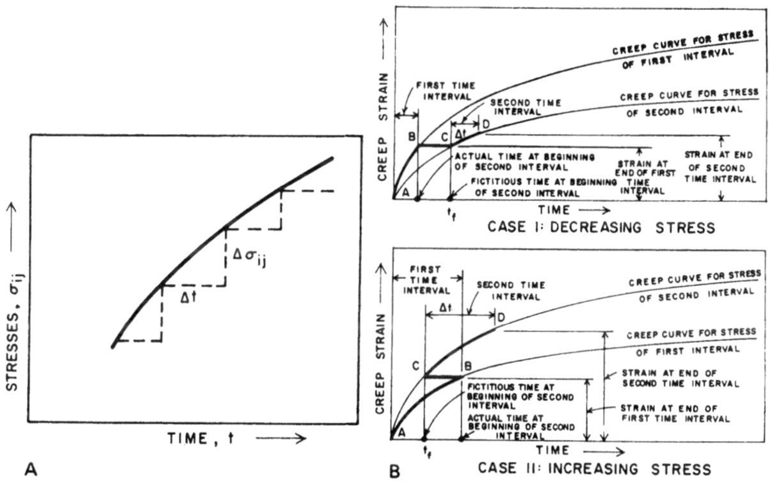

Fig. 1 Schematic representation of the incremental initial strain procedure. A. Approximation of smooth stress time curve. B. Increment of creep strain for changing stress. [1]

While the matrix formulation is lengthy [3,4] the incremental initial strain procedure essentially involves computing flow increments for each time step, and introducing them as initial strains through an equivalent load vector in the standard stiffness method (i.e. explicit approach suitable for small deformations). Steady state velocities and ice pressure development on walls for various slopes and typical flow relationships for polycrystalline ice were considered [1]. A more important application involved simulation of time dependent basal sliding, crack propagation, and basal shear strain or stress failure. The importance of critically assessing available ice flow relationships and properties was shown throughout.

Basal Sliding and Discontinuities

A simple method of simulating time dependent basal sliding was developed by introducing a thin, "soft" ice layer with a high flow rate in the basal zone. For the example shown in Figure 2, the following data were adopted:

mass ice - Meier's multiaxial flow relationship for temperate glaciers [5]

$$\dot{\varepsilon}_{oct} = 0.018\tau_{oct} + 0.13\tau_{oct}^{4.5} \quad (1)$$

where $\dot{\varepsilon}_{oct}$ is the octahedral strain rate (a^{-1}) and τ_{oct} is the octahedral shear stress (0.1 MPa)

basal ice - Colbeck and Evans' uniaxial flow relationship for ice near pressure melting, which gives effective strain rates an order of magnitude greater than Meier's for the same effective stress [6]

$$\dot{\varepsilon} = 0.21\sigma + 0.14\sigma^3 + 0.055\sigma^5 \quad (2)$$

where $\dot{\varepsilon}$ is the uniaxial strain rate (a^{-1}) and σ is the uniaxial compressive stress (0.1 MPa)

ice properties - $E = 9.1 \times 10^3$ MPa, $\nu = 0.34$ and $\gamma = 0.91$ Mg/m^3 (Noting that $\nu \simeq 0.4$ is probably more appropriate.)

rock properties - $E = 5.9 \times 10^4$ MPa, $\nu = 0.25$ and $\gamma = 2.56$ Mg/m^3

} used for all ice slopes

The basal sliding velocity in this case was about 56 percent of the surface velocity, comparable to a number of field observations. A basal failure criterion (i.e. sliding movement) can be developed in terms of a limiting cumulative shear strain for predicting time to large movements.

To simulate crack propagation and basal failure in terms of a limit-

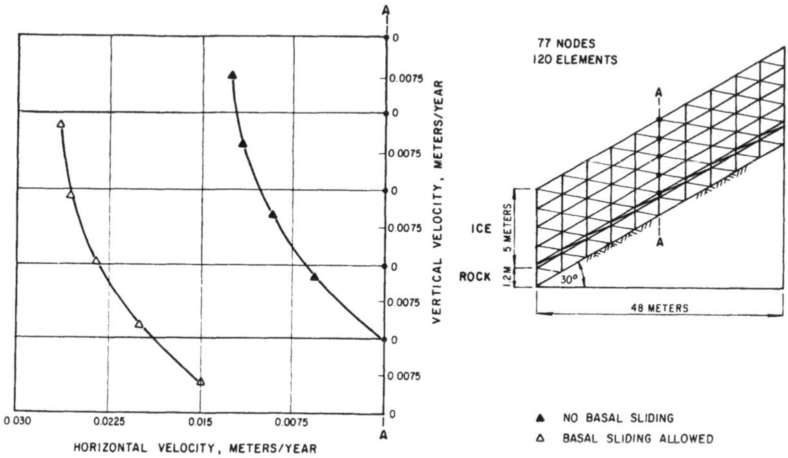

Fig. 2 Steady state velocity distributions for an ice slope with basal sliding [4]

ing cumulative shear stress, Goodman type joint elements were introduced at potential tension zones and ice-rock interfaces [7]. The normal and tangential joint stiffnesses (both 3.0 x 10^4 MPa/m) were developed through a parametric study based on maintaining the ice mass stress distribution that existed prior to joint introduction. This approach is shown in Figure 3, where elastic (time = 0), progressive basal shear (>0.1 MPa) and tensile (>0.2 MPa) failures were simulated by reducing the tangential stiffness to a residual value (15.7 MPa/m) or normal stiffness to zero, respectively. For instance, after the sixth cycle, all ten basal joint elements have failed in shear and eight vertical joint elements in tension so that the lower ice mass is free to slide downslope. Ice flow and basal sliding have been introduced for other slopes by checking for joint failures after each time increment (including time = 0) until a collapse mechanism, if any, develops. The key problem of how an ice slope stops moving after collapse occurs (i.e. surging) was not addressed at this stage.

Feasibility Study - Isothermal Steady-State Flow of the Barnes Ice Cap

The incremental, initial strain, finite element approach was used to consider the isothermal elastic and steady state flow behaviour

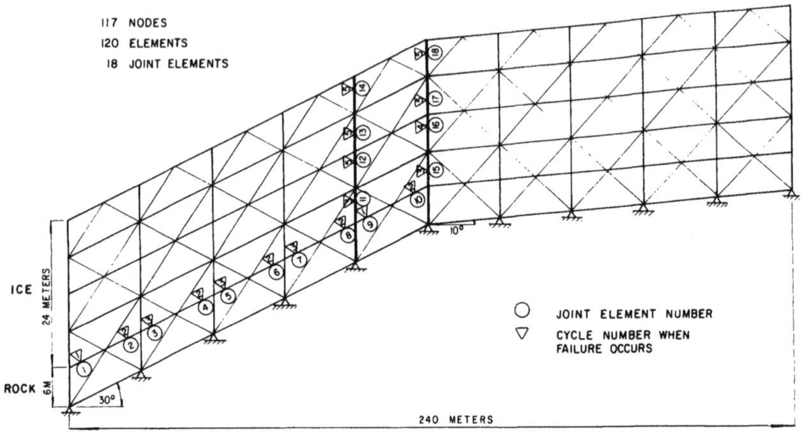

Fig. 3 Progressive basal shear and tensile failures for ice slope using joint elements [4]

of sections NO and NOS of the Barnes Ice Cap shown in Figure 4, which includes one of the five surge scars described by Holdsworth [2]. This feasibility study was completed in conjunction with the Glaciology Division of Environment Canada to provide background for incorporating required refinements to simulate surging.

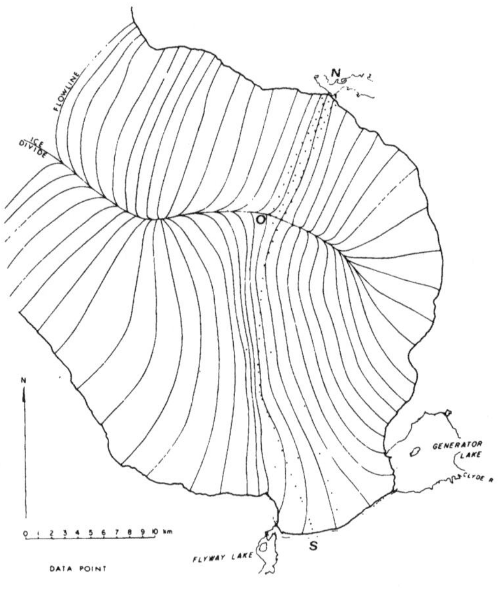

Fig. 4 South dome of the Barnes Ice Cap [2]

In addition to elastic solutions ($E = 9.1 \times 10^3$ MPa, $\nu = 0.34$ and $\gamma = 0.91$ Mg/m^3), steady state flow rates were found for a number of flow relationships including:

Meier, Equation 1

Glen, incorporating Hooke's parameters from measurements near the margin of the Barnes Ice Cap [8]

$$\dot{\varepsilon}_{eff} = 0.070\tau_{eff}^{1.65} \tag{3}$$

Nye [9]

$$\dot{\varepsilon}_{eff} = 0.175\tau_{eff}^{3.07} \tag{4}$$

where $\dot{\varepsilon}_{eff}$ is $\sqrt{3/2}\,\dot{\varepsilon}_{oct}$ (a^{-1}) and

τ_{eff} is $\sqrt{3/2}\,\tau_{oct}$ (0.1 MPa).

It was found that the use of the dynamic divide as a boundary (free vertically, fixed horizontally) to consider only section NO (Figure 4) was restrictive, and the total section NOS was generally used (see Figure 7 for the section and part of the element mesh). Large aspect ratio elements did not introduce numerical problems.

The results were most promising mainly from a qualitative viewpoint - the dynamic divide was found to be about 200 metres north of the topographic divide as compared to a measured shift of 150 metres for instance - and reasonable in a qualitative sense - surface flow rate predictions for instance. Solutions were found to be very sensitive to basal topography as indicated in Figure 5 for the steady state basal shear stresses for Meier's flow relationship.

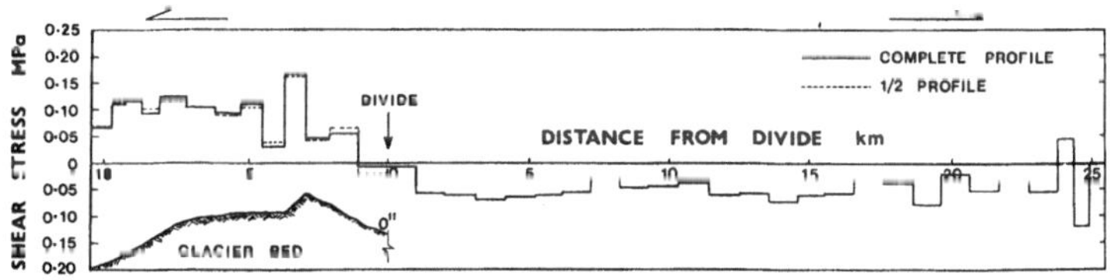

Fig. 5 Steady state basal shear stresses for section NOS of Barnes Ice Cap simulation, Meier's flow relationship [10]

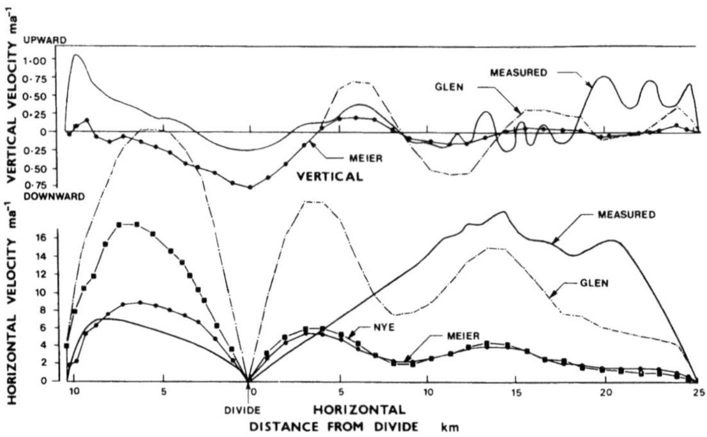

Fig. 6 Steady state surface flow rates for section NOS of Barnes Ice Cap simulation [10]

A comparison of steady state surface flow rates for various relationships (Equations 1, 3 and 4) with measured values are given in Figure 6. Close agreement was found for various relationships at different parts of the section (Glen's relationship with Hooke's parameters at the north for instance), indicating that development of an appropriate stress-strain rate-temperature relationship (functional form) would yield the desired accuracy. This study defined the refinements necessary to simulate large deformation features like surging that are described below.

Implicit Solution Method and Transient Thermal Analysis

In order to predict instability movements of large ice masses due to flow and/or temperature, the explicit finite element method based on small deformation theory is not sufficient. Therefore, an implicit plane strain finite element approach for nonlinear creep analysis, based on large deformation theory, has been adopted [11]. To incorporate the thermal regime of large ice masses, which are subject to long-term changes in temperature and geothermal flux, a transient heat flow finite element program has also been developed. By stepwise uncoupling, it is possible to input the new temperature distribution at discrete time intervals to the large deformation solution.

In contrast to the explicit finite element method, in the implicit method the incremental flow strains, using the Crank-Nicolson rate, are first expressed in terms of strain rates at times t_n and t_{n+1}. The unknown creep strain rate at t_{n+1} is then estimated by using a limited Taylor series expansion, thus incorporating stress and/or temperature dependence of flow laws through the derivative terms. Finally, a closed form expression for the stress increments is obtained as a function of total strain increments for Δt_n and creep strain rates at time t_n, and used directly in the large deformation finite element formulation. Other important features of the implicit approach include much longer time increments and equilibrium checks at each time interval. The program has been applied to some preliminary examples, and methods to ensure stability for large time intervals have been developed so that ice mass histories can be considered.

The general transient heat transfer equation to be solved for the appropriate boundary conditions is:

$$\frac{\partial}{\partial x}\left(k_x \frac{\partial \phi}{\partial x}\right) + \frac{\partial}{\partial y}\left(k_y \frac{\partial \phi}{\partial y}\right) + \frac{\partial}{\partial z}\left(k_z \frac{\partial \phi}{\partial z}\right) + \hat{Q} - C_p \frac{\partial \phi}{\partial t} = 0 \qquad (5)$$

where ϕ is the temperature, k_x, k_y and k_z are the thermal conductivities, $\hat{Q}$ is the heat flow (friction, radiation, etc.), and C_p is the specific heat. While a number of factors are involved, the geothermal flux (1HFU = 1μcal cm^{-2}sec^{-1}) and surface temperatures (typically taken at a shallow depth below seasonal variations) are of greatest importance.

As the thermal regime of the Barnes Ice Cap is important to simulating surge behaviour, section NOS has been considered for the geothermal fluxes and surface temperatures (at 17 metres), provided by the Glaciology Division, shown in Figure 7 (k = 2.1 Wm^{-1}^{o}C^{-1}). The significant difference in basal geothermal flux is reflected by a distinct upward central zone of warmer ice. Long-term changes in geothermal flux and surface temperature can be readily incorporated to give a full transient analysis.

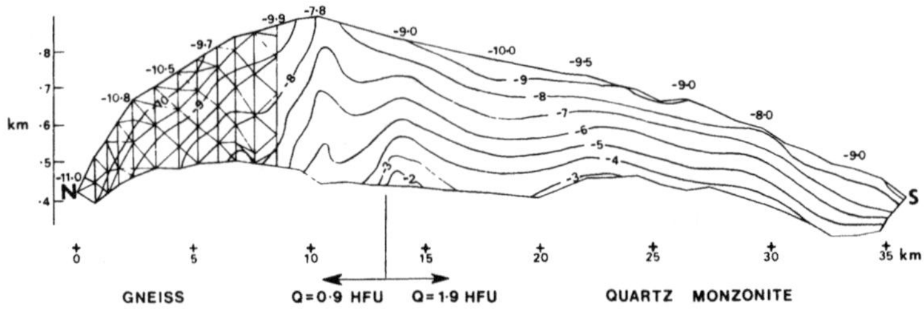

Fig. 7 Steady state temperature distribution for section NOS of the Barnes Ice Cap simulation

Functional Form Flow Relationships

Functional form flow relationships are important in any simulation to account for temperature, stress level and structure influences. Several functional form flow relationships shown in Figure 8 have been developed from "isothermal base laws" (Equations 1 and 3) by assuming thermal activation (i.e. rate process theory through Arrhenius equation with appropriate energies), or the micro-mechanistic approach of Ashby, Frost and Goodman (D,T) [12,13]. For the "Hooke base law" (Equation 3, HOOKE), both the Barnes, Tabor and Walker (H_{BTW}) [14] and Weertman (H_W) [15] activation energies were considered, noting the important change at -8^{0}C. For the "Meier base law" (Equation 1, MEIER), the Barnes, Tabor and Walker activation energies were considered (M_1) in addition to linearization and thermal activation. Dorn's definitions of effective shear stress ($\tau_D = 3/\sqrt{2}\ \tau_{oct}$) and strain rate ($\dot{\varepsilon}_D = \sqrt{2}\ \dot{\varepsilon}_{oct}$) have been used as they preserve the form of uniaxial flow relationships when generalized to the multiaxial case [1]. It is clear that there are order of magnitude differences in flow rates involved, and this is now being evaluated along with ice structure influences.

Conclusion

Several other important refinements are in progress: accumulation/ablation simulation using layers of elements that can expand or compress as required; three-dimensional (transverse strain) influences along the lines suggested by Budd [16]; velocity boundary conditions to establish current stress conditions (i.e. time = 0

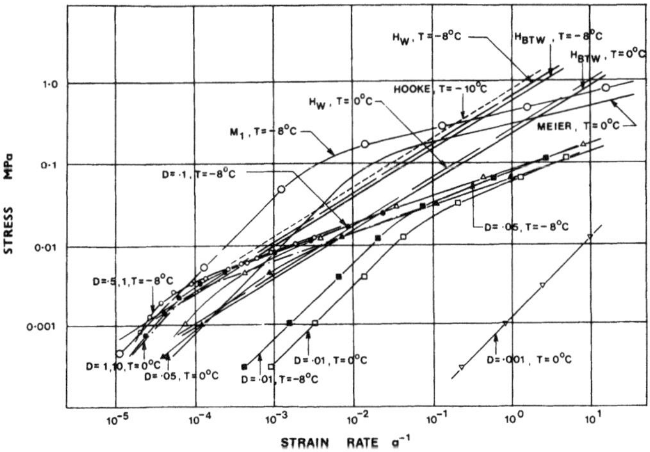

Fig. 8 Comparison of various functional form laws in terms of τ_D and ε_D (see text for definitions of flow relationships)

condition not generally known and insitu stresses difficult to monitor compared to surface flow rates); and contact representation at the ice margins. The finite element method offers the necessary flexibility to incorporate these refinements for the realistic simulation of ice mass flow behaviour. It is hoped that the concepts outlined will assist others intending to use the finite element method in ice mechanics.

Acknowledgements

A number of colleagues have made significant technical contributions to the ice flow studies described: T.Q. Nguyen, E.A. Hanafy, and J. Van Egmond through graduate research; and G. Holdsworth of the Glaciology Division, Environment Canada, through provision of important Barnes Ice Cap data, and continuing review and advice. The financial support provided by Environment Canada and the Natural Sciences and Engineering Research Council of Canada is also gratefully acknowledged.

References

1 Emery, J.J.: Simulation of slope creep. Dev. Geotech. Eng., Elsevier, 14A (1978) 669-691.

2 Holdsworth, G.: Evidence of a surge on Barnes Ice Cap, Baffin Island. Can. J. Earth Sci. 10 (1973) 1565-1574.

3 Greenbaum, G.A.: Creep analysis of axi-symmetric bodies. Ph.D. Thesis, Univ. California, Los Angeles, 1966.

4. Nguyen, T.A.: Simulation of ice flow using the finite element method. M.Eng. Thesis, McMaster Univ., Hamilton, 1976.

5. Meier, M.F.: Mode of flow of Saskatchewan Glacier, Alberta, Canada. U.S. Geol. Surv. Prof. Paper 351 (1960).

6. Colbeck, S.C. and Evans, R.J.: A flow law for temperate glacier ice. J. Glaciol. 12 (1973) 71-86.

7. Goodman, R.E. et al: A model for the mechanics of jointed rock. J. Soil Mech. Found. Div. ASCE 93 (1968) 637-659.

8. Hooke, R.L.: Structure and flow in the margin of Barnes Ice Cap, Baffin Island, N.W.T., Canada. J. Glaciol. 12 (1973) 423-438.

9. Nye, J.F.: The flow law of ice from measurements in glacier tunnels, laboratory experiments and the Jungfraufirn borehole experiment. Proc. R. Soc. Lond. Ser. A, 219 (1953) 477-489.

10. Hanafy, E.A.: Ice flow studies. EC 1630/35/M4, McMaster Univ., Hamilton, 1977.

11. Kanchi, M.B. et al: The visco-plastic approach to problems of plasticity and creep involving geometric nonlinear effects. Int. J. Num. Methods in Eng. 12 (1978) 169-181.

12. Van Egmond, J.: Simulation of large ice mass flow. M.Eng. Thesis, McMaster Univ., Hamilton, 1978.

13. Goodman, D.J.: The effect of impurities on the creep of ice and its illustration by the construction of deformation maps, IASH (1977).

14. Barnes, P.: The friction and creep of polycrystalline ice. Proc. R. Soc. Lond. Ser. A, 324 (1971) 127-155.

15. Weertman, J.: Theory of steady state creep based on dislocation climb. J. App. Physics 26 (1955) 1213-1217.

16. Budd, W.F.: The dynamics of ice masses. Anare Sci. Rep. Ser. A, 108, Melbourne, 1969.

Experience Gained by Use of Extensive Ice Laboratory Facilities in Solving Ice Problems

G.E. FRANKENSTEIN

U.S. Army Cold Regions Research and Engineering Laboratory, Hanover, New Hampshire, U.S.A.

Abstract

Ice Engineering is a relatively new field of endeavor which seeks to provide technical analysis, techniques, and designs for the solution of ice problems. An ice problem can be defined as any instance where ice or the effects of ice increases operational and/or maintenance requirements, reduces reliability, or impairs man's efforts.

Winter navigation problems due to ice are numerous. Ice builds up on navigation structures to such an extent that they become non-operational. Moving ice, broken up by ship's passage, has caused much damage to shore structures.

The discovery of offshore oil in ice-infested waters has caused major concern to the design engineers. Some of the problems associated with offshore structures are ice forces, icing, and pile-up.

Laboratory facilities have and will continue to solve many of the ice problems. The ice problem at navigation locks, for example, has been solved primarily due to laboratory studies. Also, the results of ice forces due to ice uplift have been virtually eliminated by controlled studies.

Laboratories are becoming larger and more sophisticated. This should result in an increase in laboratory studies and a decrease in field studies. Solutions will come faster because conditions can be precisely controlled.

Introduction

This paper discusses the solutions, full and partial, that have been accomplished through laboratory investigatons. A discussion is also included on the present research needs and how laboratories can assist on their solutions.

There are many institutions that have small refrigerated facilities. This type of laboratory is typically located at a university and studies the physical and mechanical properties of ice. These usually develop because one or more professors at the institution has a keen interest in ice and its properties. With increasing interest and engineering problems associated with ice some of these laboratories increase in size and staff (Kennedy [1]). All of these university laboratories provide a valuable contribution to the solution of many ice problems.

The development of oil from the Arctic regions has been responsible for the development of towing tanks that measure ice resistance of vessels, commercial carriers and ice breakers. Some of these laboratories use ice as the model material while others use a simulated material. Most of these laboratories have also investigated the problem of ice forces on marine structures.

Hydraulic models have been used for many years to study ice conditions. All of these models used synthetic materials to simulate ice. The first known hydraulic model where ice is being used is presently being constructed at CRREL (Fig. 1). Because of the size and uniqueness of the CRREL facility, it will be emphasized in the remaining part of the paper.

Ice problems can be divided into three sub-headings, namely, Navigation, Flooding and Ice Forces. Each will be discussed below.

Navigation

Continued exploration and discovery of petroleum, gas and mineral deposits has increased the need for vessels that are capable of operating in ice-infested waters.

Figure 1. The St. Clair river model in the Refrigerated Research area of the CRREL Ice Engineering Facility.

This requirement was responsible for the development of research facilities and towing tanks that could model ship resistance such that full scale phenomena could be predicted. The first facilities used a synthetic ice but the newest ones are refrigerated and use saline ice as the model material.

The model tests carried out in the above-mentioned facilities vary considerably. No two facilities can agree on method and especially on the resulting prediction equations. The primary reason for the differences is in the scaling of the elasticity, E, and strength, σ, of the ice. According to the scaling laws the ratio of E/σ should be between 2000-5000. This is the ratio found in full scale. However, this ratio is difficult, if not impossible, to achieve for model ice. The model material is either too plastic or stiff. Therefore, the prediction equation usually includes a correction factor to make it useful.

There is a great need (ITTC [2]) for conducting research that will lead to the standardization of model tests. Efforts are now underway to accomplish this through the ITTC Ice Committee. The CRREL Ice Engineering Facility is planning on conducting tests to determine the scale effect, characterize mush ice and determine its viscous flow properties, and investigate the effect of ice particle size and shape on viscosity.

The inland waterways have problems associated with extending the navigation season other than the development of ice resistant vessels. Some of these problems are the ice accumulation on the walls and gates of navigation locks, the floating ice above these locks, and the presence of mush ice in the navigation channel.

Laboratory tests were conducted on the ice problems at the locks at CRREL. The two most successful results were the development of a coating (Frankenstein, et al., [3]) and an air screen (Hanamoto, [4]).

The coating reduces the adhesive force between the coated surface and ice that forms on it. The coating is a long chain copolymer compound made up of polycarbonates and polysiloxanes. It was applied to a lock wall and the results were successful. Research will continue in the development of chemical coatings to find the ideal one that will prevent ice adhesion.

The floating ice problem at navigation locks is usually the broken-up (brash) ice pieces that a vessel moving downstream pushes in front of it. This floating ice mass hinders gate opening and creates vessel passage problems. It is often necessary to lock the accumulated ice through the lock separately. This obviously causes costly delays. To correct this problem laboratory tests were conducted on the use of a high flow, high velocity air screen. After successful laboratory tests a system was installed in front of an operating lock during the ice season. The demonstration was so successful that the system has been installed in other locations.

Flooding Caused By Ice

Any accumulation of ice pieces, sheet or frazil, restricts the flow of water. If this restriction is large enough then flooding will occur. At the right locations the flooding will cause extensive damage.

Frazil ice is responsible for most of the river ice problems. Frazil forms in supercooled turbulent river reaches and then collects in the areas where the stream velocity has decreased. The frazil particles continue to collect and have known to block a complete channel. Crude estimates can presently be made of the amount of frazil production but it is impossible to predict the threshold velocities associated with entrainment or deposition and consequent blockage effects on river flows. The most common practice used to prevent frazil production is to induce early ice growth to prevent heat loss or the introduction of artificial heat.

Presently there is limited information on the properties of frazil ice, including the mechanism of nucleation. A number of laboratories have installed hydraulic flumes (Fig. 2) in cold chambers to study frazil ice. In addition to studying frazil ice properties, some laboratories are investigating control techniques. Studies should continue and hopefully the engineers

Figure 2. A Refrigerated hydraulic flume.

will someday be able to control frazil formation.

River ice jams are associated with an increased water discharge caused by snowmelt of rain. Ice jam locations can often be predicted but not when initiated or the flood stage associated with it. This is contrary to non-ice jam flooding.

Studies are currently being conducted to better understand the formation of ice jams. The studies include the use of hydraulic models to study the formation and breakup of ice jams. These studies are being conducted in a cold environment under actual conditions. There is currently considerable controversy over whether or not distorted or non-distorted models are most appropriate for the prediction of ice movement, control, and forces. A study will soon be underway to compare an undistorted river model with a similar distorted model of the same reach, and for which extensive prototype data has been obtained.

In the past one of the major limitations of ice research was the determination of ice thickness remotely. The recent development of the pulsed-radar system (Dean, [5]) for determining ice thickness has provided for the rapid acquisition of complete vertical profiles. Refinement of the system is continuing for eventual operational use.

Mechanical Properties of Floating Ice Covers

Studies to determine ice properties have been conducted for many years but greatly accelerated since World War II. Tests have also been conducted on floating ice covers in the field as well as in laboratories. These tests did supply many practical results but revealed many shortcomings on which theory to use (Kerr [6]).

The most notable feature of ice strength data is the very wide scatter in the reported results. It is not uncommon to have results vary as much as an order of magnitude. Factors which affect the strength of ice and account for the scatter in test results are ice temperature, loading rate, ice type and grain structure, direction of loading, confining force, sample size, the presence of impurities such as salt, and testing technique. There is a great need for good controlled laboratory testing of ice properties, especially sea ice.

Kerr [7] conducted an extensive survey of the literature on the bearing capacity of ice covers. This study revealed the urgent need for establishing proper failure criteria for floating ice covers.

Available test data, suggest that an ice cover that is subjected to a vertical load deforms differently when the ambient air temperature is slightly below 0°C as compared to temperatures much below 0°C. It appears that the instantaneous response of a floating ice cover at very low temperatures follows the elastic bending theory of plates, whereas at near zero temperatures, especially for lake ice, shear deformation may be significant. It is also necessary to establish the contribution of the bending and shearing deformations for static loads of long duration (when ice covers continue to deform in creep).

To clarify this situation it is suggested to subject a "large" floating cover to a uniform load over a circular area and to measure the resulting deformations at the time of instant load application and at fixed times thereafter. The test should be conducted at different air temperature and for ice formed in still and moving water.

For loads of short duration it is generally assumed that for temperatures below freezing an ice cover responds elastically. Therefore the analyses of such problems have been based on the available bending theory of thin plates that are both elastic and homogeneous. To clarify this problem it is recommended to conduct laboratory tests (1) to establish if the first crack which occurs in a plate may be predicted with a reasonable accuracy by a plate analysis, with a σ_f-value obtained from a cantilever beam test in situ (loaded upward), (2) to establish how much larger is the break through load to the one which causes the first crack, and (3) to establish if a deflection failure criterion is applicable for loads of short duration.

The largest force that an ice cover may exert on a structure is limited by the failure strength of the ice (Michel [8]). For the dynamic case ice structure interaction appears when either the deformation of the structure or ice affect the ice failure process. This type of failure often occurs on slender flexible structures, lighthouses and piers for example, when a moving ice sheet crushes against them. This is due to the fact that the resistance of the ice is dependent on the loading rate, initial velocity, and the elasticity of the structure. Studies designed to measure the dynamic ice-structure interaction are limited. Maattanen [9] has conducted tests for high velocities and plans to continue his studies to include the lower range.

The maximum force is developed when the ice fails by crushing. This occurs when the ice makes contact with a vertical structure. The present design practice is to incline the structure from the vertical such that the ice will fail in bending. A number of laboratory tests, using both natural and artifical ice (Zabilansky [10], Tryde [11]) have been conducted to determine the angle of inclination that is necessary to cause a bending rather than a crushing type failure. Preliminary results indicate that the inclination should be greater than 15° and preferably 30°.

When a water body freezes over, the formed ice cover often attaches itself to structures. Subsequently, any change to the position of the ice plate relative to the structures, in the contact region, creates forces between them. Water fluctuations are the major cause of the uplift forces. The change in water elevation is usually created by tidal action or vessel passage. Vertical movement of piles has been observed to be several feet which can completely destroy a structure.

Laboratory investigations have been conducted (Fig. 3) to determine the magnitude of these uplift forces and to develop solutions that would reduce the damage caused by them. The most promising solution to date has been to reduce the bond of ice to pile by an epoxy sleeve. However, Haynes [12], has discussed the use of an air thermosyphon pile to prevent uplift. The thermosyphon would create a frozen bulb at the buried portion of the pile which would require a much greater uplift force to move it. Laboratory studies to determine the effect of a thermosyphon pile on reducing uplift forces are planned for the very near future.

Figure 3. A Laboratory study of uplifting forces.

References

1. Kennedy, J.F.: The Iowa low temperature flow facility. Proc. IAHR Ice Symposium (1970) 1.2, 1-13.

2. Proc. ITTC 15 (1978) 157-179.

3. Frankenstein, G.; Wuebben, J.; Jellinek, H.H.G.; Yokota, R.: Ice removal from the walls of navigation locks. ASCE Rivers 76 (1976) 1487-1496.

4. Hanamoto, B.: Ice control at navigation locks. Corps of Engineers ETL 1110-2-237 (1978) 1-6.

5. Dean, A.: An investigation of automatic data collection equipment for oceanographic applications. Proc. POAC (1977) 1111-1121.

6. Kerr, A.; Palmer, W.: The deformations and stresses in floating ice plates. Acta Mechanica, Vol. 15 (1972).

7. Kerr, A.: The bearing capacity of floating ice plates subjected to static and quasi-static loads. CRREL RR333 (1975).

8. Michel, B.: Ice Mechanics. Les Presses De L'Universite Laval, Quebec (1978).

9. Maattanen, M.: Laboratory tests for dynamic ice-structure interaction. Proc. POAC (1979).

10. Zabilansky, L.: Ice forces on model structures. Proc. Canadian Hydrotechnical Conference (1975) 400-407.

11. Tryde, P.: Intermittent ice forces acting on inclined structures. CRREL 77-26 (1977).

12. Haynes, D.: Use of an air thermosyphon pile for marine structures. CRREL Tech Note (1979).

Dynamic Ice Forces on an Inclined Structure

R. Frederking

Division of Building Research,
National Research Council of Canada.

Summary

A qualitative model of the breaking behaviour of an ice cover against an inclined structure and subsequent movement of the broken ice slabs is presented. The various force components resulting from the assumed ice behaviour are identified, physical models developed, and quantitative expressions formulated. Particular attention is given to all the assumptions. An example of the variation of the force components with time throughout the breaking cycle is presented. The results of the analytical model compared well with the results of a new series of model tests.

Introduction

The dynamic interaction of a floating ice cover and an inclined structure is complex. It involves forces due to crushing and flexural behaviour, inertial effects of the broken pieces of ice, hydrodynamic drag between the water and ice, friction between the ice and structure, and gravity. These various forces act not only while the ice is deforming up to the point of failure, but subsequently as the broken ice pieces move up and around the structure.

Various investigators have approached the problem of defining ice forces on structures with inclined surfaces analytically and experimentally, e.g., Tryde [1], Sørensen [2], Croasdale, et al [3], Bercha and Danys [4], Ralston [5], Dolgopolov, et al [6], Schwarz, et al [7].

This paper deals specifically with the case of a simple inclined plane of finite width which breaks the ice upwards. Only the general approach and assumptions in developing equations for the variation of the force components with time throughout the breaking cycle will be given, details of the derivations will be published later.

General Formulation of Problem

The following approach was used to determine the various force components acting on a structure. First a qualitative account of the breaking behaviour of the ice cover and subsequent movement of the broken ice pieces was made. Then the various force components resulting from the assumed ice behaviour were identified and physical models developed. The final step was to formulate quantitative expressions for each force component.

Observations of ice behaviour reveal a typical cycle of ice-structure interaction (Fig. 1). Let the cycle start at the point where the structure is contacted by the edge of the unbroken ice cover:

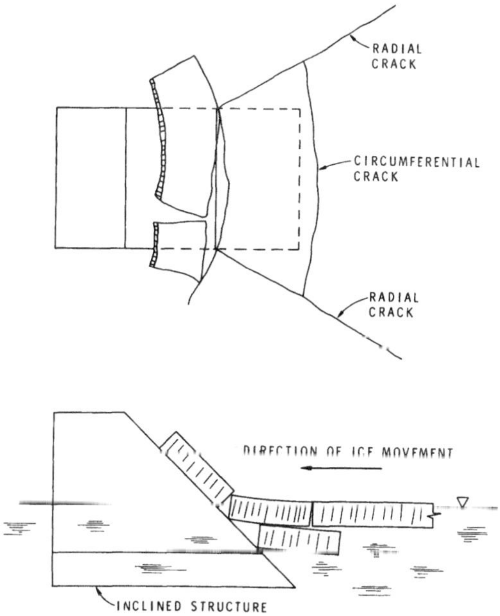

Fig. 1 Schematic of ice behaviour on inclined plane structure

(i) The inclined plane develops vertical and horizontal forces on the ice edge thereby bending it upwards. Two radial cracks propagate outwards from the edges of the structure as these forces increase. The peak forces are associated with the formation of a circumferential crack after which forces drop to zero until contact is again made with the unbroken ice cover. The forces developed on the structure in this phase are a function of incline angle, ice bending and crushing strength, coefficient of friction between ice and structure, effective modulus of the ice, and buoyant effect of the water.

(ii) The broken ice slab rotates until it is parallel to the surface of the incline. As the piece rotates, water drag forces proportional to the rotational velocity of the ice piece and inertial forces proportional to its rotational acceleration are developed on the structure. These forces vary with time and are a function of velocity, incline angle, coefficient of friction between ice and structure, surface roughness of ice underside and buoyancy.

(iii) The broken ice slab slides up the incline developing frictional and gravity forces on the surface.

The total ice force acting on the structure at any one time is the sum at that time of all the intermittent force components.

Ice Force Components

Basic expressions will now be developed for each of the force components.

1. Forces in bending and breaking ice

From the interaction mechanism illustrated schematically in Fig. 2 and assuming classical Coulomb friction the following expressions for the horizontal and vertical forces on the ice edge are defined

$$F_x = N \sin \alpha + \mu N \cos \alpha$$
$$F_z = N \cos \alpha - \mu N \sin \alpha \qquad (1)$$

where μ = coefficient of friction between ice and structure, α = angle of inclination from horizontal, and N = force normal to the surface of the plane. (Note that the forces used here are total forces on the structure, not forces per unit width.) The ratio of the horizontal to vertical force, ξ, is defined as

$$\xi = \frac{F_x}{F_z} = \frac{\sin\alpha + \mu\cos\alpha}{\cos\alpha - \mu\sin\alpha} \tag{2}$$

It is now necessary to make some assumptions about the deformation behaviour of ice. Gold and Sinha [8] presented a rational procedure for defining the region where ice behaviour can be approximated as linearly elastic and represented by an effective modulus which is a function of stress and time. This provides a sound basis for using the many elastic solutions which exist in the literature. What is required now is appropriate experimental data on ice in nature and in model basins.

Observations of the cracking pattern, (Fig. 1) suggest that the ice sheet can be approximated as a truncated wedge on an elastic foundation. This idealization is shown in the plan view of Fig. 2. Nevel [9] solved this problem analytically assuming elastic behaviour for the ice. For most field situations and model tests, where a/ℓ lies between 0.5 and 2 (a = truncation distance; ℓ = characteristic length), the following approximate expressions for tip deflection,

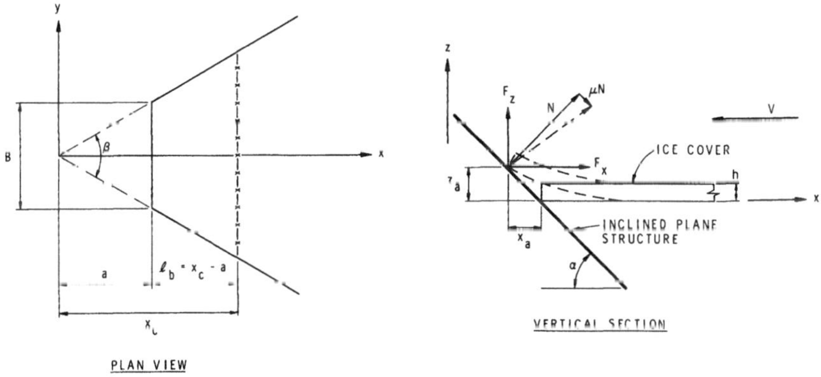

Fig. 2 Ice cover idealized as truncated wedge

z_a; circumferential crack position, x_c; and vertical load, F_z, can be formulated

$$\frac{z_a}{F_z} 2 \tan \frac{\beta}{2} k\ell^2 = \frac{0.85}{(a/\ell)^{0.6}} \tag{3}$$

$$x_c/\ell = 0.76 + 1.08\, a/\ell \tag{4}$$

$$\frac{3F_z}{\tan \frac{\beta}{2} \sigma_f h^2} = 2.3\, a/\ell + 1.1 \tag{5}$$

Truncation distance a, wedge angle β, ice thickness h, and structure width B are defined in Fig. 2. Wedge angle β is determined from observations. The characteristic length, ℓ, is given by $(D'/k)^{\frac{1}{4}}$ where D' is the modified flexural rigidity of a plate with effective modulus, E_{ef}, varying through the thickness (Kerr, [10]) and k is foundation modulus. From the wedge geometry, $a = \frac{B}{2 \tan \beta/2}$.

In addition to the vertical force, F_z, a horizontal force, F_x, is also developed on the truncated wedge as a result of the interaction with the structure. This horizontal force develops a moment in the wedge. As the tip deflection at failure is small (about $\frac{1}{4}$ of the ice thickness) and the magnitude of F_x is at most 3 times F_z (from equation (2)), the flexural stress in the wedge is assumed to be a function of only the vertical force, F_z. The horizontal force, F_x, superimposes a compressive stress in the wedge, which when combined with the flexural stress leads to the following approximate expression for a non-dimensional vertical force

$$f_z = \frac{3F_z}{\sigma_f h^2 \tan \frac{\beta}{2}} = (1.16 + 0.0325\xi) + (2.26 + 0.0825\xi) a/\ell \tag{6}$$

for $0.5 < a/\ell < 2.0$. The forces at failure, F^*, are calculated using an assumed flexural strength, σ_f^*. (The symbol *, will be used to refer to values at failure.) The non-dimensional horizontal force is obtained via Eq. (2). Once the ice cover has failed by the formation of a circumferential crack, it advances a distance $\ell_b = x_c - a$ before contacting the structure again and restarting the load cycle.

2. Static forces in lifting ice slabs

As the ice cover continues to move towards the structure, the broken slab is lifted out of the water and rotated until it is parallel with the inclined surface of the structure. Frictional, gravitational and buoyant forces are involved in this movement. The actual process is shown schematically in Fig. 3. The rotation of the slab, ϕ, is given by

$$\phi = \alpha - \text{arc sin } \{\sin \alpha \, (1 - Vt/\ell_b)\} \tag{7}$$

Equation (7) indicates that ϕ increases from 0 to α as non-dimensional time Vt/ℓ_b increases from 0 to 1. The actual form of the variation for $\alpha = \pi/4$ is illustrated in Fig. 4. The gravity force, F_g, which acts vertically downwards through the centre of gravity of the slab is independent of rotation ϕ. The buoyancy force, F_b, which acts in an upwards direction, and its point of action are involved functions of rotation ϕ. To simplify the solution, buoyancy effects will be ignored. This is a conservative assumption which results in slightly higher ice force predictions. Two reaction forces, N_r and μN_r, normal and parallel to the incline, act on the tip of the slab to

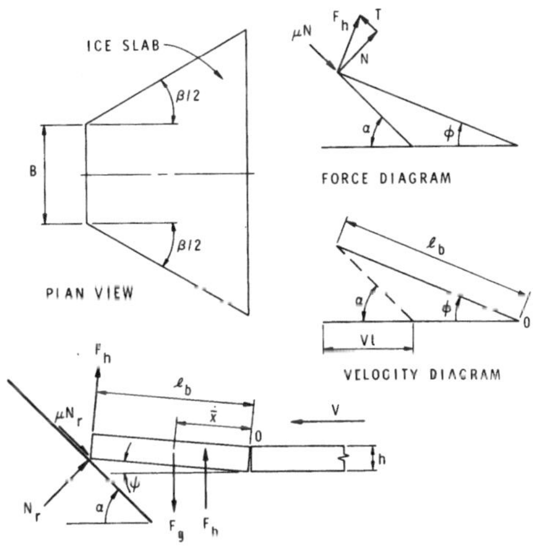

Fig. 3 Schematic of geometric relations and forces acting on rotating ice slab

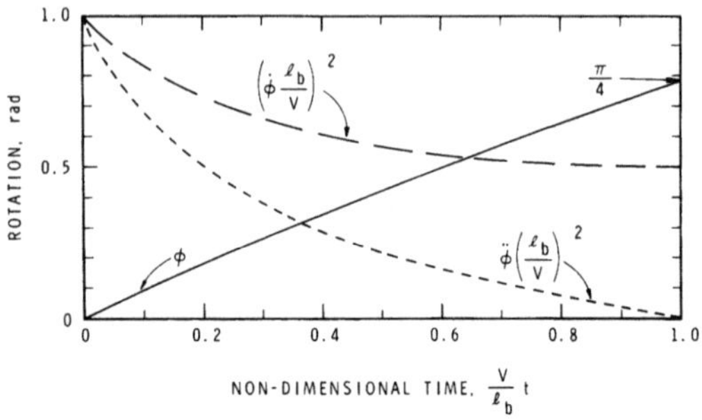

Fig. 4 Variation in rotational velocity and rotational acceleration with non-dimensional time

rotate it. Summing the moments about 0 the following simplified expression for, N_r, is obtained,

$$N_r = \frac{\bar{x}\ F_g}{\ell_b\ f(\phi)} \tag{8}$$

where

$$\bar{x} = \frac{\ell_b}{6}\ \frac{3B/\ell_b + 2\tan\beta/2}{B/\ell_b + \tan\beta/2} \tag{9}$$

$$F_g = \gamma_i\ h\ A = \gamma_i\ h\ \ell_b^{\,2}\ (B/\ell_b + \tan\beta/2) \tag{10}$$

$$(\gamma_i = \text{mass density of ice})$$

The variation of N_r as the slab rotates from the horizontal to an incline position parallel to the slope is a function of ϕ and hence time. Typical curves of $1/f(\phi)$ versus ϕ are given in Fig. 5 for $\alpha = \pi/3$ and $\pi/4$. From this variation both maximum and average forces can be determined. Horizontal and vertical forces developed on the structure due to slab rotation are obtained by substituting Eq. (8) into Eq. (1). In general it can be seen that the maximum force occurs at contact, i.e., time zero, and subsequently decreases with increasing time.

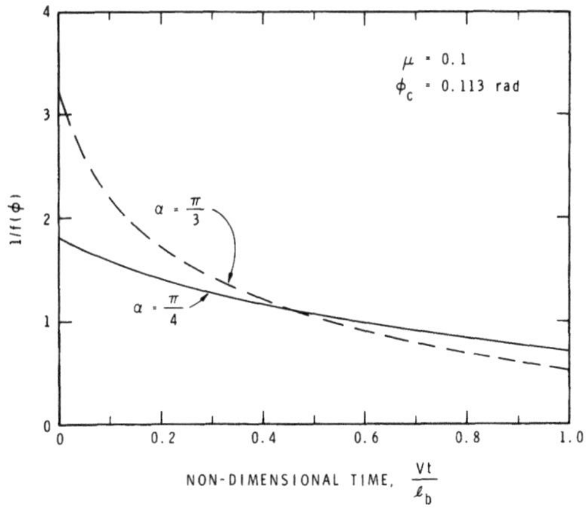

Fig. 5 Variation of static rotation force factor $1/f(\phi)$

Once the broken slab is parallel to the incline surface it slides up the slope and generates a normal force and corresponding frictional force on the structure. The normal force is approximated by

$$N_s = F_g \cos \alpha \tag{11}$$

where F_g is as defined in Eq. (10). This force will be assumed to be a constant since, after the first slab is on the incline, there will always be ice on the structure. Dynamic effect will now be considered.

3. Hydrodynamic force in rotating ice

Besides the static forces set up by rotation of broken ice slabs and movement of them up the incline, hydrodynamic forces are developed as a result of relative movement between the ice and the water. These are momentum drag forces caused by the rotation of the broken ice slabs (Brix [11]). The action of these forces is assumed to be represented by a tip force, F_h, which acts normal to the axis of the slab (Fig. 3). This force is given by

$$F_h = \frac{C}{2} \rho_w A \left(\frac{\dot{\phi}\, \ell_b}{2}\right)^2 \tag{12}$$

where $\dot{\phi}$ is the rotational velocity obtained by taking the time derivation of Eq. (7), A is the cross-sectional area swept by the rotating ice slab (Eq. (7)), ρ_w is density of water, and C is a drag coefficient associated with the shape and roughness of the ice slab. For the trapezoidal-shaped slab, C will be assumed equal to 2. The time variation of $\dot{\phi}^2$ is illustrated in Fig. 4. This tip force, F_h, can be resolved into components, $N = F_h \cos(\alpha - \phi)$ normal, and $T = F_h \sin(\alpha - \phi)$ tangential to the surface of the inclined plane. The normal force will also develop a frictional force μN. The direction of these forces is illustrated in the force diagram of Fig. 3. Horizontal and vertical forces on the structure due to hydrodynamic forces are given by

$$F_x^h = N \sin \alpha - (T - \mu N) \cos \alpha$$
$$F_z^h = N \cos \alpha + (T - \mu N) \sin \alpha \tag{13}$$

For rotations greater than $\phi_c = \tan^{-1} h/\ell_b$, the contact area, A, between the ice slab and the water begins to decrease. For computational simplicity it will be assumed that contact area remains constant up to a rotation $\phi_h = \tan^{-1} 2h/\ell_b$ and then goes immediately to zero.

4. Inertial forces in rotating ice

In addition to the static and hydrodynamic forces in rotating the broken ice slabs, inertial forces are developed. The mass of the ice slabs plus the added mass of the water accelerated along with them affects these inertial forces.

The inertial tip force, F_i, which acts normal to the slab axis, ϕ, is given by

$$F_i = \frac{I}{\ell_b} \ddot{\phi} \tag{14}$$

where ℓ_b is the breaking distance as defined before, I is the mass moment of inertia of the slab about 0, and $\ddot{\phi}$ is the second derivation with respect to time of Eq. (9). The mass of the ice slab is determined from Eq. (10), and the added mass of the water, from the empirical expression developed by Yu [12] is

$$m_w = \rho_w \left[\frac{0.79\ B\ \ell_b}{\left(B^2 + \ell_b^2\right)^{\frac{1}{2}}} + 0.62\ \sqrt{h} \right] A \qquad (15)$$

(Length dimensions must be expressed in centimetres.) The tip force, F_i, due to inertial force can be resolved into vertical and horizontal force components in the same manner as in the previous sub-section. To simplify the computations it is assumed that the added mass of the water can be neglected for ϕ greater than $\tan^{-1} 2h/\ell_b$.

5. Example problem

The preceding development has been left in dimensional form to present, as clearly as possible, the relation between the physical and mathematical models. To illustrate further the analytical model the relative size of the individual force components and their typical variations with time are given in Fig. 6. In this particular case

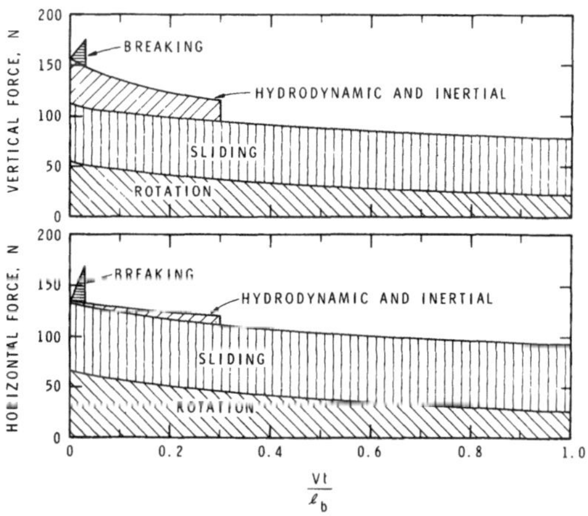

Fig. 6 Force variations for the case h = 40 mm, B = 0.75 m, $V = 0.5\ m{\cdot}s^{-1}$, $\alpha = \pi/4$, E = 40 MPa, $\sigma_f^* = 20$ kPa, $\mu = 0.1$, $\beta = \pi/3$

forces associated with breaking, rotating and sliding dominate; dynamic forces add less than 10% to the total vertical force. Increasing the velocity by about a factor of three would increase the dynamic forces by about a factor of 10. There is a periodicity to the forces characterized by an initial peak force which then decreases.

Discussion of Analytical Model

It is now possible to discuss the influence of geometric factors, e.g., structure width, ice thickness and slope inclination, and ice properties, e.g., flexural strength, effective modulus and friction. The primary effect of increasing the structure width is to increase ice slab width and breaking force F_z^*. Increasing the slab size increases the static, hydrodynamic and inertial forces. Thicker ice or greater effective modulus results in a larger characteristic length and therefore a larger slab size. Steeper slope angle, α, and thus ξ (Eq. (2)), increases the vertical force required to break the ice and also proportionately increases the corresponding horizontal force. Steeper slope angle also increases the static, hydrodynamic and inertial forces. Higher velocity brings up the hydrodynamic and inertial forces and shortens the loading time and therefore the effective modulus is increased (Gold and Sinha [8]). Ice strength only influences the forces associated with ice failure. The ice properties of effective modulus and friction however, have a very important influence.

Comparison with Model Test Results

A series of model tests was performed under contract to DBR/NRC (Abdelnour [13]) to obtain data to verify the model. Simple inclined plane structures of 0.3 m slope length were used. The independent variables included structure widths of 0.25 and 0.75 m; ice thicknesses of 25 and 40 mm; slope angles of $\pi/4$ and $\pi/3$; and velocities between 0.01 and 0.20 $m \cdot s^{-1}$. The complete test data are too numerous to present here; the results from one test series are presented in Table I. It is apparent that there is scatter in the data. For the velocities shown dynamic forces predicted by the analytical model.

Table I Model test results for inclined structure (Arctec Canada, [13]) $\alpha = \pi/4$, $\mu = 0.14$

	$h = 25$ mm, $\sigma_f^* = 14$ kPa, $E_{ef} = 14$ MPa				$h = 47$ mm, $\sigma_f^* = 24$ kPa, $E_{ef} = 83$ MPa			
	B = 750 mm		B = 250 mm		B = 750 mm		B = 250 mm	
Force, N / Velocity	$\bar{F}_x$	$\bar{F}_z$	$\bar{F}_x$	$\bar{F}_z$	$\bar{F}_x$	$\bar{F}_z$	$\bar{F}_x$	$\bar{F}_z$
$mm \cdot s^{-1}$								
20	93	60	28	29	150	116	88	92
100	96	98	28	26	246	271	144	137
200	72	113	34	35	204	226	130	116
Predicted Force, N	61	46	20	17	250	190	113	86

were not found to be significant. Only single force predictions, therefore, are given in the Table. For the 25 mm thick ice sheet the predicted values are about 1/3 less than the measured values; values for the 47 mm ice sheet agree closely. The high vertical forces on the 0.75 mm wide structure are likely due to ice piling on the model.

Acknowledgements

The many stimulating discussions with Joachim Schwarz while the author was on sabbatical leave at the Hamburg Ice Modelling Basin, and which led to the development of the basic concepts underlying the analytical model, are gratefully acknowledged. The cooperation of Razek Abdelnour, Arctec Canada, in carrying out the model tests is also acknowledged.

References

1 Tryde, P.: Intermittent ice forces acting on inclined wedges. CRREL Report 77.26, (1977).

2 Sørensen, C.: Interaction between floating ice sheets and sloping structures. Institute of Hydrodynamics and Hydraulic Engineering, Technical University of Denmark, Series Paper No. 19, (1978).

3 Croasdale, K.R., Metge, M., Verity, P.H.: Factors governing ice ride-up on sloping beaches. Proceedings IAHR 4th International Symposium on Ice Problems, Luleå, (1978), Pt. 1, 405-420.

4 Bercha, F.G. and Danys, J.V.: Prediction of ice forces on conical offshore structures. Proceedings IAHR Third International Symposium on Ice Problems, Hanover, N.H., (1975), 447-458.

5 Ralston, T.D.: Ice force design considerations for conical offshore structures. In Proceedings 4th International Conference on Port and Ocean Engineering under Arctic Conditions, (1977), St. John's, Newfoundland, 741-752.

6 Dolgopolov, Yu.V., Afanasiev, V.P., Koren'kov, V.A., Panfilov, D.F.: Effect of hummocked ice on the piers of marine hydraulic structures, IAHR Third International Symposium on Ice Problems, Hanover, New Hampshire, (1975), 469-477.

7 Schwarz, J., Frederking, R., Wessels, E., Hoffman, L.: Theoretische Untersuchungen und Durchführung stützender Eistank-Modellversuche zur Ermittlung der Eisdruckkräfte auf Rohrstützen-Tragwerke von Offshore-Plattformen, Hamburgische Schiffbau - Versuchsanstalt GmbH. Bericht Nr, 1512, (1978).

8 Gold, L.W., and Sinha, N.K.: The rheological behaviour of ice at small strains. IUTAM Symposium on the Physics and Mechanics of Ice, (1979).

9 Nevel, D.E.: The ultimate failure of a floating ice sheet. IAHR Symposium on Ice and its Action on Hydraulic Structures, Leningrad, (1972), 17-22.

10 Kerr, A.D.: The bearing capacity of floating ice plates subjected to static or quasi-static loads. Journal of Glaciology, Vol. 17, (1976), No. 76, 229-268.

11 Brix, J.: Zur Bemessung der Größe des Querstrahlsteuers, Vorsonderdruck, Schiff & Hafen, Jahr. 28, (1976), Heft. 5, 6-9.

12 Yu, Y.-T.: Virtual masses of rectangular plates and parallelepipeds in water. Journal of Applied Physics, Vol. 16, (1945), 724-729.

13 Abdelnour, R.: Model tests of an icebreaking inclined plane. Report No. FR 289C-2, Arctec Canada Limited, Ottawa, (1979).

This paper is a contribution from the Division of Building Research, National Research Council of Canada and is published with the approval of the Director of the Division.

The Rheological Behaviour of Ice at Small Strains

L.W. Gold and N.K. Sinha

Division of Building Research,
National Research Council of Canada.

Summary

Many important ice engineering problems are associated with relatively small strain, dynamic loads and stress up to the failure value. The deformation behaviour of ice can be assumed to be made up of three components: fully elastic, delayed elastic and viscous. Based on this assumed rheological model and the results of laboratory and field investigations, the temperature, time, strain and stress conditions are defined for which ice can be assumed to behave in an elastic manner for engineering purposes.

Introduction

A very significant number of applied mechanics problems for ice involve only elastic and primary creep behaviour. They include most ice pressure problems, all bearing capacity problems for moving loads, and static bearing capacity problems for which the maximum deflection is limited to the free board [1]. Few attempts have been made to date to find suitable analytical descriptions for the initial elastic and primary creep stage and, thereby, to provide an appropriate basis for the solution of this class of problem. The following describes work being carried out at the National Research Council of Canada.

Traetteberg et al

Traetteberg et al [2] made a study of the elastic modulus of ice over the nominal strain rate range of 10^{-8} to $5 \times 10^{-3}s^{-1}$ and temperature range of -10 to -39.5 °C. Stress was limited to a maximum of 0.5 $MN \cdot m^{-2}$. Analysis of these observations [3] showed that for a significant time following the initiation of loading the strain imposed on ice is essentially elastic plus delayed-elastic (recoverable), and that the time required for viscous or permanent strain to become significant depends on rate of application and duration of stress, temperature, and type of ice.

These results were analysed initially assuming that ice is a linear viscoelastic solid [4]. Observations on the time dependence of the strain recovery after load was removed, however, showed that ice does not have a simple relaxation behaviour. The apparent relaxation time calculated from the strain recovery curves was found to be proportional to the duration of the load raised to the power 2/3. Substitution of this apparent time dependence in the equation presented by Zener [4] gave:

$$\sigma + \alpha_\varepsilon t^{2/3} \dot{\sigma} = M_r \left(\varepsilon + \alpha_\sigma t^{2/3} \dot{\varepsilon} \right) \tag{1}$$

where t = time, σ = stress, ε = strain, M_r = value of the fully relaxed elastic modulus, and α_ε and α_σ were assumed to be constants.

It could be assumed in these experiments that the load was applied at a constant rate. For this condition it was possible to obtain the following solution for equation (1):

$$\varepsilon_t = \frac{\dot{\sigma}_o}{M_r} \left\{ t + (\alpha_\sigma - \alpha_\varepsilon) \left[\frac{2}{3} \alpha_\sigma t^{1/3} - t^{2/3} + \frac{2}{9} \alpha_\sigma^2 \left(\exp\left[- \frac{3t^{1/3}}{\alpha_\sigma} \right] - 1 \right) \right] \right\} \tag{2}$$

where ε_t = strain at time t, and $\dot{\sigma}_o$ = constant rate of application of stress.

Good agreement was found between the measured elastic modulus and that given by equation (2), i.e., $E(t) = \dot{\sigma}_o/\dot{\varepsilon}_t$, for the duration of load less than 10^3 s for granular ice and less than 10^2 s for columnar-grained S2 ice [5]. The analysis has been carried out to date only on observations made at -10°C.

Sinha

Sinha [6, 7] undertook a study of the elastic/delayed-elastic behaviour of ice under conditions of constant stress. He proposed the following empirical equation:

$$\varepsilon_t^\sigma = \frac{\sigma}{E_o} + c\left(\frac{\sigma}{E_o}\right)^s \left\{1 - \exp\left(-\left[a_T t\right]^b\right)\right\} + \dot{\varepsilon}_1\left(\frac{\sigma}{\sigma_1}\right)^n t \tag{3}$$

where ε_t^σ = strain for stress, σ, and time, t; E_o = Young's modulus (time independent and, for ice, only slightly temperature dependent); a_T = temperature dependent factor = $a_o \exp\left(-\frac{Q}{RT}\right)$; Q = activation energy; R = gas constant; T = temperature; $\dot{\varepsilon}_1$ = viscous creep rate for unit stress; and a_o, c, s, b, n are factors initially assumed to be constants.

The first term on the right hand side is the initial elastic deflection; the second term describes the delayed-elastic (recoverable) strain; the third term is the viscous flow. Sinha found equation (3) to be a good description of experimental results for n = 3, b = 0.34 ≃ 1/n and s = 1.

It is of interest that for b = 1/3 the first two terms of equation (3) are a solution to equation (1) for the constant load condition, i.e.,

$$\alpha_\sigma t^{2/3}\dot{\varepsilon} + \varepsilon - \frac{\sigma}{M_r} = 0 \tag{4}$$

Although based on different approaches, the equations obtained by Sinha [6, 7] and by Gold and Traetteberg [3] are solutions for the same general equation. Work is now in progress to establish whether experimental results are in agreement with this implied equivalence.

Strain Behaviour

This success provides an encouraging basis for exploring the dependence of the elastic/delayed-elastic behaviour of ice on time, stress, strain and temperature. Considering the variability that is normally present in ice engineering problems, it is reasonable to assume that deformation is essentially elastic/delayed-elastic as long as the permanent viscous strain is small, say less than 5% of the total strain. Equations (2) and (3) indicate that this condition will depend on the load and how it is imposed on the ice. For the purpose of the present discussion only the condition of constant stress will be considered.

Sinha [6] has compared equation (3) with experimental results. This

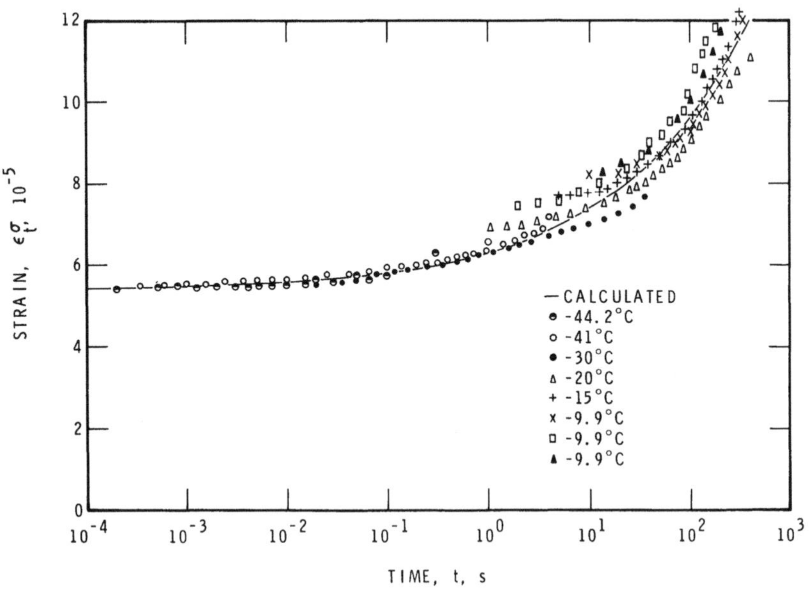

Fig. 1 Creep of S2 ice reduced to a reference temperature of -10°C, σ = 0.49 MN·m^{-2}

comparison is presented in Fig. 1 for inclusion-free type S2 ice subject to a stress of 0.49 $MN{\cdot}m^{-2}$; all results are shifted to the reference temperature of -10°C using the appropriate value for a_T. The analysis indicates that ice is thermo-rheologically simple; the activation energy for the delayed-elastic term has the same value as that determined for viscous creep, i.e., Q = 16 K cal/mol (67 KJ/mol).

Figure 2 shows the time dependence of ε_t^σ calculated from equation (3) for various applied constant stresses for S2 ice at -10°C and average grain size 3 mm. The values of the factors in equation (3) were assumed to be E_o = 9.5 $GN{\cdot}m^{-2}$; Q = 67 KJ/mol; c = 3; n = 3;

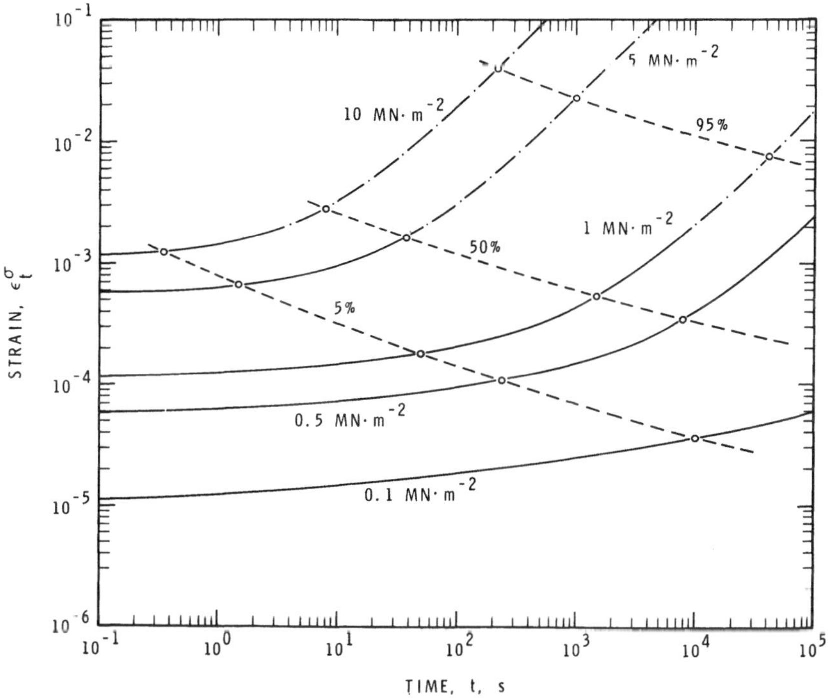

Fig. 2 Creep of S2 ice of average grain size of 3 mm at -10°C for several stress levels. Dot-dash part of each creep curve indicates region for which equation (3) is assumed to be invalid owing to neglect of crack formation. Broken lines indicate time and strain at which viscous strains are 5, 50 and 95% of total strain

$b = 0.34$; $s = 1$; $a_o = 2.5 \times 10^{-4}\ s^{-1}$; $\dot{\varepsilon}_1 = 1.76 \times 10^{-7}\ s^{-1}$ for $\sigma = 1.0\ MN{\cdot}m^{-2}$, $T = 263°K$. The dashed lines give the time for each stress for which the viscous strain is 5, 50 and 95% of the total strain, assuming the specimen has not failed or yielded prior to that time.

The effects of cracking are not taken into consideration in Fig. 2. Observations have shown that for a constant stress greater than about $0.6\ MN{\cdot}m^{-2}$ internal cracks are formed that affect the conditions of yield and brittle behaviour [8]. The strain to the formation of the first crack for a stress of $1\ MN{\cdot}m^{-2}$ is about 5×10^{-4}. For stress greater than about $8\ MN{\cdot}m^{-2}$ a brittle type of behaviour can be expected, with premature failure occurring at strains of about 1.5×10^{-3} [9]. For ductile behaviour, yield for S2 ice occurs at a strain of from 5×10^{-4} to 25×10^{-4} for the range of stresses 1 to $8\ MN{\cdot}m^{-2}$ under conditions of constant rate of increase of stress [9]. It is clear that beyond a certain time and strain for each stress the curves shown in Fig. 2 are invalid. This range is indicated by the dot-dash part of each curve. Information to date suggests that the proposed elastic/delayed-elastic model is a reasonable approximation of the response of ice to a uniaxial load for stress up to $10\ MN{\cdot}m^{-2}$ and viscous strain less than 5% of the total strain.

Elastic Modulus

Sinha [6, 7] has shown that a convenient way to look at the delayed-elastic behaviour of ice is to consider the creep compliance function

$$D_t^\sigma = \frac{1}{E_t^\sigma}$$

where D_t^σ = creep compliance function for stress, σ, and time, t; E_t^σ = effective elastic modulus for stress, σ, and time, t. As ice is nonlinear, the values of D and E at a given time depend on stress. For purposes of presentation and discussion it is useful to consider the normalized form of the compliance function

$$\frac{D_o^\sigma}{D_t^\sigma} = \frac{E_t^\sigma}{E_o} = \frac{\varepsilon_o^\sigma}{\varepsilon_t^\sigma}$$

where D_o^σ is the compliance function at t = o; E_o is Young's modulus (elastic modulus at t = o); ε_o^σ is the initial elastic strain; $E_t^\sigma = \sigma/\varepsilon_t^\sigma$; ε_t^σ is given by equation (3).

D_o^σ/D_t^σ is plotted against time in Fig. 3 for various stresses. The dashed line is the elastic compliance function calculated from the first two terms of equation (3), i.e., elastic/delayed-elastic response only. Either Fig. 2 or Fig. 3 shows that the elastic compliance function is applicable, at -10°C, for a stress of 0.5 $MN{\cdot}m^{-2}$ for 250 s, for a stress of 1.0 $MN{\cdot}m^{-2}$ for 50 s, and for a stress of 5.0 $MN{\cdot}m^{-2}$ for 15 s. The time for which ice can be assumed to be "elastic" for various stresses at other temperatures can be obtained by shifting the curves in Fig. 3 along the time axis, using

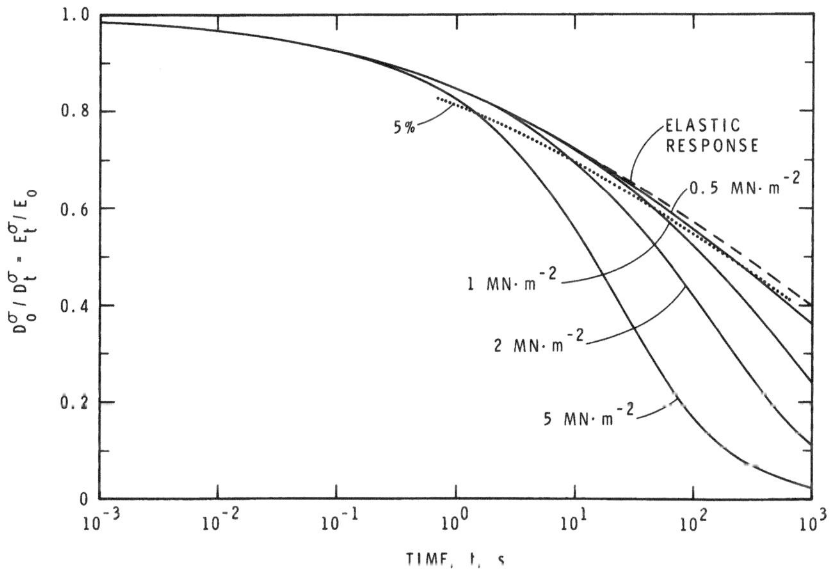

Fig. 3 Normalized creep compliance function for S2 ice, average grain size of 3 mm, at -10°C for different loads. Dotted line indicates condition at which viscous strain is 5% of total strain

the appropriate value for a_T. For example, permanent strain for stress of 0.5 $MN{\cdot}m^{-2}$ would be less than 5% of the total strain at -45°C for up to about 25,000 s. At -10°C and stress less than 0.1 $MN{\cdot}m^{-2}$ ice can be assumed to be elastic for about the first 10^4 s following the application of load.

The temperature dependence of the compliance function is shown for stresses of 1 and 2 $MN{\cdot}m^{-2}$ in Fig. 4 for 1, 5 and 30 s following application of load. A reasonable time of measurement of laboratory tests is 5 s and it is significant that the experimental results of Gold [10] on coarse-grained ice agree well with the 5 s curve in Fig. 4.

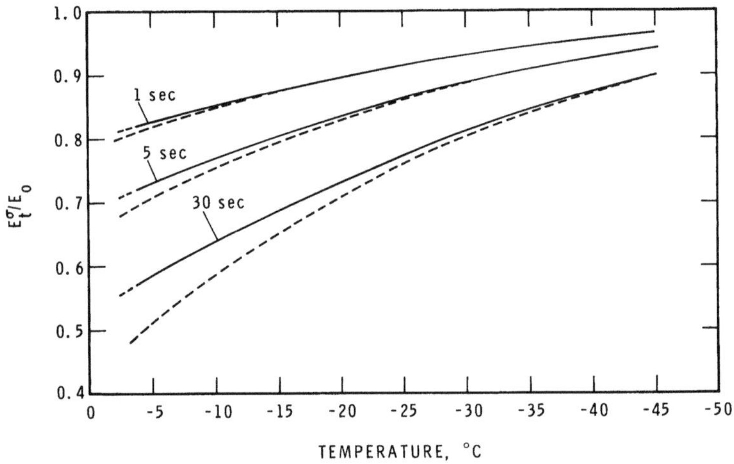

Fig. 4 Temperature-dependence of normalized effective modulus of S2 ice for loads of 1 $MN{\cdot}m^{-2}$ (solid line) and 2 $MN{\cdot}m^{-2}$ (broken line) at three different times

Dynamic Modulus

Sinha [7] has approximated a cyclical load of frequency f by a step function in which the time of loading, t, is assumed equal to 1/2 f. The modulus calculated using equation (3) and σ = 0.3 $MN{\cdot}m^{-2}$ for 3 mm grain size is compared in Fig. 5 with values compiled by

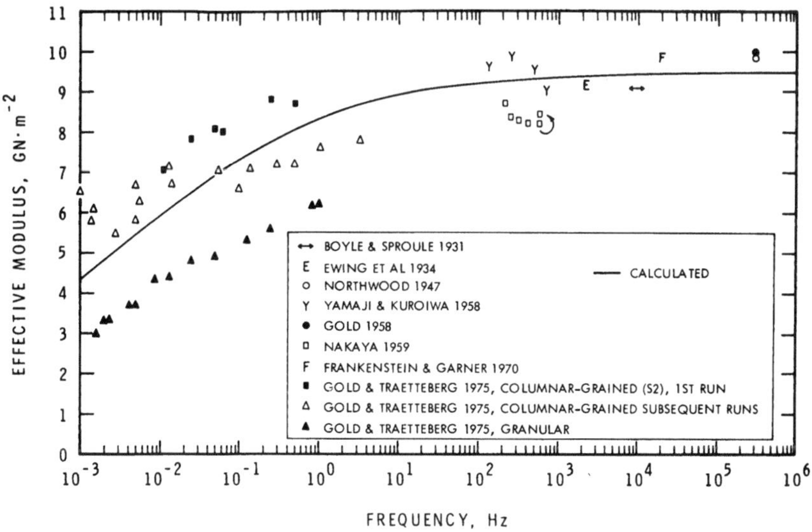

Fig. 5 Frequency dependence of effective modulus for polycrystalline ice at -10°C. Experimental data are taken from Ref. 1, which includes the other references cited in this figure. Calculated results are shown by the solid line for stress amplitude of 0.3 $MN \cdot m^{-2}$, E = 9.5 $GN \cdot m^{-2}$

Gold [1]. The agreement for S2 ice is encouraging and indicates that dynamic moduli calculated from the elastic and delayed-elastic terms only are valid for stress less than 0.5 $MN \cdot m^{-2}$ for frequencies down to at least 10^{-3} Hz at T = -10°C. The range of validity for higher stresses is still to be determined. Sinha is currently investigating grain size effects [11].

Discussion

The viscoelastic nature of ice has interesting implications with respect to the analysis of its deformation behaviour. Fig. 2 shows that the duration of the primary or transient creep period is very greatly dependent on stress. For stress less than 0.1 $MN \cdot m^{-2}$ the time required for viscous strain to reach 95% of the total strain is about 1 year at -10°C. It is rare that creep tests are carried out for such periods of time at low stress, although they are often analysed as if the secondary creep stage had been developed. Sinha [7] has considered the implications of this; and the result of his analysis is presented in Fig. 6 for S2 ice at -10°C. The general trend for stress less than 0.1 $MN{\cdot}m^{-2}$ and time less than 10^5 s is similar to that reported by several other investigators. Evidence from the studies being conducted at the National Research Council of Canada indicates that the power law dependence of the viscous strain on stress extends well down into the low stress range if proper account is taken of the delayed-elastic effect [11].

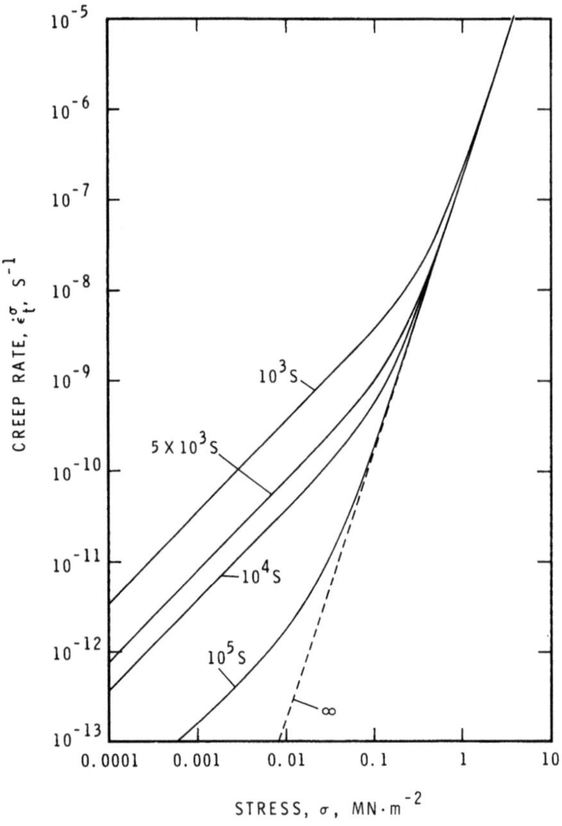

Fig. 6 Creep rate of S2 ice in uniaxial compression against applied stress at -10°C for several loading times. Curves are calculated on the basis of proposed rheological relation

The assumed elastic/delayed-elastic model for ice explains the wide range of reported values of the elastic modulus. Laboratory measurements of the "static" elastic modulus can be made within 5 s of application of the load. In-situ field measurements usually take longer, and often insufficient attention is given to close control and recording of the time and rate of loading. Field values are subject to additional causes of variation such as temperature gradients and impurities. It is not unreasonable, therefore, that reported values measured in the field by static methods should be lower than laboratory values. With greater appreciation of the factors controlling elastic response of ice it should be possible to design both field and laboratory methods for measuring the elastic moduli that will allow a proper, rational interpretation of the results. The need to standardize methods of measuring elastic moduli so as to have a proper basis for comparing results has been recognized in the field of plastics, for example by ASTM [12]. Similar steps must be taken for ice, so that gradually it will be possible to define conditions for which elastic/delayed-elastic behaviour can be assumed for applied mechanics problems involving ice.

Conclusion

Using standard methods of analysis established for the field of rheology, analytical expressions can be found for the elastic/delayed-elastic behaviour of ice under various boundary conditions. Through these analytical expressions and the one for viscous strain it is possible to define the limits of elastic/delayed-elastic behaviour of ice in terms of time, stress, strain and temperature.

References

1 Gold, L.W.: Engineering properties of fresh water ice. J. Glaciol., 19, No. 81 (1977) 197-211.

2 Traetteberg, A.; Gold, L.W.; Frederking, R.M.: The strain rate and temperature dependence of Young's modulus of ice. Proc. 3rd Int. Symp. on Ice Problems, IAHR, Hanover, N.H. (1975) 479-486.

3 Gold, L.W.; Traetteberg, A.: Young's modulus of ice and ice engineering problems. Proc. 2nd Symp. on Applications of Solid Mechanics, June 17, 18, 1974, Hamilton, Canada. Ed. by Weaver, D.S., Faculty of Engineering, McMaster. (1975) 1-16.

4 Zener, C.: Elasticity and anelasticity of metals, Univ. Chicago Press (1948)

5 Michel, B.; Ramseier, R.O.: Classification of lake and river ice. Can. Geotech. J., 8, No. 1 (1971) 36-45.

6 Sinha, N.K.: Rheology of columnar-grained ice. Exper. Mech., 18, No. 12 (1978) 464-470.

7 Sinha, N.K.: Short-term rheology of polycrystalline ice. J. Glaciol., 21, No. 85 (1978) 457-474.

8 Gold, L.W.: The process of failure of columnar-grained ice. Phil. Mag., 26, (1972) 311-328.

9 Sinha, N.K.: Rate sensitivity of compressive strength of columnar-grained ice. Presented at 1979 Spring Meeting, Society of Experimental Stress Analysis, San Francisco, California, May 20-25 (1979).

10 Gold, L.W.: Some observations on the dependence of strain on stress for ice. Can. J. Phys. 36, No. 10 (1958) 1265-75.

11 Sinha, N.K.: Grain size influence on effective modulus of ice. Proc. of Workshop on the Bearing Capacity of Ice Covers, Winnipeg, Canada, 16-17 Oct 1978. National Research Council of Canada; Associate Committee on Geotechnical Research, TM No. 123 (1979) 65-79.

12 ASTM, D674-56: Standard recommended practices for testing long-time creep and stress relaxation of plastics under tension or compression loads at various temperatures. (Reapproved, 1969).

This paper is a contribution from the Division of Building Research, National Research Council of Canada and is published with the approval of the Director of the Division.

Critical Stress Intensity Factor (K_{Ic}) Measurements at High Loading Rates for Polycrystalline Ice

D.J. GOODMAN
Physics and Chemistry of Solids
Cavendish Laboratory
Madingley Road
Cambridge CB3 OHE
UK

Abstract

In this paper results are presented from experiments to measure K_{Ic} the critical stress intensity factor, at -4, -11, and -24°C. Rectangular bars with a slot ending in a sharp crack in one side were broken in four point bending. The mean values were 118 ± 32 kN $m^{-3/2}$ (44 samples) at -4°C, 119 ± 34 kN $m^{-3/2}$ (40 samples) at -11°C, and 108 ± 21 kN $m^{-3/2}$ (44 samples) at -24°C. These agree well with the results of other studies, and show that G, the strain energy release rate is close to twice the surface energy, Γ (overall mean for K_{Ic} is 115 kN $m^{-3/2}$, implies G equals 1.38 Jm^{-2} compared to 0.2 Jm^{-2} for 2Γ).

Some preliminary results from the measurement of crack opening displacement are also presented.

1 INTRODUCTION

In this paper we report the results of measurements of the critical stress intensity factor K_{Ic} of polycrystalline ice at -4, -11 and -24°C at relatively high rates of strain. This work was prompted by an interest in the way in which small swells break up sea ice. The sea ice is cyclically strained (with a typical period of 12s and strain amplitude of 5 x 10^{-6}) as the swell passes through the pack ice. The loading rate is correspondingly high.

The first study of fracture toughness appears to have been carried out by Gold (1963) [1] using a thermal shock technique (for his review see Gold, 1972 [2]). The first investigation of K_{Ic}, using the methods of more modern fracture-mechanics was due to Liu and Loop (1974) [3]. They employed a compact tensile specimen (CTS) and a crack arrest method using a wedge. Since then indentation methods and 4-point bending tests have been used in a study by Goodman (1977) [4] and by Goodman and Tabor (1978) [5]. In 1977 Vaudrey [6] reported some preliminary results on fracture using a 4-point bending test to measure K_{Ic} as a function of brine volume. The most recent paper is that of Liu and Miller (1979) [7]; their work is a study of the fracture of ice using the methods described by Liu and Loop. They discuss the effect of temperature and strain rate.

The present paper deals with a 4-point bending test in which a standard crack is first inserted into one side of a bar of ice. It involves measurements of three parameters - the stress intensity factor K_{Ic}, the crack opening displacement (COD) at failure and the rate of change of K_{Ic}.

2 FRACTURE-TOUGHNESS PARAMETERS

Miller [8] elsewhere in this volume discusses the concepts of linear elastic fracture mechanics (K_{Ic}, COD, J and G). Knott (1973) [9], Lawn and Wilshaw (1975) [10] and issue No. 4, vol. 10 of the Journal of Strain Analysis, also provide useful summaries. Briefly, K_{Ic} is a material parameter which describes the stress required to make a crack of a known size propagate. It is directly related to the strain energy release rate, G, by the expression (for plane strain):

$$K^2_{Ic} = \frac{GE}{1-\nu^2} \tag{1}$$

where E is the Young's modulus, and ν Poisson's ratio. The "I" refers to Mode I failure; the crack surfaces are opening in directions normal to the crack faces. For a purely brittle solid G equals 2Γ, where Γ is the surface energy. For different crack geometries K varies, for the case of interest here, 4-point bending, for a specimen with dimensions shown in figure 1, and thickness b, K is given by the expression:

$$K_{Ic} = \frac{3P\ (\ell_1-\ell_2)\ a^{\frac{1}{2}}}{2bh^2}\ f(a/h) \tag{2}$$

where f(a/h) is:

$$f(a/h) = 1.99 - 2.47(a/h) + 12.97(a/h)^2 - 23.17(a/h)^3 + 24.80(a/h)^4 \tag{3}$$

(Brown and Srawley, 1966 [11])

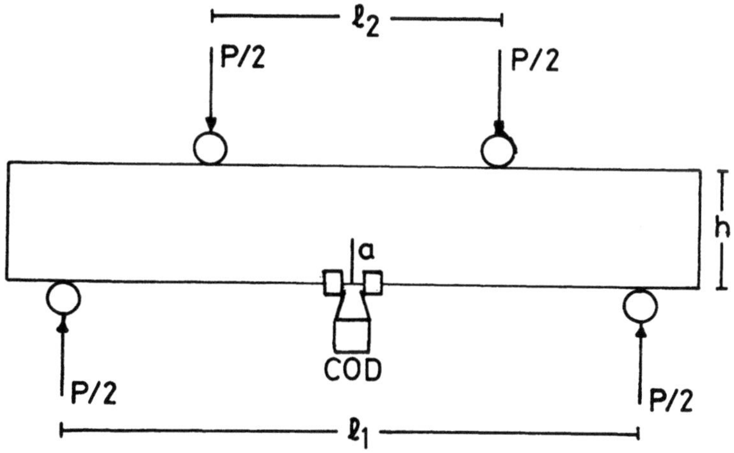

Figure 1: The 4-point bend test geometry.

K_{Ic} is only a useful fracture parameter if certain conditions are met. Namely, that the region of plasticity in the specimen is confined to a small zone near the crack tip and that the crack is sharp. The specimen thickness should be wide enough to ensure that the centre of the crack remains in the plane strain.

Linear elastic fracture mechanics incorporates the small zone of plasticity that must occur near the crack tip (to prevent an infinite stress at the crack tip) by extending the crack length from a to $(a + a_p)$, where a_p is a correction to allow for the plastic zone. In the Dugdale model this correction is given by:

$$a_p = \frac{\pi^2 K^2}{8\sigma_y^2} \tag{4}$$

where σ_y is the yield stress. For ice, at temperatures close to its melting temperature, a yield stress cannot be uniquely defined. A strain rate must be specified. Consider figure 2, a deformation mechanism map for ice, which summarises the creep properties.

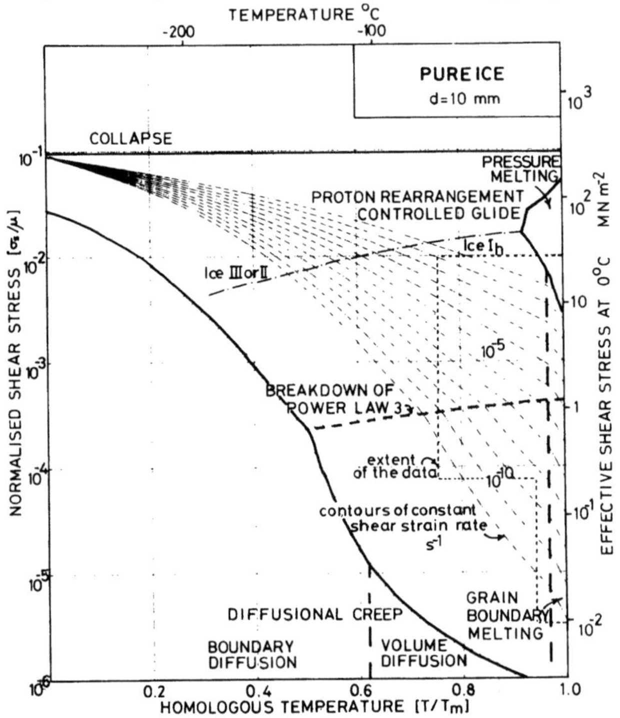

Figure 2: The creep properties of ice. The lines crossing the diagram are contours of constant shear strain rate.

If the ice were at a homologous temperature of 0.3 T_m (82°K), as the stress was increased on the ice no effect would be observed until a stress of 0.1 μ was reached. However at a homologous temperature of 0.95, the ice creeps with a finite strain-rate at all stresses within the range 3×10^{-5} to 3×10^{-3} (0.1 MN m^{-2} to 10 MN m^{-2}).

If a crack is loaded, the stresses around the crack tip will continuously be relaxed by creep. To make an estimate of a_p, we can choose a representative strain rate of say $10^{-4} s^{-1}$ and find the creep stress at 0.95 Tm from figure 2 ($\sim \sigma_{e/\mu} = 10^{-3}$), and insert this into equation 4 (using a value of $K = 115$ kN m^{-2} - see section 212).

Since the size of the plastic zone increases with the value of K, by choosing the final value of K deduced from the experiments we ensure that our value of a_p is not an underestimate. The value of a_p so deduced (for the region within which the creep rate is greater than $10^{-4}s^{-1}$) is of order 0.6 mm and therefore very much smaller than the crack length.

The displacement at the crack tip (or COD) is related to the 'yield stress' in the plastic zone by the expression:

$$\delta = K^2 \frac{(1-\nu^2)}{E} \cdot \frac{1}{\sigma_y{}^*} = \frac{G}{\sigma_y{}^*} \tag{5}$$

(the * indicates the stress is the constrained yield stress - in a normal material with a well defined σ_y in plane strain $\sigma_y \sim 3\sigma_y$). In a 4-point bend test, the crack opening displacement is related to the displacement measured on the face of the specimen, δ_m, by the expression:

$$\delta = \frac{\delta_m}{1 + \dfrac{2.7(a+z)}{h-a}}$$

where z is the height of the knife edge supporting the gauge above the ice surface, a the crack length, and h the specimen width.

In the experiments described in this paper, the problems associated with creep near the crack tip have been side-stepped by using a high loading rate (in excess of 500 N s^{-1}). The K_{Ic} values derived are lower bound. At lower strain rates the extra amount of plastic work done at the crack tip will always tend to increase K_{Ic}.

3 EXPERIMENTS

3.1 Experimental Method

Polycrystalline ice samples were manufactured from distilled water by upward freezing of water in a tray. The tray was insulated across the top and around the sides, and placed on a metal block inside a deep freeze maintained at -13°C. The ice grew from the bottom to produce bubble free blocks of ice. To ease specimen manufacture, the tray was subdivided by perspex plates so that twenty rectangular ice blocks could be manufactured at once. The grain size was 5 - 10 mm, with a columnar orientation, and horizontal c-axis. The blocks were then transfered to a small deep freeze, whose temperature could be set to $1^{o}C$, and allowed to reach thermal equilibrium over several hours. After this period each block was broken in 4-point bending after a sharp crack had been nucleated at the end of a slot (cut using a band saw) with a scalpel blade (this ensured the crack radius was sharp). The crack and slot length, and specimen dimensions were measured with vernier calipers. The load was applied by a hand driven hydraulic ram with the arrangement shown in figure 3. For some of the tests a crack opening displacement gauge (COD) was attached to the block with a clamp, and the COD as a function of the load measured.

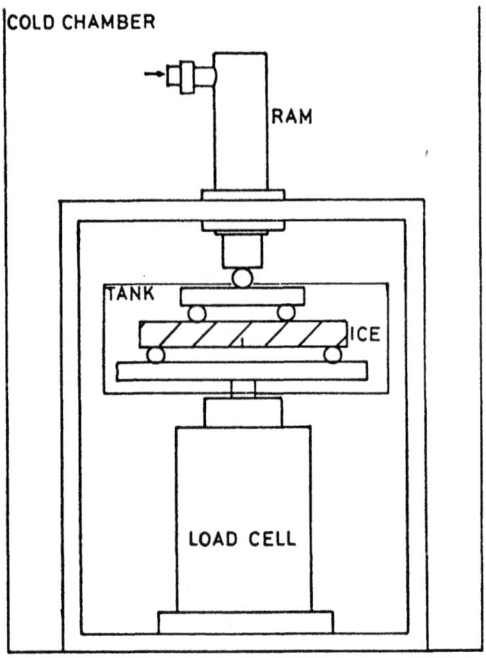

Figure 3: The loading arrangement.

The 4-point bend test applies a uniform bending moment to the crack which need not be located directly under the ram but only somewhere between the upper loading points. The specimen dimensions (width 20 mm, thickness 30 mm and length 140 mm) were chosen so that the crack was in plane strain in the centre of the block and satisfied the criteria for the correct measurement of K_{Ic} (K_{Ic} reaches a minimum value as the specimen thickness is increased, and crack tip radius reduced, Knott, 1973 [9]). Figure 4 shows an example of of the specimens; the sharp crack initiated with the scalpel blade is visible at the end of the slot.

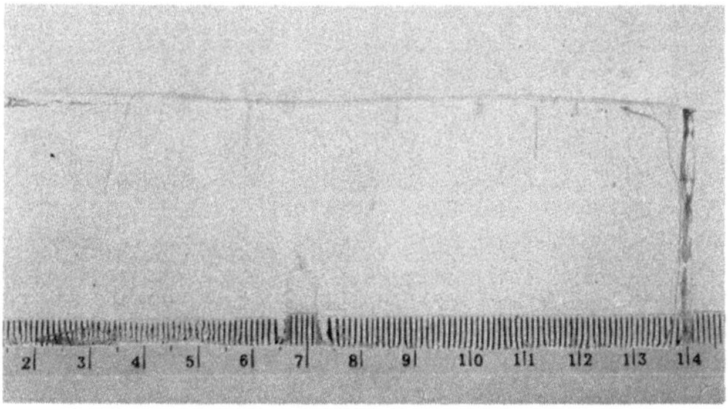

Figure 4: A typical ice specimen showing the sharp crack at the end of the slot.

The signals from the load cell and COD gauge were recorded on a FM tape recording system and replayed through either a UV chart recorder or, in the case of the COD measurements, on an X-Y chart recorder. This arrangement had the advantage that signals could be replayed at a slower speed than that at which they were recorded. From the slope of the replayed signal the loading rate and hence K could be calculated.

4 RESULTS AND DISCUSSION

Figure 6 is a set of histograms which show the results from the 4-point bend tests, and figure 5 plots the data as a function of temperature. (The loading rate data appear in the appendix.)

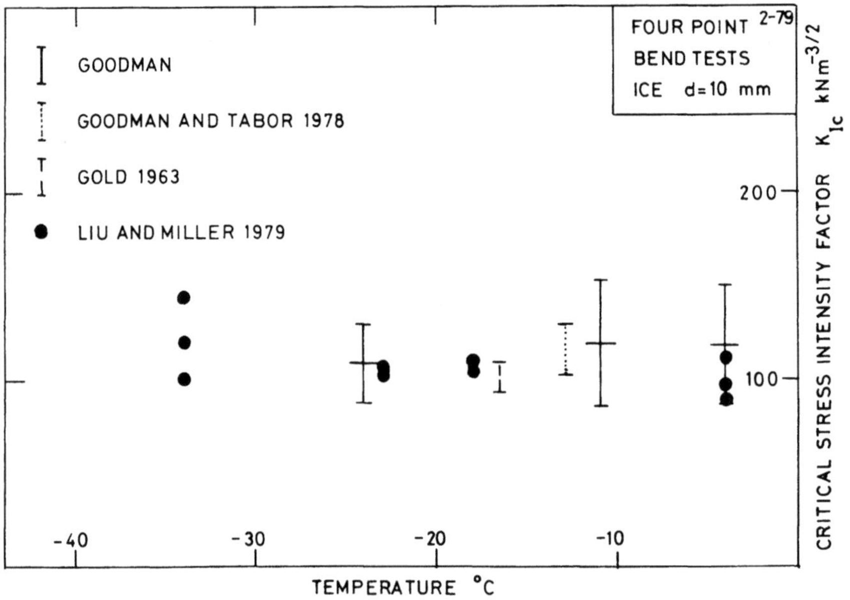

Figure 5: The fracture toughness values as a function of temperature

The results show that the critical stress intensity factor K_{Ic} probably does not vary significantly with temperature for the loading rates of these tests ($\dot{K} \sim 1000$ kN m$^{-3/2}$ s^{-1}). The mean values and corresponding G values are tabulated in table I together with the work of other authors. (The Young's modulus used to calculate G was taken from Dantl (1968) [12]. E = 8.716 (1-0.00125T) GN m^{-2} and ν = 0.3.

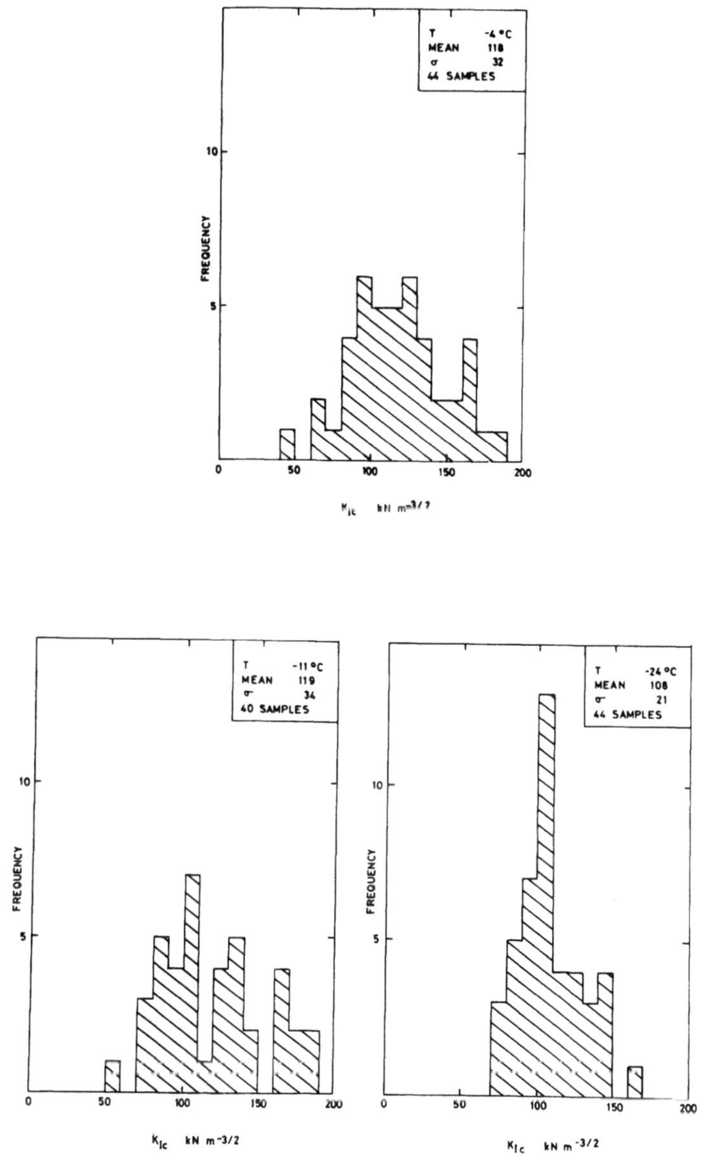

Figure 6: Histograms to show the spread of K_{Ic} values for -4, -11 and -24°C.

Table I: K_{Ic} and G values for polycrystalline ice

	T ^{o}C	K_{Ic} $kN\ m^{-3/2}$	G Jm^{-2}
This work	-4	118 ± 32	1.46 ± 0.79
	-11	119 ± 34	1.45 ± 0.83
	-24	108 ± 21	1.30 ± 0.51
Goodman and Tabor (1978) (3-point bend test)	-13	116 ± 13	1.41 ± 0.31
Gold (1963) (Thermal Shock)	-16.7	94 - 110	0.923 - 1.53
Liu and Miller (1979) (average of high loading rate tests between -4° and -46°)	-	108 ± 10	1.47 ± 0.27

Jones (1973) [13] finds the surface energy (air-solid) to be of the order of 0.1 Jm^{-2}. In a purely brittle solid G would equal twice this energy.

Our results are in agreement with the previously observed values.

The results indicate that K_{Ic} probably does not vary with temperature at the loading rates used here. Preliminary results obtained from further measurements at $-38^{o}C$ indicate a similar mean to the measurements at -4, -11 and $-24^{o}C$. As we discussed in our previous paper, the values of G are very close to twice the surface energy. This indicates that little energy is lost in plastic work as the crack advances (for steel 2Γ is about 10 Jm^{-2}, while G is of the order of 40,000 Jm^{-2}).

Two examples of the records obtained from the COD measurements are shown in figure 7. The COD appears to rise linearly with the load right up to the failure point. Using equation 4 an approximate value for the apparent 'yield stress' at the crack tip can be deduced. Table II gives the results for two examples.

Figure 7:
Two examples of COD versus load for ice.

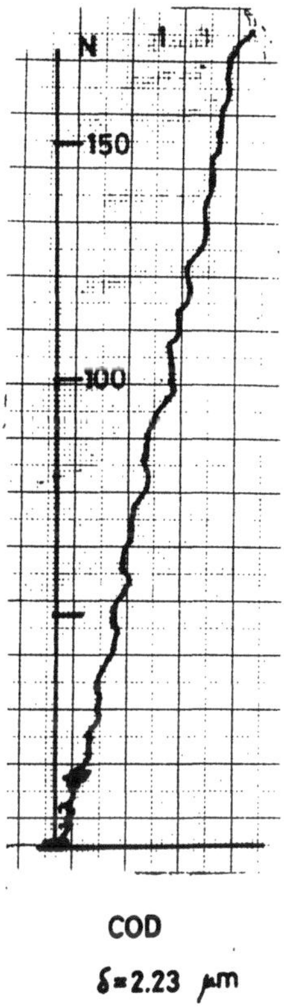

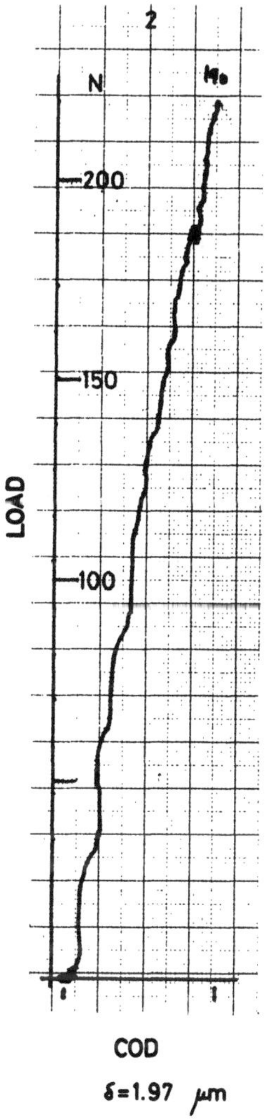

Table II: COD measurements

	δ μm	K_{Ic} kN $m^{-3/2}$	G Jm^{-2}	'σ_y' MN m^{-2}
Example 1	2.23	112	1.30	0.58
Example 2	1.97	130	1.76	0.89

Further experiments are in progress, but the preliminary results suggest that the COD should be measured in each fracture test to serve as a guide to the extent of plasticity at the crack tip, and in particular the rate of change of the COD indicates the amount of creep occurring at the crack tip. The apparent yield stresses in Table II, are at a level where gross yielding of the ice would be expected. The results further suggest that the assumed size of the plastic zone is reasonable and that linear elastic fracture mechanics can be applied to ice problems.

5 SUMMARY

Ice samples have been loaded to failure in a 4-point bend apparatus to measure the critical stress intensity factor K_{Ic} at three temperatures. The results indicate, that at these loading rates, the fracture toughness is independent of temperature. This is expected if creep at the crack tip is not an important term in the work required to make the crack advance (the creep rate for a given stress depends exponentially on the reciprocal of the temperature). Preliminary results from the COD measurements suggest that COD can provide a useful indication of the extent of plasticity at the crack tip.

These experiments are continuing, and it is hoped soon to publish further results. Microscope mechanisms of fracture have not been discussed here but experiments are in progress to study the effect of grain size on K_{Ic} for a fixed $\dot{K}_{Ic}$, and to replicate the fracture surface to determine whether the crack runs across grains or around grain boundaries.

6 ACKNOWLEDGEMENTS

I am very grateful to L. Holt for carrying out much of the experimental work, and PA Goodman for reducing some of the data; and to Professors D. Tabor and M.F. Ashby, and Dr. Brown, Hagan and Chaudhri for stimulating discussions. The SRC, Royal Society, and Girton College, Cambridge kindly provided funds to support the work.

7 REFERENCES

1 Gold, L.W.: Crack formation in ice plates by thermal shock. Canadian Journal of Physics 41 (1963) 1712-1138.

2 Gold, L.W.: Engineering properties of fresh water ice. Journal of Glaciology 19(1977) 197-212

3 Liu, H.W.; Loop, L.W.C.: Fracture toughness of fresh water ice. Personal Communication.

4 Goodman, D.J.: Creep and fracture of ice, and surface strain measurements on glaciers and sea ice. Ph. D thesis. University of Cambridge (1977).

5 Goodman, D.J.; Tabor, D. : Fracture toughness of ice: a preliminary account of some new experiments. Journal of Glaciology 21 (1978) 651-660.

6 Vaudrey, K.D.: Ice Engineering - study of related properties of floating sea-ice sheets and summary of elastic and visco-elastic analyses. Technical report R860 Civil Engineering Laboratory, Port Hueneme, California. (1977).

7 Liu, H.W.; Miller, K.J.: Fracture toughness of fresh water ice. Journal of Glaciology 22 (1979) 135-143.

8 Miller, K.J.: The application of fracture mechanics to ice problems. Proceedings of the IUTAM Symposium on the Physics and Mechanics of Ice. Copenhagen, August 6-10 1979. This Volume.

9 Knott, J.F.: Fundamentals of fracture mechanics. London. Butterworth.

10 Lawn, B.R.; Wishaw, T.R.: Fracture of brittle solids. Cambridge, University Press.

11 Brown, W.F.; Srawley, J.E.: Plane strain crack toughness of high strength metallic materials. American Society for Testing and Materials (ASTM Special Technical Publication No. 410). (1966).

12 Dantl, G.: Die elastischen Moduln von Eis-Einkristallen. Physik der Kondensierten Materie, 7(1968) 390-397.

APPENDIX

Temperature $-4^{\circ}C$

	K_{Ic} (kN $m^{-3/2}$)	K_{Ic} (kN $m^{-3/2}$ s^{-1})
1	113.9	335
2	98.5	5668
3	115.5	1218
4	65.2	750
5	144.0	729
6	157.1	1243
7	159.3	2017
8	135.7	1323
9	111.0	299
10	166.3	2105
11	88.8	702
12	164.5	992
13	163.3	899
14	139.4	1260
15	83.9	885
16	95.6	3025
17	122.6	817
18	127.1	555
19	122.2	595
20	85.4	451
21	95.3	1508
22	148.2	1563
23	107.1	437
24	85.8	1811
25	120.6	2545
26	132.7	730
27	124.1	1208
28	174.9	1230
29	184.9	687
30	109.9	662

Temperature -4°C (cont'd)

	K_{Ic} (kN $m^{-3/2}$)	K_{Ic} (kN $m^{-3/2}$ s^{-1})
31	102.5	1082
32	168.3	1066
33	99.4	434
34	139.2	1356
35	94.5	1197
36	100.8	5105
37	96.9	876
38	66.4	525
39	79.8	388
40	122.2	1368
41	118.7	1503
42	43.6	394
43	105.7	836
44	123.8	978

Temperature -11°C

	K_{Ic} (kN $M^{-3/2}$)	K_{Ic} (kN $m^{-3/2}$ s^{-1})
1	87.8	231.4
2	124.6	830
3	166.3	1403
4	72.6	575
5	130.8	613
6	100.2	1268
7	101.9	1434
8	77.7	579
9	129.2	244
10	89.3	452
11	93.5	1497
12	94.4	2391

Temperature $-11^{\circ}C$ (cont'd)

	K_{Ic} ($kN\ m^{-3/2}$)	$\dot{K}_{Ic}$ ($kN\ m^{-3/2}\ s^{-1}$)
13	109.3	659
14	186.8	2958
15	160.4	4060
16	99.4	898
17	71.1	648
18	185.0	1561
19	170.0	1957
20	169.9	1654
21	142.4	1002
22	104.6	2649
23	59.1	468
24	89.1	806.1
25	175.6	1852
26	134.9	813
27	138.8	977
28	114.8	807
29	100.7	708
30	129.2	1258
31	162.9	666
32	145.1	633
33	108.0	1367
34	130.1	1373
35	130.0	1496
36	89.5	1259
37	80.5	1699
38	125.2	600
39	92.8	1173
40	100.5	749

Temperature $-24^{o}C$

	K_{Ic} (kN $m^{-3/2}$)	$\dot{K}_{Ic}$ (kN $m^{-3/2}$ s^{-1})
1	107.7	799
2	102.3	806.6
3	142.4	743.7
4	103.5	421.2
5	161.4	926
6	148.0	1167
7	116.9	615
8	125.4	879
9	100.2	2727
10	95.8	1209
11	90.5	519
12	83.7	660
13	72.3	761
14	124.0	1565
15	140.3	1106
16	107.5	617
17	102.3	1614
18	98.3	775
19	132.8	1676
20	102.7	720
21	139.2	1254
22	101.9	989
23	118.1	1243
24	101.2	982
25	101.3	1597
26	93.3	535
27	98.9	1249
28	136.0	1073
29	98.6	1245
30	105.2	1020
31	78.7	1240

Temperature $-24^{o}C$ (cont'd)

	K_{Ic} ($kN\ m^{-3/2}$)	$\dot{K}_{Ic}$ ($kN\ m^{-3/2}\ s^{-1}$)
32	82.7	1159
33	93.5	437
34	123.8	459
35	115.0	1462
36	128.1	951
37	116.1	1332
38	107.4	904
39	103.6	817
40	85.3	1345
41	73.4	926
42	140.8	1616
43	85.3	1537
44	80.4	922

Theory of Floating Ice Sheets

K. HUTTER and F.M. WILLIAMS*

Laboratory of Hydraulics, Hydrology and Glaciology,

Federal Institute of Technology

8092 Zürich, Switzerland

Summary

In ice sheets, extensions in horizontal direction are very much larger than perpendicular to it, suggesting a two-dimensional approximate analysis. It is indicated how the usual plate models for stress analyses can be derived from general approximating techniques, and how two-dimensional descriptions for thermo-mechanical interactions in floating ice sheets are obtained. A "plate model" for the thermal response of an ice shelf is thus derived. The governing equations include accumulation at the top and phase change at the bottom and take convection and dissipation into account. An analysis of the steady state response is presented, and the response of the shelf to climatic conditions is determined. Generalizations to time-dependent situations are indicated.

Introduction

With common plate theories stress analyses are performed by reducing the three-dimensional stress problem to a simpler two-dimensional one. In the classical Kirchhoff-Love theory reduction was achieved by introducing certain restrictions on the deformation and by replacing the true stress by thickness averages and stress moments. Whereas large deflection (von Kármán) and shear deformation (Reissner) could be taken into account by this

---

* Permanent address: P.O. Box 9251, Ottawa, K1G 3T9 Canada

ad-hoc technique, there was a need for a systematic derivation which would incorporate more complex situations, for example composite plates. The first paper where such a derivation was attempted is perhaps Mindlin and Medick [1]; today there is an extensive literature on extensions and alternatives of this technique (see Naghdi [2], and [3]).

Engineering treatments of floating ice plates date back to Hertz [4]. In current research, ice plates are usually modeled by a thin isotropic plate with elastic or spring-dashpot constitutive laws [5][6]. To account for depth variations of the material constants Assur [7] asserted that the classical plate theory is applicable provided the plate constant is expressed by an appropriate thickness average of Young's modulus and Poisson's ratio. A first attempt at proving this was made by Kerr and Palmer [8], a second and more rigorous one by Hutter [9]. Hutter's analysis justified the assumption that Poisson's ratio is temperature independent. The proof is sufficiently general that it allows for the inclusion of materials which are not elastic. The approach was taken up by Williams [10] in a linearly viscoelastic model and later by Hutter [11] to model creep in plates. In these derivations, the constitutive relations of the plate are expressed in terms of the local three-dimensional relations.

In glaciers and ice shelves the stress and temperature problems are interdependent, but are usually treated separately by assuming a certain state for the companion problem. The deformations of ice shelves are studied by Weertman [12] and Thomas [13]. Temperature distributions are considered by Robin [14], Weertman [15] and Zotikov [16]. The equations for the fully coupled problems, in a glacier, are presented by Grigoryan et al.[17]. A simultaneous solution, for the case of an ice shelf, is attempted by Shumskiy and Krass [18], without considering, however, the influences of climate on shelf thickness. Averageing techniques are particularly suitable for a reanalysis of Shumskiy's problem, and we will demonstrate, by applying them to the energy equation, that the flow and thermal parts of the problem are both dependent on the climate.

Sections 2 and 3 give a description of the approximating technique and a selection of possible applications. Section 4 contains the analysis of the ice shelf problem. Extensions and future possibilities are discussed in Section 5.

2. A General Method for Plate Problems

Floating ice sheets are frequently suitable for plate analysis, because it is not necessary to know the precise distributions of stress and temperature through the thickness, provided resultant stresses and energy levels are obtainable.

The theoretical foundation for the general method for plate problems is the fundamental lemma of the Calculus of Variations, together with the techniques of shape function representation, such as orthogonal function expansions. The method of weighted residuals can be developed from these two principles and this method can, in turn, be regarded as the basis for numerical approximating techniques such as Finite Difference techniques, spectral methods and the Kantorovitch method [19].

In each of these techniques, the objective is to reduce the dimension of the domain in which the boundary value problem must be solved.

The theoretical foundation generates two general methods for reducing the dimension of the boundary value problem. If an unknown function f is expressed as a series in some appropriate set of shape functions $\{\phi_j\}$,

$$f(z) = \sum_j f^{(j)} \phi_j(z) \quad , \tag{1}$$

then the boundary value problem for f is reduced to a set of new problems for the functions $\{f^{(j)}\}$. Alternatively, the boundary value problem for the function f may be replaced by a set of problems for weighted thickness averages of the function,

$$g^{(j)} = \int_L^U \phi_j(z) f(z) dz \quad , \tag{2}$$

where the integration indicated is from the lower to upper surface of the plate. The relationship between the two methods, suggested by the notation, is made precise in the theory of orthogonal functions. In general, the coefficients $f^{(j)}$ of the expansion (1) are given by

$$f^{(j)} = \frac{1}{N} \int_L^U f(z) \; w(z) \phi_j(z) dz \tag{3}$$

where N is a normalizing factor and $w(z)$ is a weight function. To illustrate, in the case where the $\{\phi_j\}$ are trigonometric functions, (3) and (1) are the finite Fourier transform over the thickness, and its inverse.

The choice of shape functions may be motivated by an eigenvalue problem, a characteristic behavior, or computational convenience. In practice the summation (1) is usually approximated by a finite number of terms. The relationship between the $\{\phi_j\}$ and the $f^{(j)}$ may then be made precise through a set of linear algebraic equations. The usual choice of shape functions is the set $\{z^n\}$, as in [3,10], and the most familiar application is in stress boundary value problems. The capabilities of the general method are, however, large. It may, for instance, also be applied to cases in which energy and momentum equations are coupled, or to plates with changing thickness.

3. Examples of Plate Problems

3.1 Forced Flexural Vibration of an Ice Sheet

In a situation where an ice sheet is deforming in response to wave action, the effect of the deformations on the temperature in the ice may be significant.

A mathematical formulation of this problem consists of the momentum and energy equations, the boundary conditions associated with each field equation, and appropriate constitutive postulates for stress, entropy, and heat flux. An elastic constitutive relation would neglect dissipation of mechanical energy, and hence a thermoviscoelastic model, say of the Kelvin type, might be selected.

The coupled field equations may be averaged over the thickness of the plate. As shape functions $\{1, z, z^2\}$ prove to be suitable. The corresponding series expansions may be reduced to two terms by considering symmetry. For harmonic excitation, the net heat production per cycle within the ice, and the corresponding temperature rise, may then be calculated [20].

3.2 Prolonged Loading of an Ice Cover

In many engineering applications, an ice cover is subjected to pressures with duration between one hour and several days. Over this period, the temperature distribution in the ice cover will be relatively stable. Consequently, temperature will not appear explicitly in the field equations, but it makes the coefficients dependent on depth. Furthermore, in the vicinity of shore structures, variations in thickness are significant.

Considering the time scale described, a linear viscoelastic response in the ice is anticipated. An illustration will indicate how the plate methods may be applied in this case. The stress σ is related to the history of the strain ε by

$$\sigma(x,z,t) = G(z,t,t)\varepsilon(x,z,t) - \int_0^t \frac{\partial G(z,t,\tau)}{\partial \tau} \varepsilon(x,z,\tau)d\tau \qquad (4)$$

and such relations appear in the field equations. Define

$$f^{(m)}(x,t) = \int_{L(x)}^{U(x)} z^m f(x,z,t)dz \qquad (5)$$

and let

$$\varepsilon(x,z,t) = \sum_{n=0}^{M} z^n \varepsilon_n(x,t) \ . \qquad (6)$$

Then the macroscopic stress $\sigma^{(m)}$ is related to macroscopic strain ε_n through

$$\sigma^{(m)}(x,t) = \sum_{n=0}^{M} (G^{(m+n)}(x,t,t)\varepsilon_n(x,t) - \int_0^t \frac{\partial G^{(m+n)}(x,t,\tau)}{\partial \tau} \varepsilon_n(x,\tau)d\tau) \qquad (7)$$

The variation in thickness of the ice cover over its extent appears in equation (5) as an x-dependence of the limits of integration, and consequently in the moment

$G^{(m+n)}(x,t,\tau)$ of the function $G(z,t,\tau)$. Thus the field equations over (x,z,t) are reduced to a set of M field equations over (x,t), see [10].

3.3 Velocity and Temperature Distributions in a Polar Ice Shelf

The behavior of glacier ice is complicated by a nonlinear flow law, complex boundary conditions, and by the interaction of thermal and mechanical processes in the ice. The flow law is highly temperature dependent, and the temperature profile, in turn, is affected by the velocity distribution through convection and dissipation. When glacier ice forms a floating shelf, both shape and boundary conditions suggest the application of plate methods.

For the mechanical problem, the field equations are the flow law and the momentum equation. The full kinematic boundary conditions incorporate the influences of surface irregularities and change in shape and position of the surface due to phase changes and mass accumulation.

The temperature distribution is determined by the balance of energy equation, which includes heat convection, conduction, and mechanical dissipation of energy. Energy exchanges at the upper surface of the ice are governed by atmospheric conditions and by the amount of energy available in the ice for transfer. At the ice water interface, the difference between the heat conducted into the ice and that available from the ocean is absorbed in phase change.

The coupled thermo-mechanical problem is greatly simplified by the assumption that small variations in velocity along the length of the shelf will not substantially affect temperature profiles. Then the solutions to the stress and velocity problem are immediately available to feed into the energy equation.

One interesting feature in the ice shelf problem is that quantities which may be regarded as parameters in the mechanical problem become essential variables in the thermal problem. These quantities are shelf thickness, the ratio of velocity to depth, and vertical velocity at the surface. Furthermore, the question of whether or not the vertical velocity changes sign in the ice shelf becomes critical in the thermal problem.

The thermal problem is particularly suitable for plate analysis.

The total thermal energy in the plate is directly proportional to the integral of the temperature distribution; the dissipation energy is accounted for by the integral of the relevant term in the energy equation. Any changes in total energy level of the plate are due to energy, gained or lost at the surface. These exchanges are precisely accounted for by distinct terms in the integrated energy equation, and in the boundary conditions. Thus no information pertaining to the energy balance in the plate is lost, or approximated, by the averaging processes.

4. On the Temperature Regime of Floating Ice Shelves

As an illustration of the concepts we have discussed we present an abbreviated analysis for a floating ice shelf or a tongue. The analysis involves velocity, stress and temperatur distributions within the shelf and gives information on the response of the shelf to environmental conditions.

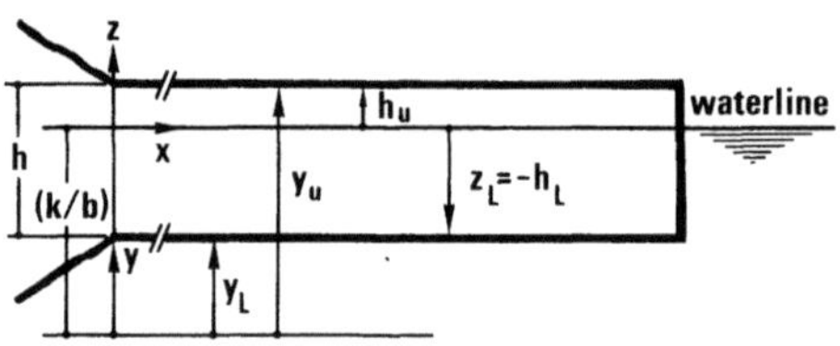

Fig 1

The geometry is depicted in figure 1. With the usual definitions of stress and velocity (u is horizontal, w is vertical) Weertman's solution [12] reads

$$(\sigma_{xx} - \sigma_{zz}) = 2b^{1/n}B, \quad \sigma_{zz} = -\rho_i g(h_u - z), \quad \sigma_{xz} = 0$$

$$u = bx + \beta \ , \quad w = -bz + k$$

$$h_u = \frac{1}{1+\lambda}\, h \ , \quad h_L = \frac{\lambda}{1+\lambda}\, h \ , \quad \lambda = \frac{\rho_i}{\rho_w - \rho_i} \ , \tag{8}$$

$$b = \frac{\rho_i g h^2}{4(1+\lambda)\bar{B}} \ , \quad \bar{B} = \int_{-h_L}^{h_u} B\, dz \ ,$$

The symbols B and n are parameters of Glen's flow law, which in pure shear reads $\dot{\gamma} = \tau / B^n$; β is the longitudinal speed at the hinge of the tongue and k the vertical velocity at the water line. All quantities may depend parametrically on time.

The above solution must be complemented by the solution to the thermal problem, since this determines h . The energy equation for temperature T is

$$\rho_i c_p \left[\dot{T} + wT,_z \right] = \kappa T,_{zz} + \frac{\sigma_{xx} - \sigma_{zz}}{2} (u,_x - w,_z) \tag{9}$$

and the boundary conditions are

$$\dot{h}_u - w = a_s^* \ , \quad Q_A^* = -\kappa T,_z \quad , \quad \text{at} \quad z = h_u \ , \tag{10}$$

$$\dot{h}_L + w = \frac{1}{L} (Q_w - \kappa T,_z) \qquad , \quad \text{at} \quad z = -h_L \ . \tag{11}$$

Accumulation a_s^*, heat flux into the atmosphere Q_A^*, and water temperature T_w^* are assumed known. The heat flux from the water into the ice, Q_w, is unknown.

The first step in the solution consists in approximating T by a series of shape functions. Both physical and mathematical considerations, detailed elsewhere [21], lead to the representation

$$T(y) = \frac{T_1}{|y|} - \frac{T_1}{|y_L|} + T_w^* , \qquad y = z - \frac{k}{b} , \qquad w = -by . \tag{12}$$

In selecting (12), it is recognized that the function possesses a singularity at the origin, and that the approximation (12) cannot be considered to represent T in the neighbourhood of the singularity.

A suitable smoothing for equation (9) is the first moment over the thickness, since this avoids singularities. Neglecting conduction terms one obtain

$$\dot{\theta} - \dot{y}y\,T \Big|_{y_L}^{y_U} - by^2 T \Big|_{y_L}^{y_U} + 2b\theta = d \tag{13}$$

$$\theta = \int_{y_L}^{y_U} y\,T\,dy , \qquad f \Big|_{y_L}^{y_U} = f(y_U) - f(y_L) ,$$

$$d = \frac{2}{\rho c_p} b^{\frac{(n+1)}{n}} \int_{y_L}^{y_U} y\,B\,dy \simeq \frac{1}{\rho c_p} b^{\frac{n+1}{n}} \bar{B}(y_U{}^2 - y_L{}^2) . \tag{14}$$

Equations (8) to (14) may be reduced to a set of differential equations involving the unknowns θ, h, k and Q_w. The system is highly nonlinear, but the steady state case separates conveniently, and provides a basis for a perturbation treatment of the time dependent problem [21].

In the following we summarize the results of the steady state analysis. These results are expressed in terms of the <u>thickness balance parameter</u> x, and the dimensionless heat flux $\mathbb{Q}$, defined by

$$x = \frac{bh}{a_s{}^*} , \qquad \mathbb{Q} = \frac{4(1+\lambda)c_p}{g\kappa} Q_A{}^* \tag{15}$$

It may be shown that

$$\mathbb{Q} = \begin{cases} x(2-x), & x \leq 1 , \\ x^2 , & x > 1 . \end{cases} \tag{16}$$

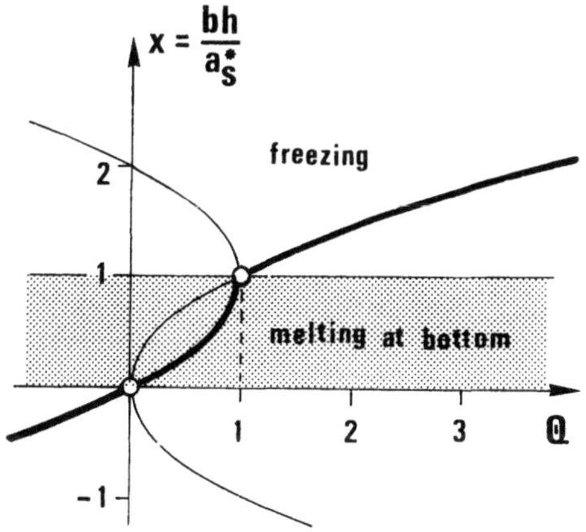

Fig. 2

The inverse relation is shown in figure 2. As long as heat flux at the upper boundary is out of the ice x is positive and a single valued function of $\mathbb{Q}$. The value of k in equation (12) is given by

$$k = \frac{a_s^*}{1+\lambda} \cdot \begin{cases} -\lambda - \sqrt{1-\mathbb{Q}}\,, & \mathbb{Q} \leq 1 \\ -(\lambda+1) + \sqrt{\mathbb{Q}}\,, & \mathbb{Q} > 1 \end{cases} \tag{17}$$

and the melting or freezing at the lower surface by

$$\bar{Q} = \frac{1}{L}\,(Q_w + \frac{\kappa}{y_L^2}\,(\text{sgn}\; y_L)T_1) = bh - a_s^* \; . \tag{18}$$

Thus the steady state thickness of the ice depends on the accumulation, the horizontal strain rate, and the thickness balance parameter x , which is determined by the energy exchange between ice and atmosphere. It does not depend on β, the horizontal velocity at the hinge, nor does it depend explicitly on the flow law parameters B and n , though these influence the value of b. The vertical velocity at the water line, k , is determined by the same quantities which govern h .
The two disjoint domains of the thickness balance parameter x separate two distinct physical behaviour. If $x < 1$

the origin of y is below the ice plate and the temperature increases monotonically with depth. The vertical velocity is downward throughout the ice. Fig. 3b shows typical temperature profile and flow lines. Accumulation is

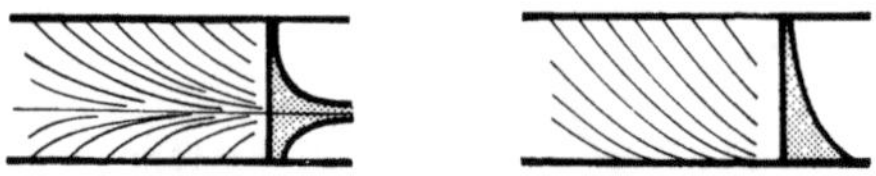

a b

Fig 3

greater than thinning due to spreading and thickness is maintained by melting at the lower surface ($\bar{Q} < 0$). When $x > 1$ thinning is more rapid than accumulation and thickness is maintained by freezing at the ice water boundary, ($Q > 0$). The origin of y, and the singularity of T_1 are within the ice plate. Ice formed at the bottom flows slowly upwards, fig 3a. Due to dissipation the temperature of this added material increases as it rises. In practice the temperature profile is regularized in the neighbourhood of its singularity by conduction of heat away from this region. Shumskiy suggests that the solution for $x > 1$ corresponds to a physically unstable situation. However the steady state analysis provides no information on stability.

The results of the steady state analysis may be used to relate various parameters of the velocity and temperature distributions to available data. For this purpose we select $\rho_i = 900\ \mathrm{kgm^{-3}}$, $c_p = 4.0\ \mathrm{Jkg^{-1}deg^{-1}}$, $a^*_{so} = 0.03\ \mathrm{my^{-1}}$, where a^*_{so} is a reference accumulation rate. Weertman [12] reports temperatures of $-17^{\circ}C$ and $-1.5^{\circ}C$ at the top and bottom, respectively of an ice shelf with thickness $h = 185$ m. With this data, and the above results, it may be shown that $(x, Q) = (0.48, 0.73)$ if $\chi < 1$, or $(x, Q) = (1.52, 2.3)$ if $x > 1$. If $b = 1.8 \times 10^{-3}\ \mathrm{yr}^{-1}$, then $a_s^* = 0.69$ or $0.22\ \mathrm{myr}^{-1}$. These calculations were possible without resort to values of B and n . The value of h, however, depends on these, as

$$h = h_o x^{1/(n+1)}, \quad h_o := \left(\frac{4(1+\lambda)B}{\rho_i g}\right)^{n/(n+1)} a_s^{*\,1/(n+1)} \tag{19}$$

For $a_s^* = a_{so}^*$ and $B = \bar{B}$ h_o is plotted against $\bar{B}$ in figure 4 along with data for $(h_o, \bar{B})$ from a study of ice shelves by Thomas [22].

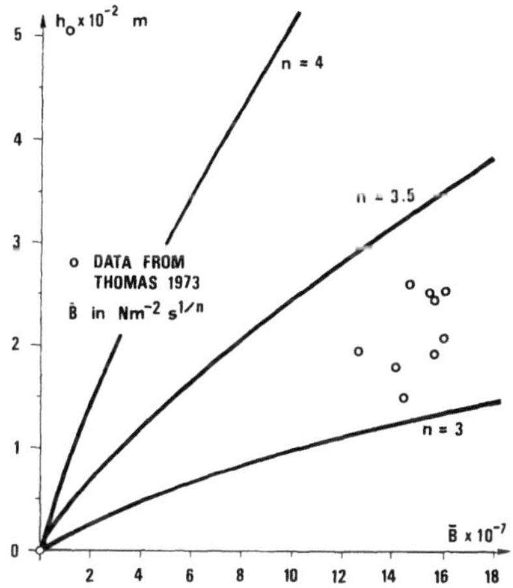

Fig 4

Equations (19) may be written

$$\log\frac{h}{h_o} = \frac{n}{n+1}\log\frac{B}{\bar{B}} + \frac{1}{n+1}\log\frac{a_s^* x}{a_{so}^*},$$

and using $n = 3.5$, $\bar{B} = 1.5 \times 10^{-8}\ \mathrm{Nm^{-2}\ sec^{2/7}}$ it can be shown that Thomas's data fall within the limits $h/h_o < \frac{1}{2}$, $a_s^*/a_{so}^* < 0.04$. Unless accumulation is very small this suggests that for Thomas's ice shelves $x < 1$.

Further conclusions are possible for which the reader is referred to [21].

5. Conclusions

We have presented a review of plate techniques for floating ice sheets, showing how classical plate theory can be extended to accommodate coupled thermal and mechanical fields. A system of plate equations for the thermal response of an ice shelf to climatic conditions was presented. The analysis showed that in steady state, the velocity and temperature distributions in the ice shelf depend on the surface energy flux Q and the thickness balance parameter x . The theroy also indicates how average values of B and n may be verified.

The methods presented here may be applied to the analysis of the time dependent response of the shelf, thus permitting a stability analysis. They may also be applied in wave interaction problems with thermomechanical coupling. Space limitations prevent elaboration, but we hope that the above considerations have provided sufficient material to demonstrate the usefulness of the technique.

References

1 Mindlin, R.D.; Medick, M.A.: Extensional vibrations of elastic plates. J. Applied Mechanics. 26 (1959).

2 Naghdi, P.M.: The Theory of Shells and Plates. Handbuch der Physik, Vol VIa/2. Berlin, Göttingen, Heidelberg: Springer 1972.

3 Hutter, K.: On the fundamental equations of floating ice. Mitteilung der Versuchsanstalt für Wasserbau, Hydrologie und Glaciology, ETH Zürich. 8 (1973) 1-150.

4 Hertz, H. Ueber das Gleichgewicht schwimmender elastischer Platten. Widemanns Annalen der Physik und Chemie. 22 (1884) 449-455.

5 Nevel, D.E.: Creep theory for a floating ice sheet. Special Report 76-4. Cold Regions Research and Engineering Laboratory, Hanover, New Hampshire (1976) 1-98.

6 Kerr, A.D.: The bearing capacity of flcating ice plates subjected to static or quasistatic loads. Journal of Glaciology, 17(1976) 229-268.

7 Assur, A.: Flexural and other properties of sea ice sheets. Proc. Conf. Inst. Low. Temp. Sc. Hokkaido Univ. (1966).

8 Kerr, A.D.; Palmer, W.T.: The deformations of stresses in floating ice plates. Acta Mechanica, 15 (1972) 57-72.

9 Hutter, K.: Floating sea ice plates and the significance of the dependence of the Poisson ratio on brine content. Proc. Royal. Soc. A 343 (1975) 85-105.

10 Williams, F.M.: Time dependent deflections of nonhomogeneous ice plates. Acta Mechanica. 25 (1976) 29-44.

11 Hutter, K: On the mechanics of floating ice sheets. Mitteilung der Versuchsanstalt für Wasserbau, Hydrologie und Glaziologie, ETH Zürich. 28 (1978) 1-103.

12 Weertman, J.: Deformation of floating ice shelves. Journal of Glaciology, 3. (1957) 38-42.

13 Thomas, R.H.: The creep of ice shelves: Theory. Journal of Glaciology. 12 (1973) 45-53.

14 Robin, G.de Q.: Ice movement and temperature distribution in glaciers and ice sheets. Journal of Glaciology, 2 (1955) 523-532.

15 Weertman, J.: Mechanism for the formation of inner moraines found near the edge of cold ice caps and ice sheets. Journal of Glaciology, 3 (1961) 975-978.

16 Zotikov, I.A.: Bottom melting in the central zone of the shield on the antarctic continent and its influence upon the present balance of the ice mass. Bulletin Int. Ass. Sc. Hydr. 8 (1963) 36-44.

17 Grigoryan, S.S.; Krass, M.S.; Shumskiy, P.A.: Mathematical model of a three-dimensional non-isothermal glacier. Journal of Glaciology, 17 (1976) 401-417.

18 Shumskiy, P.A.; Krass, M.S.: Mathematical models of ice shelves. Journal of Glaciology 17 (1976) 419-432.

19 Rektorys, R. (editor): Survey in Applicable Mathematics. Translation from the Russian. Boston: The MIT Press 1969.

20 Williams, F.M.; Hutter, K.: On the temperature distribution in harmonically excited thermoviscoelastic plates. In preparation.

21 Williams, F.M.; Hutter, K.: Thermal response of ice shelves to climatic conditions. In preparation.

22 Thomas, R.H.: The creep of ice shelves: Interpretation of observed behaviour. Journal of Glaciology. 12 (1973) 55-70.

On the Buckling Force of Floating Ice Plates

ARNOLD D. KERR
Department of Civil Engineering
University of Delaware
Newark, Delaware
U.S.A.

Summary

The calculation of the largest horizontal force a relatively thin floating ice plate may exert on a structure requires the knowledge of the buckling load for this floating plate. In the published literature on the stability of continuously supported beams and plates it is usually assumed that this buckling force corresponds to the lowest bifurcation force p_{cr}. However, recent studies indicate that, generally, this is not the case. The purpose of the present paper is to clarify this situation for floating ice plates.

At first this problem is studied on a simple model that exhibits the buckling mechanism of a floating ice plate but is amenable to an exact nonlinear analysis. This study shows that, depending on the ratio of the rigidities of the "liquid" and "plate", the post-buckling branch may rise or drop away from the bifurcation point. Thus, p_{cr} may or may not be the actual buckling load. It is also shown that when lift-off of "plate" from the "liquid" takes place, the actual buckling load may drop substantially. This study is followed by an analysis of a floating compressed semi-infinite plate with a straight free edge, assuming that there is no lift off. It was found that for this case there always exists a buckling load that is lower than p_{cr}.

According to the obtained results, the value p_{cr} should be used with caution as a buckling load for floating ice plates. It is suggested that the buckling load be determined using the post-buckling equilibrium branch, taking into consideration the possibility of lift-off of the ice cover from the liquid base.

1. Introduction

The analytical determination of the largest horizontal force a floating ice cover may exert on a structure is based on the criterion that an *ice force cannot be larger than the force capable of breaking up the ice cover.*

In many publications, as discussed by K.N. Korzhavin [1] and B. Michel [2], this failure criterion is related to a "crushing strength" of the ice cover. There it is assumed that the failure mechanism consists of the crushing or/and splitting of the floating ice plate in the immediate vicinity of the structure.

However, in a number of laboratory and field tests it was observed that for relatively thin ice covers (compared to the width of contact of plate and structure) the ice cover failed by buckling in the vicinity of the contact area. Thus, for these cases, the buckling force is smaller than the crushing force and the determination of the largest force should be based on the determination of the force at which the floating ice cover is expected to buckle.

The related buckling analyses and tests were recently reviewed and discussed by A.D. Kerr [3]. In all these analyses it was assumed that the buckling load corresponds to the first bifurcation load, p_{cr}; an assumption justified when the post-buckling equilibrium branch rises monotonically with increasing displacements. However, according to the recent studies by J.G. Lekkerkerker [4], A.D. Kerr [5], and R.H. Plaut [6], this is not the case in general. The purpose of the present paper is to clarify this situation for the floating ice plate problem.

2. Simple Model for Ice Cover Buckling

To study the basic features of the title problem we consider first the simple model, shown in Fig. 1, which

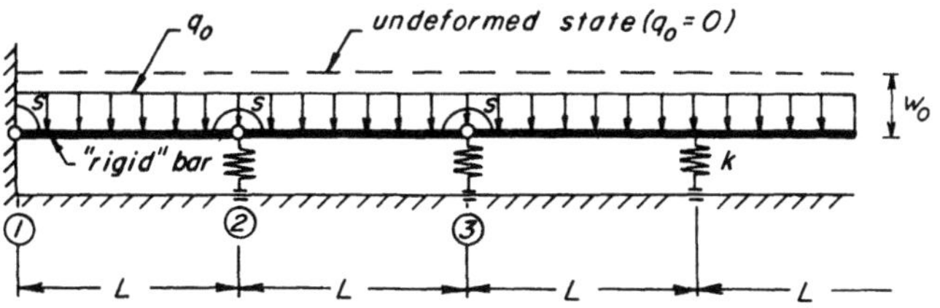

Fig. 1. Simple Buckling Model

consists of "rigid"bars inter-connected by elastic springs. This model contains the observed buckling mechanism of a compressed floating ice cover and is amenable to a simple nonlinear exact analysis. It is assumed that the rigid bars are horizontal when subjected to the uniform load distribution (own weight) q_o. In this position of equilibrium the spiral springs are stress free and each of the straight vertical springs is compressed by a force kw_o. Thus,

$$w_o = q_oL/k \tag{2.1}$$

Next, the bars are subjected to an axial force P. For increasing P, deformed states of equilibrium may exist, as shown in Fig. 2(a). Assuming that the bars at joint 2 will not separate from the base spring, thus that

$$w_o \geqq L \sin\theta \quad \text{or} \quad q_o/k \geqq \sin\theta, \tag{2.2}$$

the equilibrium equation for the deformed state is

$$P^* \sin\theta = \theta + k^* \sin\theta \cos\theta \tag{2.3}$$

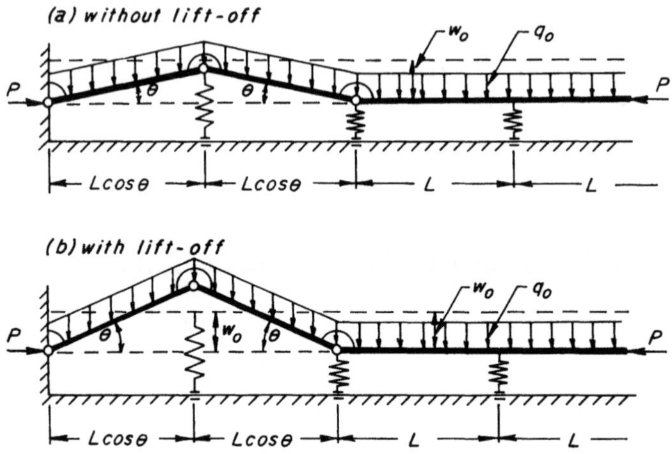

Fig. 2. Deformed Model

where, denoting by s the parameter of the spiral springs,

$$P^* = PL/3s \; ; \qquad k^* = kL^2/6s \qquad (2.4)$$

Note that because of the linearity of the base springs, q_o does not appear in equilibrium equation (2.3). Also note that eq. (2.3) is satisfied for $\theta=0$. Thus, the straight state is in equilibrium (not necessarily stable) for any load P. For very small values of θ, eq. (2.3) reduces to

$$(P^* - 1 - k^*)\ \theta = 0 \qquad (2.3')$$

Thus, at the bifurcation point

$$P^*_{cr} = 1 + k^* \qquad (2.5)$$

Eq. (2.3) was evaluated numerically for various values of k^*. The results are shown in Fig. 3(a).

For a proper interpretation of the results shown in Fig. 3(a) note that, as shown in Fig. 4 (A.D. Kerr [7]),

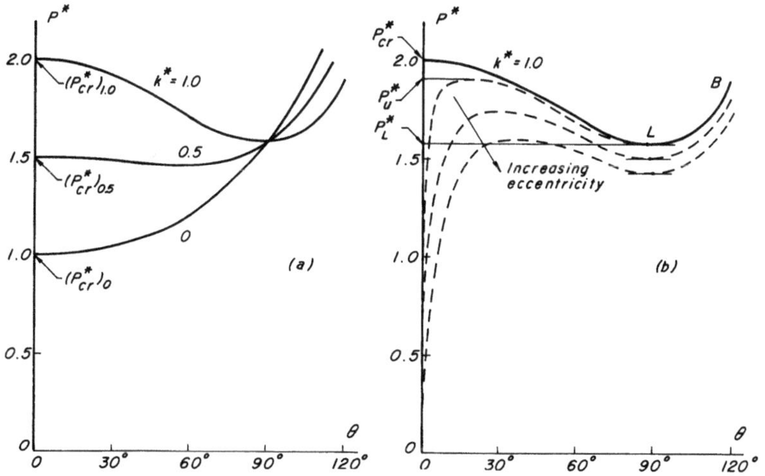

Fig. 3. Equilibrium Branches for Model Without Lift-Off. For θ<0 the branches are symmetrical to P^*-axis.

when the model bars are constrained only by a spiral spring (bending effect) then the equilibrium branch r i s e s monotonically, whereas when the model bars are constrained by a straight spring (effect of base) the equilibrium branch d r o p s. This is the reason why in Fig. 3(a) for large values of k^*, i.e. when the base stiffness predominates, the equilibrium branch drops at first and then rises exhibiting a l o w e r buck- l i n g load $P_L < P_{cr}$.

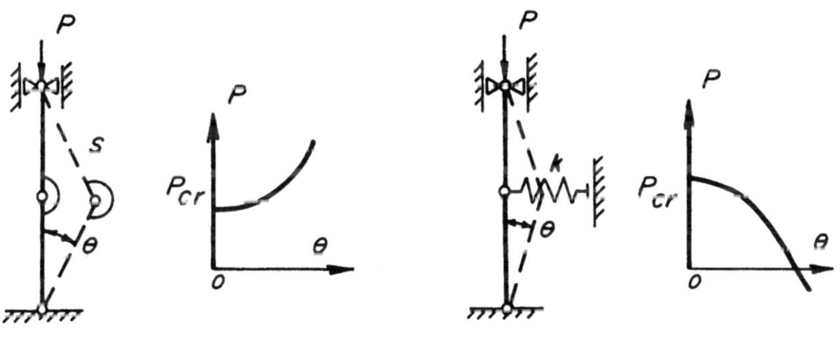

Fig.4

From the graphs in Fig. 3(a) it follows that the value of k^* which separates the different responses is located in the interval $0 < k_s^* < 0.5$. Using the perturbation method, A.D. Kerr [5] showed that for the model under consideration

$$k_s^* = \frac{1}{3} \tag{2.6}$$

Consider the model with $k^* = 1.0$. According to Fig. 3(a) or 3(b) for each $P < P_L$ there exists only one straight equilibrium state, for $P_L < P < P_{cr}$ there exist five equilibrium states, and for each $P > P_{cr}$ three states of equilibrium are possible. It may be shown, (A.D. Kerr [8]), that a straight equilibrium state is stable below P_{cr} and unstable above it, that the equilibrium states on branch $P_{cr}L$ are unstable, and that the equilibrium states on branch LB are stable.

When the axial load P has a small vertical eccentricity the equilibrium branches are as shown by the dashed lines in Fig. 3(b). The effect of vertical shape imperfections is similar. Note that with increasing small eccentricities, or shape imperfections, the upper buckling load P_u drops rapidly. Structures of this type are referred to in the literature as *imperfection sensitive*.

Because of the existence of load and geometrical imperfections (and dynamic inputs) in an actual situation, buckling will take place for $P_L < P < P_{cr}$, where P_L is the value for the structure with imperfections. Thus, for the cases that exhibit a P_L value, P_{cr} is not the true buckling load.

Next consider the case when at joint 2 the bars are resting on the base spring, i.e. they may *lift-off* the spring base, as shown in Fig. 2(b). This may occur in an actual situation with floating ice covers. For this case, until lift-off takes place, the governing equilibrium equation is eq. (2.3). After lift-off, namely when

$$w_o < L \sin\theta \qquad \text{or} \qquad q_o/k < \sin\theta, \tag{2.7}$$

the equilibrium equation is

$$P^* \sin\theta = \theta + q^* \cos\theta \tag{2.8}$$

where

$$P^* = PL/3s \;; \quad q^* = q_o L^2/6s \tag{2.9}$$

The corresponding equilibrium branches for $k^* = 0.2,\ 1.0$ and a range of q^* values are shown in Fig. 5. The reason

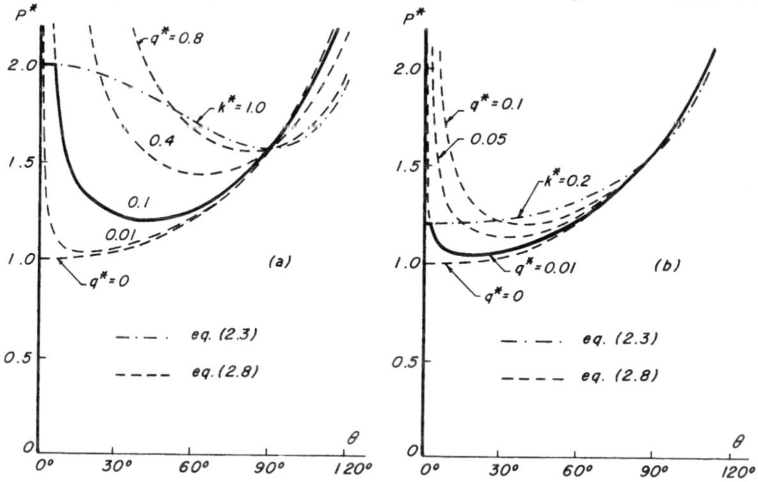

Fig. 5. Equilibrium Branches for Model With Lift-Off

for the large post-buckling deformations in Fig. 3 and Fig. 5 is the assumption that an axial force P remains the same also at the post-buckling equilibrium points. However, it was shown by A.D. Kerr [9] that when the axial compression force is induced by a temperature raise, the axial force drops and the post-buckling deformations are much smaller.

According to Fig. 5(a) the admissibility of lift-off reduces substantially the P_L-value, especially for small values of q^*. The resulting equilibrium branch for $q^* = 0.1$ is shown as a solid line. Note, that the corresponding P_L^* equals 1.2; thus about 25% lower than the P_L^* for the case when lift-off is not allowed.

Fig. 5(b) shows that the admissibility of lift-off creates a $P_L < P_{cr}$ even for an equilibrium branch that is monotonically increasing with no lift-off. This is an example where a stability analysis of the bifurcation point on the undeformed branch (for example, the Koiter method) is not suitable for predicting whether P_L exists.

The findings of this section suggest that for compressed floating ice plates, even in the absence of lift-off between plate and liquid base, a $p_L < p_{cr}$ may exist, as shown in Fig. 3(a). The results in Fig. 5 suggest that the possibility of lift-off may greatly affect the post-buckling equilibrium branches and thus also the lower buckling loads p_L. Occurrance of lift-off may even create a p_L value where one does not exist when lift-off is not allowed.

To establish whether floating ice plates do respond that way requires the solution of the nonlinear equations for plates on a liquid base. Because of space limitation the study of all these phenomena is beyond the scope of the present paper. In the remainder of the paper only the behavior of the equilibrium branches of compressed floating ice plates in the vicinity of p_{cr} [as shown in Fig. 3(a)] is investigated in order to establish if, and under what conditions, a $p_L < p_{cr}$ will exist in the absence of lift-off.

3. Semi-Infinite Floating Plate With Free Edge

The problem under consideration is shown in Fig. 6. Because the load p (per unit length of plate boundary) is assumed to be constant the plate displacements will be functions of x only. The vertical displacement component

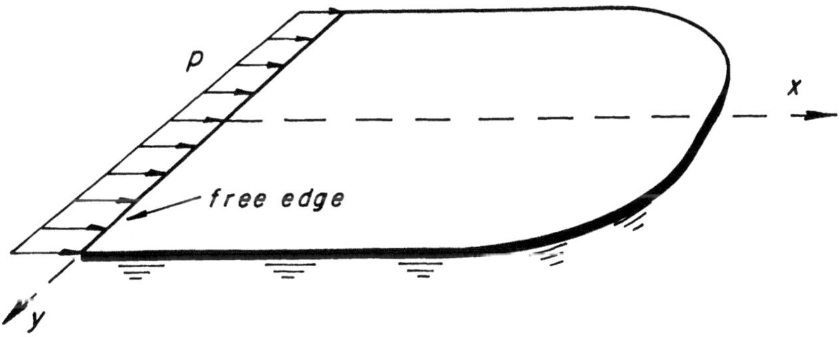

Fig. 6. Semi-Infinite Plate With Free Edge

of a point (x,y) on the middle plane is assumed in the form $w_o + w(x)$; w_o = const is the rigid body displacement due to the weight of plate q_o (per unit area) and w(x) is caused by bending deformations. Thus, $q_o = \gamma w_o$.

Following the derivations in [5][10], using Lagrange coordinates and neglecting the extensibility of the middle plane, the differential equation for w(x) is

$$\frac{w''''}{1-w'^2} + \frac{4w'w''w'''+w''^3}{(1-w'^2)^2} + \frac{4w'^2w''^3}{(1-w'^2)^3} + \frac{p}{D}\frac{w''}{(1-w'^2)^{3/2}} + \frac{\gamma}{D}w = 0 \tag{3.1}$$

where $D = Eh^3/12(1-\nu^2)$ is the flexural rigidity of the plate, γ is the base modulus, and $(\ )' = d(\ )/dx$. The boundary conditions are

$$w''(0) = 0 \tag{3.2}$$

$$\left[D\left(\frac{w''}{\sqrt{1-w'^2}}\right)' + pw'\right]_{x=0} = 0 \tag{3.3}$$

The formulation is highly nonlinear. In the mathematical literature there are no methods, as yet, for obtaining an exact closed form solution for this boundary value problem. Therefore, in the following, we analyze the post-buckling equilibrium branch in the vicinity of p_{cr}, in order to establish if p_L exists. This is done using the perturbation method [5].

Following J.B. Keller [11] the variables p and w(x) are expanded in a Taylor series

$$p(\varepsilon) = p(o) + p_\varepsilon(o)\varepsilon + p_{\varepsilon\varepsilon}(o)\frac{\varepsilon^2}{2!} + p_{\varepsilon\varepsilon\varepsilon}(o)\frac{\varepsilon^3}{3!} + \dots$$

$$w(x,\varepsilon) = w(x,o) + w_\varepsilon(x,o)\varepsilon + w_{\varepsilon\varepsilon}(x,o)\frac{\varepsilon^2}{2!} + w_{\varepsilon\varepsilon\varepsilon}(x,o)\frac{\varepsilon^3}{3!} + \dots \qquad (3.4)$$

Denoting symbolically the non-linear formulation, consisting of equations (3.1),(3.2),(3.3), by

$$F[w(x),p] = 0 \qquad (3.5)$$

it follows, because of (3.4), that $F = F(\varepsilon)$ only. Assuming that F is differentiable, we write

$$F(\varepsilon) = F(o) + F_\varepsilon(o)\varepsilon + F_{\varepsilon\varepsilon}(o)\frac{\varepsilon^2}{2!} + \dots = 0 \qquad (3.6)$$

Since ε is an arbitrary parameter, the above equation is satisfied when

$$F(o) = 0; \quad F_\varepsilon(o) = 0; \quad F_{\varepsilon\varepsilon}(o) = 0; \quad \dots \qquad (3.7)$$

These are the equations for the determination of the coefficients in (3.4)*.

It should be noted that in (3.4) the term w(x,o) represents the pre-buckling state. Hence

$$w(x,o) \equiv 0 \qquad (3.8)$$

*) Equation (3.6) may also be obtained by substituting (3.4) into (3.5) and then by grouping terms of equal powers in ε.

The formulation which corresponds to the first condition in (3.7), $F(o)=0$, is satisfied for any point on the equilibrium branch with $w(x,o)\equiv 0$.

Applying the second condition in (3.7), $F_\varepsilon(o)=0$, to the equations (3.1),(3.2),(3.3) and noting that because of eq. (3.8) the derivatives $w^{(n)}(x,o)=0$ we obtain

$$L\, w_\varepsilon(x,o) = 0 \qquad 0 \leqq x < \infty \tag{3.9}$$

where

$$L = D(\ )^{iv} + p(o)(\ )'' + \gamma \tag{3.10}$$

and the boundary conditions

$$w''_\varepsilon(o,o) = 0 \tag{3.11}$$

$$Dw'''_\varepsilon(o,o) + p(o)w'_\varepsilon(o,o) = 0 \tag{3.12}$$

This is a linear eigenvalue problem for $w_\varepsilon(x,o)$. It is identical to the formulation in the Euler method for obtaining p_{cr}.

This eigenvalue problem is satisfied by the trivial solution

$$w_\varepsilon(x,o) \equiv 0 \tag{3.13}$$

which represents the continuation of the undeformed equilibrium branch. It is also satisfied, for the equilibrium branch which corresponds to $w(x,\varepsilon) \neq 0$, by the first eigenvalue [12]

$$p(o) = p_{cr} = \sqrt{\gamma D} \tag{3.14}$$

and the corresponding eigenfunction

$$w_\varepsilon(x,o) = Be^{-\alpha x}\left[\sin(\sqrt{3}\alpha x) - \sqrt{3}\cos(\sqrt{3}\alpha x)\right] \tag{3.15}$$

where B is an arbitrary constant and $\alpha = \sqrt[4]{\gamma/16D}$.

Next, we form $F_{\varepsilon\varepsilon}(o)=0$. The resulting equations are, the differential equation for $w_{\varepsilon\varepsilon}(x,o)$

$$Lw_{\varepsilon\varepsilon}(x,o) = -2p_\varepsilon(o)w''_\varepsilon(x,o) \qquad 0 \leqq x < \infty \tag{3.16}$$

and the boundary conditions

$$w''_{\varepsilon\varepsilon}(o,o) = 0 \tag{3.17}$$

$$Dw'''_{\varepsilon\varepsilon}(o,o) + p(o)w'_{\varepsilon\varepsilon}(o,o) = -2p_\varepsilon(o)w'_\varepsilon(o,o) \tag{3.18}$$

where the differential operator L is given in (3.10).

The above boundary value problem is satisfied for

$$w_\varepsilon(x,o) \equiv 0 \quad \text{and} \quad w_{\varepsilon\varepsilon}(x,o) \equiv 0 \tag{3.19}$$

which represent the continuation of the undeformed equilibrium branch. For the equilibrium branch at the bifurcation point which corresponds to $w(x,\varepsilon) \neq 0$ the operator L in (3.10) is singular [Note that the corresponding homogeneous problem is identical to the one for $w_\varepsilon(x,o)$]. Hence, a solution $w_{\varepsilon\varepsilon}(x,o)$ will exist only if the Fredholm alternative is satisfied.

In this connection the non-homogenuity in the second boundary condition is eliminated first, by introducing a new variable $v_1(x)$, as follows:

$$w_{\varepsilon\varepsilon}(x,o) = v_1(x) - \frac{2p_\varepsilon(o)\; x}{p(o)} w'_\varepsilon(o,o) \tag{3.20}$$

With this transformation differential equation (3.16) becomes

$$L\, v_1 = 2p_\varepsilon(o)\left[\frac{\gamma x}{p(o)}\, w'_\varepsilon(o,o) - w''_\varepsilon(x,o)\right] \tag{3.16'}$$

and the boundary conditions in (3.17) and (3.18) become

$$v''_1(o) = 0 \tag{3.17'}$$

$$Dv_1'''(o) + p(o)v'_1(o) = 0 \tag{3.18'}$$

According to the Fredholm alternative a solution $v_1(x)$, and hence $w_{\varepsilon\varepsilon}(x,o)$ will exist only if

$$\int_o^\infty p_\varepsilon(o)\left[\frac{\gamma x}{p(o)}\, w'_\varepsilon(o,o) - w''_\varepsilon(x,o)\right]\psi(x)\,dx = 0 \tag{3.21}$$

where $\psi(x)$ is the non-zero solution of the homogeneous adjoint problem.

Because the operator is self-adjoint, it follows that

$$\psi(x) = w_\varepsilon(x,o) \tag{3.22}$$

and the solvability condition (3.21) reduces to

$$p_\varepsilon(o)\int_o^\infty \left[\frac{\gamma x}{p(o)}\, w'_\varepsilon(o,o) - w''_\varepsilon(x,o)\right] w_\varepsilon(x,o)\,dx = 0 \tag{3.23}$$

Noting that $w_\varepsilon(x,o)$ is given in (3.15) and performing the integrations it may be shown that the integral does not vanish. Hence eq. (3.23) is satisfied when

$$p_\varepsilon(o) = 0 \tag{3.24}$$

With condition (3.24), the boundary value problem for $w_{\varepsilon\varepsilon}(x,o)$ given by (3.16),(3.17),(3.18) reduces to the problem for $w_\varepsilon(x,o)$ given by (3.9),(3.11),(3.12). Thus $w_{\varepsilon\varepsilon}(x,o)$ and $w_\varepsilon(x,o)$ differ by an, as yet, undetermined coefficient β. Namely,

$$w_{\varepsilon\varepsilon}(x,o) = \beta w_\varepsilon(x,o) \tag{3.25}$$

To determine the next coefficients in (3.4) we form $F_{\varepsilon\varepsilon\varepsilon}(o)=0$. The resulting equations are the differential equation

$$Lw_{\varepsilon\varepsilon\varepsilon}(x,o) = f(x) \qquad o \leqq x < \infty \tag{3.26}$$

where, noting that $p_\varepsilon(o)=0$,

$$f(x) = -\Big[6\,D\,(4w'_\varepsilon w''_\varepsilon w'''_\varepsilon + w''^3_\varepsilon + w'^2_\varepsilon w^{iv}) + 9pw'^2_\varepsilon w''_{\varepsilon\varepsilon} + 3p_{\varepsilon\varepsilon} w''_\varepsilon\Big]_{\varepsilon=o} \tag{3.27}$$

and the boundary conditions

$$w''_{\varepsilon\varepsilon\varepsilon}(o,o) = 0 \tag{3.28}$$

$$Dw'''_{\varepsilon\varepsilon\varepsilon}(o,o) + p(o)w'_{\varepsilon\varepsilon\varepsilon}(o,o) = -3\Big\{p_{\varepsilon\varepsilon}(o)w'_\varepsilon(o,o) + D\Big[w'^2_\varepsilon(o,o)w'''_\varepsilon(o,o) + 2w'_\varepsilon(o,o)w''^2_\varepsilon(o,o)\Big]\Big\} \tag{3.29}$$

The above equations are satisfied for

$$w_\varepsilon(x,o) \equiv 0 \; ; \quad w_{\varepsilon\varepsilon}(x,o) \equiv 0 \; ; \quad w_{\varepsilon\varepsilon\varepsilon}(x,o) \equiv 0 \tag{3.30}$$

which represent the continuation of the undeformed equilibrium branch.

For the deformed branch at the bifurcation point, the above equations constitute a non-homogeneous boundary value problem with a singular operator. To facilitate the use of the Fredholm alternative, the non-homogenuity in the second boundary condition is eliminated, by introducing a new variable v(x), as follows:

$$w_{\varepsilon\varepsilon\varepsilon}(x,o) = v(x) - \frac{3\{\cdots\cdot\}x}{p(o)} \tag{3.31}$$

where $\{\cdots\cdot\}$ is from eq. (3.29). With this transformation, differential equation (3.26) becomes

$$Lv = f(x) + \frac{3\gamma x}{p(o)}\Big(p_{\varepsilon\varepsilon}(o)w'_\varepsilon(o,o)$$

$$+ Dw'^2_\varepsilon(o,o)w'''_\varepsilon(o,o) + 2Dw'_\varepsilon(o,o)w''^2_\varepsilon(o,o)\Big) \tag{3.26'}$$

and the boundary conditions (3.28)(3.29) become

$$v''(o) = 0 \tag{3.28'}$$

$$Dv'''(o) + p(o)v'(o) = 0 \tag{3.29'}$$

According to the Fredholm alternative a solution v(x), and hence $w_{\varepsilon\varepsilon\varepsilon}(x,o)$, will exist only when

$$\int_o^\infty \Big[f(x) + \frac{3\gamma x}{p(o)}\Big(p_{\varepsilon\varepsilon}(o)w'_\varepsilon(o,o)$$

$$+ Dw'^2_\varepsilon(o,o)w'''_\varepsilon(o,o) + 2Dw'_\varepsilon(o,o)w''^2_\varepsilon(o,o)\Big)\Big]\phi(x)\,dx = 0 \tag{3.32}$$

where

$$\phi(x) = w_\varepsilon(x,o) \tag{3.33}$$

is the non-zero solution of the homogeneous adjoint problem.

Performing the integrations indicated in (3.32), noting

that f(x) is given in (3.27) and $w_\varepsilon(x,o)$ in (3.15), we obtain

$$p_{\varepsilon\varepsilon}(o) = -\frac{9}{28}\gamma B^2 \qquad (3.34)$$

where B is an, as yet, undetermined coefficient. Thus, according to (3.4),

$$p(\varepsilon) = p_{cr} - \frac{9}{56}\gamma B^2 \varepsilon^2 \qquad (3.35)$$

4. Conclusions

Equation (3.35) proves analytically, as predicted by the model study in Section 2, that the post-buckling branch of the semi-infinite plate with a free compressed edge exhibits values of $p < p_{cr}$, even in the absence of lift-off. According to recent results by Plaut [6] this may also occur for a floating plate strip simply supported and compressed along the straight parallel boundaries.

These findings indicate that for floating ice covers the critical buckling value p_{cr} should be used with caution. In this connection note that the analysis presented in Section 3 determines only the behavior of the post-buckling equilibrium branch in the vicinity of p_{cr}. The determination of the value p_L, when one exists, requires the solution of the nonlinear formulation based on eq. (3.1).

During buckling of ice covers lift-off may take place. Therefore, the analysis for the determination of p_L should take this possibility into consideration. As shown in the model study, lift-off may strongly affect the p_L-value and hence also the buckling load.

Acknowledgments

This research was supported by the Ice Engineering Research Branch, CRREL, U.S. Army. The author wishes to thank G.E. Frankenstein, D.E. Nevel, and L.J. Zabilanski for participating in helpful discussions on the subject of ice forces.

References

1 Korzhavin, K.N.: Action of ice on engineering structures. (in Russian), Izd. Sibir. Otd. AN SSSR (1962), Translated into English by CRREL U.S. Army, DT 260 (1971).

2 Michel, B.: Ice pressure on engineering structures. Cold Regions Science and Engineering, Monograph III-Blb, CRREL, U.S. Army (1970).

3 Kerr, A.D.: On the determination of horizontal forces a floating ice plate exerts on a structure. J. of Glaciology, Vol. 20, No. 82 (1978).

4 Lekkerkerker, J.G.: On the stability of an elastically supported beam subject to its smallest buckling load. Proc. Kon. Ned. Akad. Wet., B65 (1962).

5 Kerr, A.D.: On the stability of continuously supported beams. New York University, Technical Report NYU-AA-72-35,(November, 1972).

6 Plaut, R.H.: Post-buckling behavior of continuous, nonconservative elastic systems. Acta Mechanica, Vol. 30 (1978).

7 Kerr, A.D.: Discussion of paper by W.T. Feldt, S. Nemat-Nasser, S.N. Prasad and G. Herrmann: Instability of a mechanical system induced by an impinging fluid jet. J. of Applied Mechanics, Vol. 36, Nr. 4 (1970) p. 1206.

8 Kerr, A.D.: On the stability of the railroad track in the vertical plane. Rail International, No. 2, (1974).

9 Kerr, A.D.: Model study of vertical track buckling. High Speed Ground Transportation J., Vol. 7 (1973).

10 Kappus, R.: Zur Elastizitätstheorie endlicher Verschiebungen. Zeitschrift für Angewandte Mathematik and Mechanik, Vol. 19, 1939.

11 Keller, J.B.: Perturbation theory, Printed notes of lectures presented at the Department of Mathematics, Michigan State University, 1968.

12 Rzhanitsyn, A.R.: Stability of equilibrium of elastic systems (in Russian). Gos. Izd. Fiz.-Mat. Lit. Moscow, 1955.

Implications of Structure Width for Design Ice Forces

P. R. KRY
Research Department, Esso Resources Canada Limited
Calgary, Canada T2G 2B3

Abstract

Hydrocarbon exploration activities in the Canadian Southern Beaufort Sea have provided incentives to further develop our knowledge of ice and its interactions with structures. The shallow water regions (to 13 m depth) have been explored using artificial islands as exploration drilling platforms. To have sufficient surface area for drilling operations and personnel quarters, these islands represent very wide structures compared to the maximum ice thickness. Their width is further enlarged by ice rubble fields formed by early winter ice motions.

The analysis of ice loading on wide structures in comparison to that on narrow structures requires consideration of clearing of broken ice and the possibility of non-simultaneous failure across the width. Inability to clear broken ice around a wide structure leads to the occurrence of ice ride-up or pile-up. Since significant ice movement occurs in the early winter when ice is thin, and since artificial islands have a profile including a rough steep slope, ice pile-up occurs during freeze-up forming a rubble field which provides protection against subsequent ice ride-up. However, the rubble field increases effective structure width and this amplifies forces on the structure. Additional sliding resistance due to rubble grounding only partially offsets these force increases unless the rubble is extremely extensive. The forces arise from effective stresses generated by ice failure at the periphery of the rubble field. The extensive width increases the importance of localized failures due to irregularities at the failure interface. Such non-simultaneous failure implies that effective stresses are subject to statistical variations such that design stresses for wide structures are less than peak values appropriate for narrow structures.

Introduction

Offshore exploration for hydrocarbons has been conducted from 16 artificial islands in the Canadian Southern Beaufort Sea. The islands have been constructed as exploration islands to drill single wells in water depths to 13 m. The first, Immerk, was completed in 1973; and currently construction is proceeding on Issungnak in 20 m of water. The islands have provided a drilling platform suitable for year round operations. Fig. 1 is a map of the Mackenzie Delta region showing the locations of these islands and the ocean depth contours. In this area the ice becomes landfast out to a depth of 20 m - 30 m (60 ft. - 90 ft.).

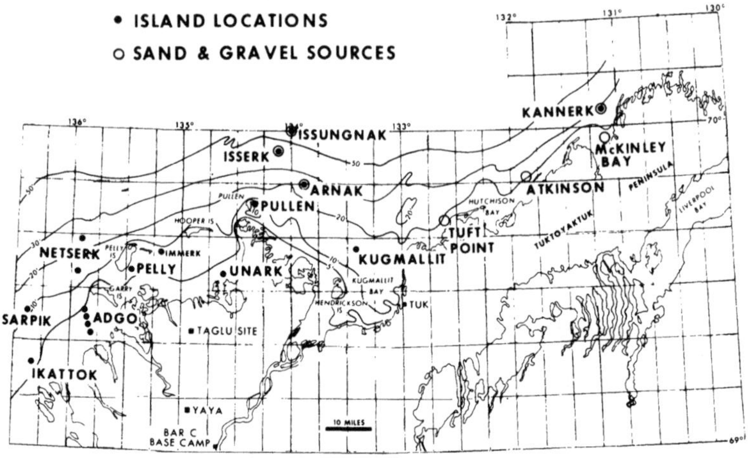

Fig. 1 Mackenzie Delta region of Beaufort Sea with artificial island locations.

The harsh environment governs the design of these exploration islands and various construction techniques have been employed to meet the design goals (1, 2). The top working area is defined by the space requirements of a drilling rig and a camp. The island profile must be designed taking into account wave height, storm surge, wave erosion, fill material, seabed foundation characteristics and the interaction with ice.

Ice in the landfast zone interacts with islands primarily as sheet ice which moves against an extremely wide structure compared to the ice thickness. Their width has important implications for this interaction compared to that for narrow structures such as piles and bridge piers. Non-simultaneous failure increases in likelihood and importance with increasing structure width and once the ice has failed the width inhibits clearing of the broken pieces. This paper is concerned with further development of the significance of these two points. The background for this discussion is established with a review of island designs and construction techniques and the observed ice regimes.

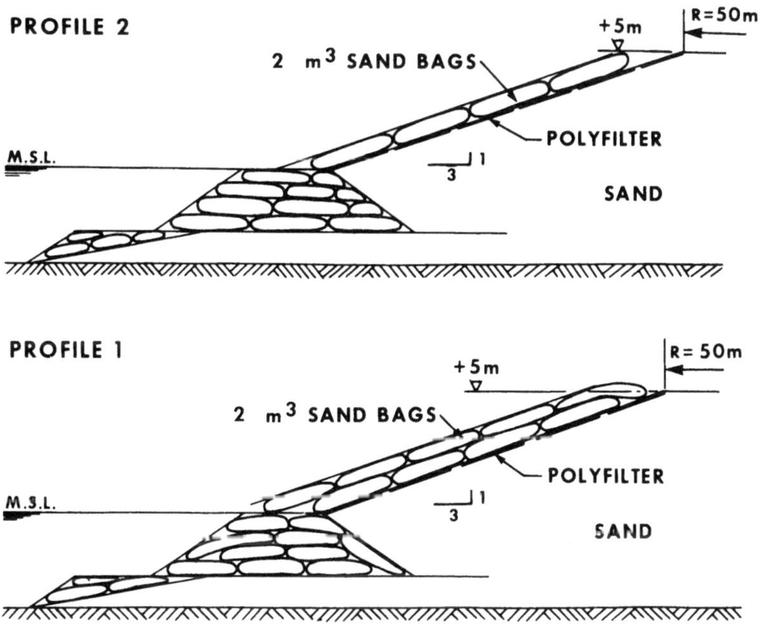

Fig. 2 Cross section of sandbag retained island beach.

Construction Techniques

Two types of artificial island have been constructed by Esso Resources Canada Limited (3). The sandbag retained island, illustrated in Fig. 2 is built by placing a berm of sandbags at the proposed island perimeter and filling the interior with dredged material. The 2 m^3 sandbags also provide slope protection against wave erosion. This island design minimizes dredged fill requirements and is used where local fill material is not plentiful. Construction of these islands has taken place both in summer during the open water season and in winter by trucking fill over the ice and dumping it into a hole cut in the ice cover.

Fig. 3 illustrates the sacrificial beach island which incorporates significantly more fill and uses the erosion of the beach as a defence against wave action. This is the preferred island design if sufficient fill exists at hand.

Both island designs have relatively steep slopes in their profiles (1:3). Croasdale and Marcellus (1) discussed the potential failure modes under horizontal ice loads and identified the most probable as a horizontal shift of the entire frozen island cap.

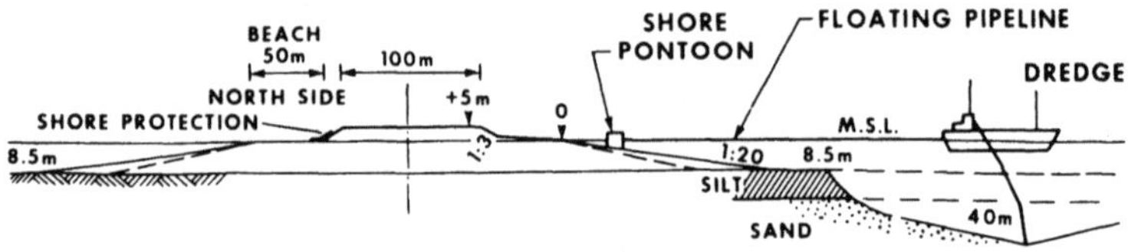

Fig. 3 Cross section of sacrificial beach island.

Ice Regime

Rubble fields of ice form around the islands in the early winter (4). Fig. 4 illustrates the extensive rubble field which formed around the sacrificial beach island, Isserk, located in 13 m of water, during the 1977-1978 winter. The rubble field formed in the early winter when ice thicknesses were generally less than 30 cm and remained unchanged from late December until break-up. Sufficient areas of the rubble field are grounded to isolate it from tidal movements and allow steady consolidation to develop a core of 2 - 3 m of solid ice acting as rigid annulus around the island. However, at the perimeter of the rubble field, small motions due to tides or horizontal motions of the landfast ice maintain an active zone. It is at this active zone that ice failure occurs in any movement of the landfast ice and thus there is an increase in interaction width due to the presence of the rubble field.

The extent of the rubble field is governed by the extent of ice movement by the island location during the winter. The growth is a direct consequence of the inability of the ice rubble formed during the failure of thin ice sheets to clear around the structure. As long as grounding of the rubble occurs the water depth appears to have only an indirect effect on the extent of the rubble field through the delay in the ice becoming landfast at deeper locations. More thin ice can potentially move by an island location in deeper waters. Sufficient sliding resistance of the rubble due to grounding appears to occur to resist the forces generated during the creation of rubble piles.

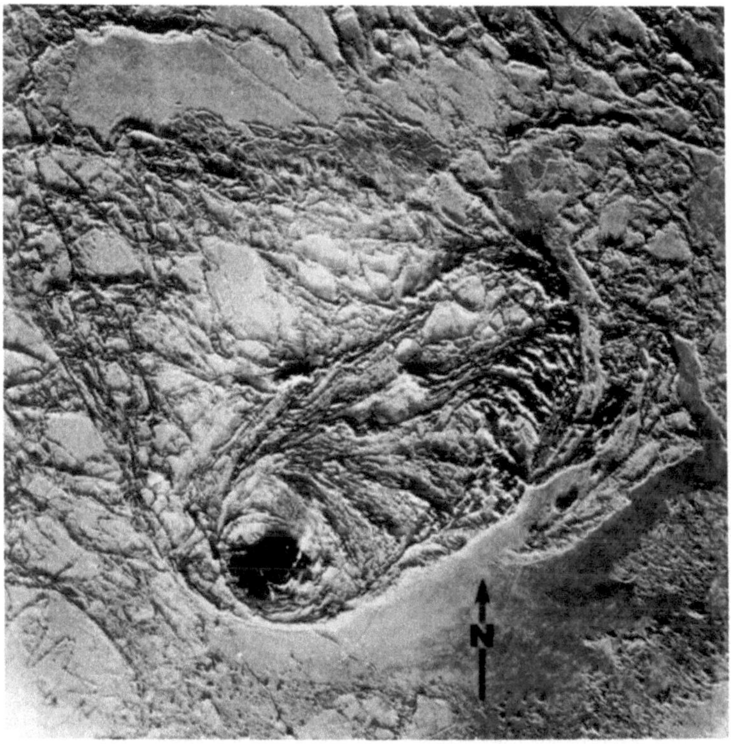

Fig. 4 Aerial photograph from March 1978 of Isserk artificial island located in 13 m water depth.

Ice Clearing

The wider a structure, the wider is the zone in front of that structure in which moving ice cannot clear around the sides. It must either ride up over the structure or pile-up in front of the structure. Experience demonstrates that pile-up is the mode which occurs for artificial islands interacting with thin ice sheets. This is also to be expected if consideration is given to the parameters influencing ride-up. The beach parameters which discourage ride-up are steep slopes, high friction and high freeboards. Similarly, specific aspects of design profiles can initiate kinematic instabilities as outlined by Croasdale et al (5). The weaker and thinner the ice the higher is the probability that ice pile-up will occur rather than ice ride-up. In particular for islands during freeze-up, the thinness of the ice is an important parameter since as Fig. 4 illustrates, pile-ups do form near the beach - water interfaces of the shallow sloped sacrificial beach. However thin ice has been reported to ride-up shallow slopes (6). The steep sloping portion of the island profiles in Fig. 2 and 3 thus provides necessary further insurance against ride-up of thin ice. In addition the large scale roughness of the sandbagging enhances the probability of pile-up.

Once a pile forms a sufficient distance in front of the structure (∿20 m) it forms a natural defense during winter against subsequent ride-up. In fact experience shows as demonstrated in Fig. 4 that seaward pile-ups form more often than enlargements of pre-existing piles. That is, rubble fields tend to expand in extent rather than height.

During the growth of a pile its volume increases to such an extent that eventually it grounds and generates sufficient sliding resistance to exceed the average horizontal forces of its own growth. This is a result of the low magnitude of forces necessary to build a pile and implies that for a wide structure, once a pile-up starts, clearing of the ice quickly becomes impossible.

The average forces involved in the growth of a pile depend on the geometric shape of the pile and its growth mechanism. As initially indicated by Parmerter and Coon (7) for ridges, the two important contributions to the force are the component to generate the potential energy and the component to overcome frictional processes of the growth mechanism.

In the case of a long uniform pile with a triangular cross section for the sail of height H and sides sloping at an angle Θ to the horizontal, Kry (4) has shown that provided the sail and keel grow horizontal at the same rate, the force per unit length along the pile to produce the potential energy in a pile grounded in water depth H_{km} is given by:

$$F_e = \tfrac{1}{2}\rho_i g\, Ht \left[\frac{\dfrac{(\rho_w-\rho_i)}{\rho_i}\dfrac{(H_{km})^2}{(H)^2} + 1}{\dfrac{H_{km}}{H} + 1} \right] \qquad (1)$$

where ρ_i and ρ_w are densities of ice and water respectively,

t is the thickness of the ice sheet from which the rubble is forming and g is the acceleration due to gravity.

In accordance with the rubble building mechanism observed by Kovacs and Sodhi (6), the frictional component of the force per unit length to build the ridge may be approximated as:

$$F_f = \mu_i\ (1-\gamma)\ \rho_i g\ f\ H^2\ \cot\Theta \qquad (2)$$

where μ_i is the coefficient of friction of ice on ice, γ is the porosity of the sail and f is the fraction of the sail mass which acts to create the frictional force. Conceptualizations by Kovacs and Sodhi (6) suggest f lies between 0.1 and 0.5.

The evaluation of a grounding force depends on an assumed profile for the keel and on the assumption that the lowest sliding resistance occurs at the seabed under the pile. The latter assumption is reasonable due to the interlocking and immediate refreezing of ice blocks in the keel and the fact that the compressive vertical load in the rubble is a minimum at the seabed. Assuming a rectangular keel of sufficient width to provide buoyancy to give a keel depth to sail height ratio of k_s, the sliding resistance of a grounded rubble pile is given by:

$$F_g = \mu_s(1-\gamma)\ \rho_i g\ H^2\ \cot\Theta \left[1 - \frac{(\rho_w-\rho_i)\ A_k}{\rho_i\ H^2\cot\Theta} \right] \qquad (3)$$

where μ_s is the coefficient of friction of the keel with respect to the seabed and the porosity of the keel and sail are assumed equal. A_k is the cross sectional area of the keel given by:

$$A_k = 2H_{km}\ \cot\Theta \left[H - \frac{H_{km}}{k_s} + \frac{\rho_i\ H_{km}}{2k_s^{\ 2}(\rho_w-\rho_i)} \right] \qquad (4)$$

The sail height at which such a rubble pile has sufficient sliding resistance to completely counter the forces of its growth is given by the solution to the equation:

$$F_e + F_f = F_g \qquad (5)$$

Table 1 lists the parameters and an estimate of their range and likely value that determines the sail height as a function of sheet ice thickness and water depth from Equation 5 using Equations 1 - 4. Fig. 5 is a plot of the solution to equation 5 for a ridge with a keel to sail ratio of 4.5. It shows the ratio of the water depth H_{km} to the sail height H as a function of the water depth divided by the thickness of the ice sheet forming the rubble pile. The cross-section of the ridge for which the solution applies is also shown in Fig. 5. The line at

$H_{km}/H = 4.5$ corresponds to the keel to sail ratio at which grounding first occurs. If a given sail were less than 1/4.5 of the water depth, that rubble pile would not be grounded. The line and associated shading is the solution to Equation 5 for the values of the parameters in Table 1.

Parameter	Range	Estimate
$\frac{\mu_s}{\mu_s - f\mu_i}$	1 - 1.5	1.1
$\frac{\tan\Theta}{(\mu_s - f\mu_i)(1-\gamma)}$	1.1 - 5.1	1.8
		$k_s = 4.5$

Table 1. Parameters and range for determining sail height at which a grounded rubble pile can resist the forces of its growth.

Figure 5 illustrates that once a pile-up is higher than half the water depth, it is almost certainly able to resist the forces of its own growth. As an example, if in 4 m of water a 0.5 m ice sheet is failing and piling-up to a sail height of 2 m, that pile-up is sufficiently grounded that it is extremely unlikely it would be cleared or moved during continuous growth. This occurs because the average forces per unit length during its growth are relatively low. In contrast, the forces per unit length which could be generated by movement of a 2 m thick ice sheet which must fail by continuous crushing against a consolidated rubble field could be an order of magnitude greater. It is these forces for which the island must be designed and it is for these forces that the extra width of a rubble field becomes detrimental.

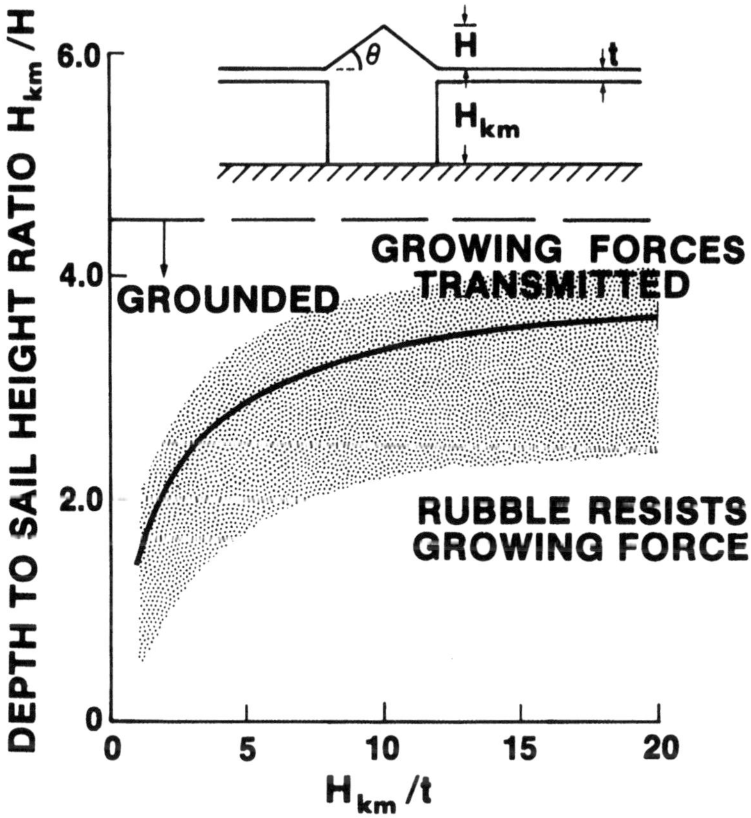

Fig. 5 Ratio of water depth and rubble sail height as a function of water depth divided by ice sheet thickness at which a rubble pile is capable of resisting the forces causing its growth.

Force Amplification

The existence of a rubble field increases the effective width along which ice must fail if it moves past an island location. The rubble field does add to the sliding resistance of the island, but since ice rubble is much less dense than island fill and the average height of the rubble field is of the same order or less than the height of an island, and the rubble field is only fractionally grounded, it does not add significantly to the sliding resistance of the island unless it is very extensive (4). Fig. 6 schematically illustrates this situation. In the case of a bare island, which is loaded by a certain force per unit width, the design includes a safety factor such that the expected peak force divided by the sliding resistance is less than one. However, as the width is increased by the presence of a rubble field, for the same force per unit

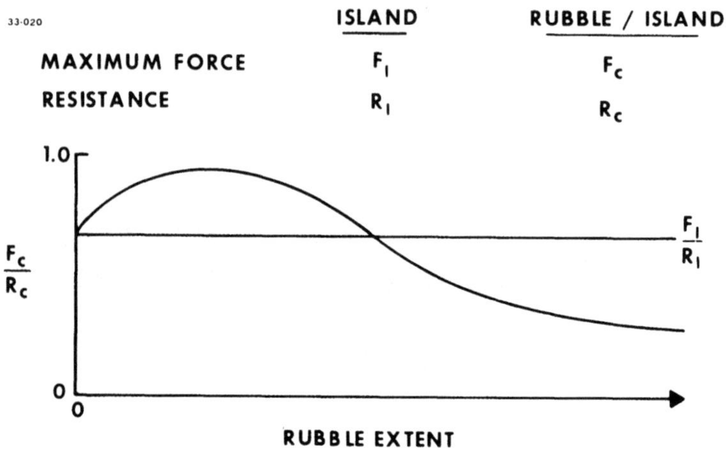

Fig. 6 Force amplification due to rubble field.

width, the ratio of total force applied to sliding resistance increases. Eventually the area of rubble field becomes sufficient to significantly contribute sliding resistance and the force ratio decreases. For the case of a circular rubble field centered on the island, and reasonable assumptions as to the nature of the rubble field the total diameter at which the curve crosses the bare island ratio is four island diameters.

Such an analysis in actual operating conditions must account for the specific grounding pattern of the rubble field and the particular geometry of the rubble field. Only such an analysis can determine if the possible forces might exceed the factor of safety of the island, necessitating defense measures (1). The possibility of failure of the rubble field must be considered along some internal weakened zone, either naturally observed or artificially created. That such failures of rubble fields occurs in nature is clear from observations of the island of grounded sea ice "Katies Floeberg" (8,9).

Stress Reduction

The forces experienced by a wide structure can be expressed as an effective stress (force per unit width of structure) times the interaction width along which the ice must fail. Wide structures differ from narrow structures in that the influences of local ice failures and local variations in contact interface do not generate a significant influence across the entire width of the structure. Therefore independent and non-simultaneous failures across the interaction width have a more significant influence on effective failure stress for a wide structure than for a narrow structure.

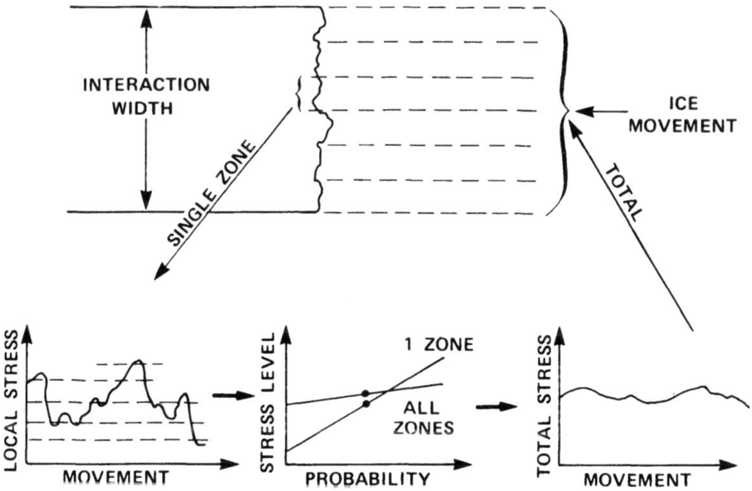

Fig. 7 Statistical influence on peak expected stress as a function of structure width.

Kry (10) proposed an engineering approximation which allows calculation of the statistical influence of non-simultaneous failures across a wide structure. The basis of this approximation is illustrated in Fig. 7 where the interaction width has been divided into a number of statistically independent and equivalent zones. Since the effective stress on the entire width is the average of the simultaneous local stress values in all zones the peak effective stress on the entire width is less than the peak local stress in any one zone for the same length of ice movement. A design stress is associated with a particular risk of its exceedance during the life of the structure, or equivalently with an expected amount of ice movement. Therefore, on a statistical basis design stresses are less for wide structures than for narrow structures.

To illustrate the magnitude of expected design stress reductions, consider the case where for one zone, the local stress has a log-normal probability distribution with median $\bar{\sigma}_1$, and geometric standard deviation σ_{g1}. A particular local stress σ_{D1} has an instantaneous probability of exceedance P and is given by:

$$\sigma_{D1} = \bar{\sigma}_1 \sigma_{g1}{}^{y} \qquad (6)$$

where y is related to P through the Probability Integral and is tabulated in Table 2.

P	10^{-1}	10^{-2}	10^{-3}	10^{-4}	10^{-5}	10^{-6}
y	1.28	2.33	3.09	3.72	4.265	4.753

Table 2. Probability P and associated parameter y.

Consider the case in which the risk of exceedance of a particular stress over the life of a structure is a unique function of the instantaneous probability of exceedance of stress and independent of the number of zones. Kry (10) demonstrated the occurrence of this case for a particular example. Then, given a value P (and y) corresponding to a particular acceptable risk, the appropriate design stress for an interaction width n zones wide is given by:

$$\frac{\sigma_{Dn}}{\bar{\sigma}_1} = \frac{\exp(\tfrac{1}{2}\ln^2\sigma_{g1})\exp(y[\ln(1-\frac{1}{n}+\frac{1}{n}\exp(\ln^2\sigma_{g1}))]^{\frac{1}{2}})}{[1-\frac{1}{n}+\frac{1}{n}\exp(\ln^2\sigma_{g1})]^{\frac{1}{2}}} \tag{7}$$

Equation (6) for an n zone structure has been used with the extrapolations from 1 to n zones for the median and geometric standard deviations as given by Kry (10).

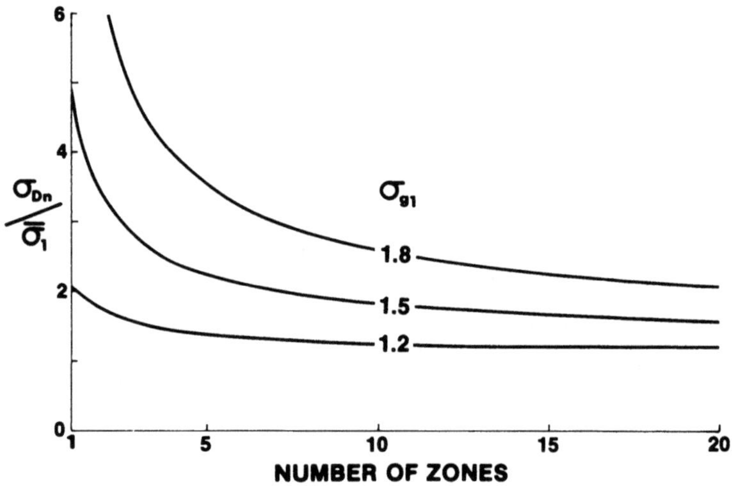

Fig. 8 Design stress reduction as a function of number of zones for three geometric standard deviations of one-zone stress histories.

Equation (7) is plotted in Fig. 8 for three values of the one zone geometric standard deviation and the arbitrary probability $P = 3 \times 10^{-5}$. The variability of the stresses in one zone is characterized by σ_{g_1} with a value 1 corresponding to no variation.

Fig. 8 demonstrates that only a few zones are necessary to achieve the majority of the statistical reduction in design stress. The existence of more than a few zones does not proportionally reduce the design stress. In addition, the magnitude of the reduction is more pronounced the greater the variability of the stresses in a single zone. The potential exists for design stress reductions of 50%.

The relationship of number of zones and structure width hinges upon the width of a single zone and the extent to which ice failures in nearby regions of the ice sheet are independent. A lower limit for this width corresponds to an aspect ratio of four to five, where experimentally it is observed (11) that the contribution to the indentation stress from the edges of the indentor becomes small. Around the islands, observations of the largest sizes of failed ice pieces correspond to this aspect ratio. This implies that the most appropriate data from which to generate one - zone statistics comes from indentation tests at appropriate conditions with an indentor of aspect ratio of 4 to 5 or greater. In this case, islands with their rubble fields represent structures at least 10 zones wide.

Summary

Artificial islands have been successfully used for year round exploratory drilling operations in the Canadian Southern Beaufort Sea.

The width of islands has two significant implications for interaction with ice compared to narrower structures. Both the inability to clear failed ice by the structure and the enhanced possibility of and significance of non-simultaneous failure across the width, influence design ice forces on a wide structure.

An inability to clear ice requires either ride-up over the top or pile-up in front of the structure. The fact that the first interaction is with thin ice coupled with additional benefits of the island beach profile ensures that pile up occurs and that a rubble field around the island is created, protecting it against subsequent ride-up during the winter. Simple calculations demonstrate that for relatively low pile-up heights, the pile can ground sufficiently to resist the average forces of its growth and prevent any clearing action.

The width of the rubble field around the structure amplifies the ice forces which may be exerted. For sufficiently large rubble fields, the additional sliding resistance due to grounding could also help protect the island from horizontal ice loads.

Non-simultaneous failure across a wide interface leads to a reduction in the design stress by potentially 50% compared to that for a narrow structure under the same conditions.

Acknowledgements

The author is grateful to Esso Resources Canada Limited for permission to present this paper.

References

1. Croasdale, K.R. and Marcellus, R.W.
   Ice and wave action on artificial islands in the Beaufort Sea. Canadian Journal of Civil Engineering, Vol. 5, No. 1 (1978) p. 98 - 113.

2. deJong, J.A.A., Stiger, C. and Steyn, B
   Design and building of temporary artificial islands in the Beaufort Sea. Proc. Third Int. Conf. Port and Ocean Engineering under Arctic Conditions, Fairbanks, Alaska, Vol. II (1975) p. 753 - 790.

3. Garratt, D.H. and Kry, P.R.
   Construction of artificial islands as Beaufort Sea drilling platforms. Journal of Canadian Petroleum Technology, April - June (1978) p. 73 - 79.

4. Kry, P.R.
   Ice rubble fields in the vicinity of artificial islands. Proc. Fourth Int. Conf. on Port and Ocean Engineering under Arctic Conditions, Vol. I (1977) p. 200 - 211.

5. Croasdale, K.R., Metge, M. and Verity, P.H.
   Factors governing ice ride-up on sloping beaches. Proc. Int. Assoc. for Hydraulic Research, Symposium on Ice Problems, Luleå, Sweden, Part 1 (1978) p. 405 - 420.

6. Kovacs, A. and Sodhi, D.S.
   Shore ice pile-up and ride-up. Workshop on Problems of the Seasonal Sea Ice Zone, Naval Post Graduate School, Monterey, California (1979).

7. Parmerter, R.R. and Coon, M.D.
   Mechanical models of ridging in the Arctic sea ice cover. Aidjex Bulletin No. 19, Arctic Ice Dynamics Joint Experiment, University of Washington, Seattle, Washington (1973), p. 59 - 112.

8. Barrett, S.A. and Stringer, W.J.
Growth mechanisms of "Katies Floeberg". Arctic and Alpine Research, Vol. 11, No. 4, (1978) p. 775 - 783.

9. Kovacs, A., Gow, A., and Dehn, W.F.
Islands of grounded sea ice. Third Int. Conf. on Port and Ocean Engineering under Arctic Conditions, Fairbanks, Alaska, Vol. I, (1975) P. 333 - 348.

10. Kry, P.R.
A statistical prediction of effective ice crushing stresses on wide structures. Proc. Int. Assoc. for Hydraulic Research, Symposium on Ice Problems, Lulea, Sweden, Part 1 (1978) p. 33 - 47.

11. Neill, C.R.
Dynamic ice forces on piers and piles. An assessment of design guidelines in the light of recent research. Canadian Journal of Civil Engineering, Vol. 3 (1976) p. 305 - 341.

A Danish Icebreaker Design

KAI KURE, BENT KOFOED JACOBSEN
Danish Ship Research Laboratory
Academy of Technical Sciences
Lyngby, Denmark

Summary

Basic ice research and icebreaking research have obtained a rational form in recent years. Design processes benefit from the results and the application of the basic findings evolves a motivation for further research. The paper evaluates the new Danish icebreaker under construction. The hull shape and its icebreaking capabilities have been studied for the continuous mode and for the ramming mode. The icebreaker is a multipurpose vessel having additional mission objectives. It appears, though, that the icebreaking capabilities are well developed in spite of the other requests to the vessel. The paper comments on recent research findings and propose aspects of research on propeller ice interactions. A trial test objective is finally outlined from a research viewpoint.

Introduction

Icebreaking has a long tradition in Denmark reflecting the particular conditions of ice problems in the navigable waters of the nation. The conditions range from a heavy brackish sea ice cover during several months to practically no ice cover at all during the entire winter. Individual winters are determined by a very unstable climatic balance between high barometer continental cold weather from Europe and the inflow of lows like beads on a string from Newfoundland carrying mild Atlantic weather. Statistics call for icebreaking operations every fourth winter on the average. Sometimes two periods go together without iceproblems, and sometimes two icewinters follow each other.

Due to economic reasons Danish icebreakers must therefore be multi-purpose vessels and the one under construction is designed to do hydrografic surveys or oil spill mitigation, when icebreaking is not needed. Other design constraints laid on the vessel are easy maintenance and mothballing in no-ice winters, when the vessel must be on station for icebreaking. This has affected the choice of prime mover to be a direct drive diesel with variable pitch propellers. The propellers are designed to match the icebreaking strategy requesting quick transfer of the icebreaker from one geografic area to another overnight via icefree zones to be on published schedule there next morning. The hydrografic particulars of the inland waters to be assisted impose additional constraints on the design such as draft limitations.

The icebreaking aspects of the design have been evaluated in this paper against recent research findings.

The Hull

The design of optimal icebreakers involves the latest findings of ice research and of naval architecture under the constraints of the mission objectives and particulars, and of requirements from regulatory bodies and international law and conventions. The design of the new icebreaker is governed by the Norske Veritas' classification rules for icebreakers leaving a wide choice of ship parameters. The very complex design process yielded a hull form and main particulars for the new icebreaker as presented in fig. 1 and complying in a balanced way with all the constraints imposed.

The bow shape is very important for the icebreaking capability and needs special consideration. The present case has been evaluated against the analytical results obtained in ref. (1) by using the constraint imposed on the bow form by eq (29):

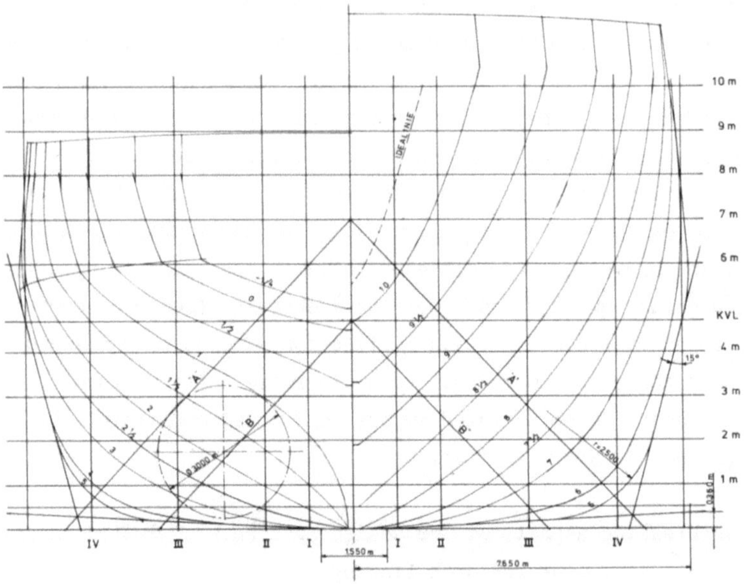

Fig. 1. The frame sections of the icebreaker shown in the figure are drawn by computer after mathematical lines fairing. The corresponding main dimensions of the vessel are a length of 60 m between perpendiculars and a max. breadth at the KVL-waterline of 15.0 m. The draft is 4.7 m. The hull contains bunker capacity for 3 weeks operation and displaces 2250 m^3 of water.

$$D = \sin\gamma \cos\beta + \mu(\sin\psi \sin\gamma \sin\beta + \cos\psi \cos\gamma) > 0 \qquad (1)$$

in order that the ship will not get stuck in the ice. Fig. 2 shows a polar plot of this inequality. The simplified form of the inequality given by (1) for the displaced backing mode with ψ = 180 deg. reads from eq (31) of (1):

$$u < \tan\gamma \cos\beta \qquad (2)$$

The polar plot in fig. 2 indicates this to be sufficient for application in design giving a limiting icefriction of 0.40 at 200 deg. against the 0.38 at 180 deg. for the present icebreaker.

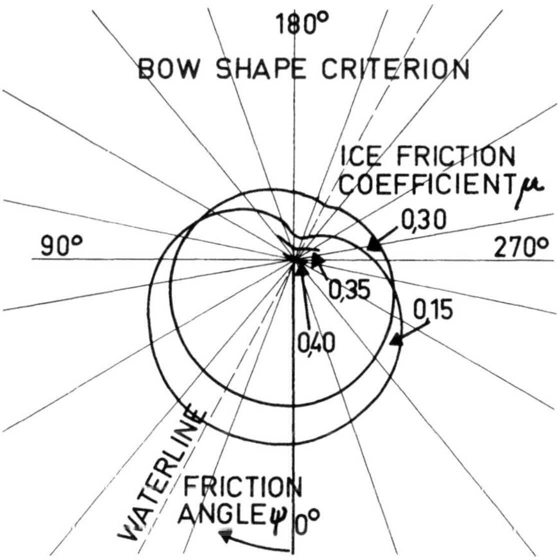

Fig. 2. Polar plot of bow shape criterion for various values of ice friction coefficient. The vessel does not get stuck in the ice when the criterion is positive for any angle of the friction force here measured from the bow line along the bow plating. A coefficient of .40 is the limiting case where reverse power is needed to free the vessel from the ice.

Icebreaking

Two icebreaking modes have been considered for the new icebreaker to evaluate its capability. The continuous breaking mode is characterized by the ice force lengthwise on the hull to be overcome by a steady propeller thrust when transiting the ice. The other mode considered is the ramming mode governed by kinetic energy at ice impact.

The ice resistance to constant ship propulsion through a uniform ice cover has been rationally expressed in (2), (3) and other recent research papers. The resistance is expressed in (2) by three terms, one for breaking the ice, one for submerging or fully emerging the broken floes below or above the remaining ice cover and a third term to account for the acceleration of the broken floes.

$R_{ice} =$	total force
$C_o \sigma_f h^2$	breaking
$+ C_1 \rho_E g B h^2$	submergence
$+ C_2 \rho_E B h v^2$	acceleration

h is ice thickness, B is ship breadth, g is gravity acceleration, ρ_E is ice density and σ_f icebreaking strength. The coefficients C_o, C_1, and C_2 are given average values in (2), they are, rounded, $C_o = 0.15$, $C_1 = 9$, $C_2 = 6$. Breaking velocity is v, and a standard value for σ_f is 15 kp/cm^2.

TABLE 1 ICE RESISTANCE

Ice Thick-ness	Force Components		Total Ice Force			
	Breaking	Submersion	V = 0 knots	V = 2 knots	V = 4 knots	V = 6 knots
m	tons	tons	tons	tons	tons	tons
0.1	.23	1.28	1.5	2.4	5.2	9.8
0.2	.90	5.13	6.0	7.9	13.4	22.7
0.3	2.02	11.54	13.6	16.3	24.7	28.5
0.4	3.60	20.52	24.1	27.8	38.9	57.4
0.5	5.63	32.06	37.7	42.3	56.2	-
0.6	8.10	46.17	54.3	59.8	-	-

The table lists the values of the two first components of the expression for ice resistance. They add up to the total iceforce in the column for v = 0 knot. The third term is added in the last three columns of the table for v = 2,4 and 6 knots. No values have been given in excess (approximately) of the max. thrust of the new icebreaker, i.e. for 4 and 6 knots in 0.5 and 0.6 m ice. The value of the force due to icebreaking as such is seen to be only abt. 20 per cent of the corresponding submergence force and at, say 4 knots, only 10 per cent of the total force in this case.

The ramming mode icebreaking involves backing of the icebreaker in the cleared channel, stopping, accelerating towards the next impact and deceleration during ice penetration until all kinetic energy at the time of the impact plus the work done by propulsion thrust during penetration has been consumed by the ice resistance. This is expressed by the same three terms as for the continuous breaking mode, though with varying velocity during penetration.

The energy equation reads:

$$\tfrac{1}{2} MV_i^2 = \int_0^P (R_{ice} - T_{prop})\, d P$$

P for penetration length, M for icebreaker effective mass and V_i for impact velocity. Assuming a deceleration velocity curve of cosine shape from impact to a definite halt,the integration yields:

$$\tfrac{1}{2} MV_i^2 = (C_o \delta h^2 + C_1 \rho_E g B h^2)P + \tfrac{1}{2}C_2 \rho_E B h V_i^2 P - T_{prop} \cdot P$$

in complete compliance with eq (24) and (68) of (3).

The ramming capability of the new icebreaker has been estimated by using the raw icebreaker mass in the above expression and the full effective propulsion thrust to calculate the ramming energy from:

$$\tfrac{1}{2} MV_i^2 = T_{prop} \cdot \text{approach distance}$$

The result appear in fig. 3.

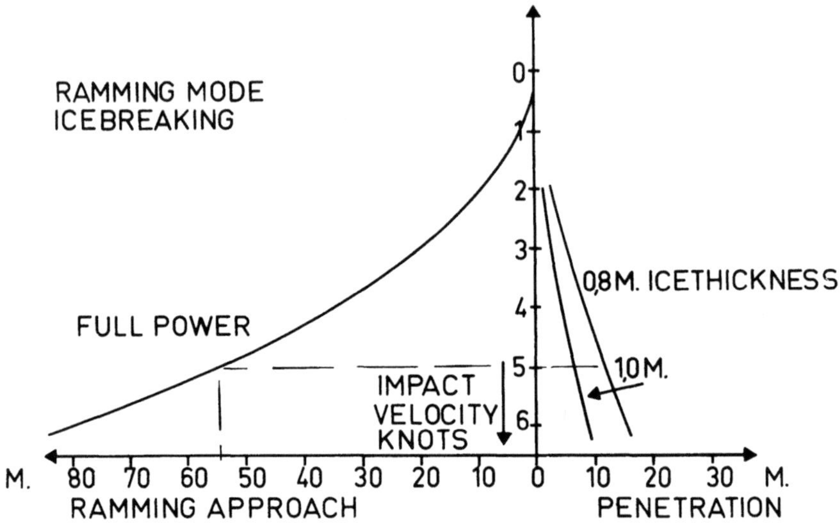

Fig. 3 shows the ramming mode capability of the new Danish icebreaker. The example given is for full power i.e. 55 tons thrust applied over abt. one ship length approach distance before impact. The ram velocity is abt. 5 knots and the penetration is 12 m in 0.8 m thick ice.

Propulsion

The prime mover of the new icebreaker has been chosen as a direct diesel engine drive system with reduction gears and elastic couplings and controllable pitch propellers. Fuel economy of this system compared to others was a main reason for the decision. Ref. (4) deals with the subject in general. Another decision parameter was the easy mothballing of the engines in non-heated engine rooms without risking corrosion damage during inoperative winter periods.

There are two main engines

B & W Alpha	type	12 U 28 LU
Max. continuous rating		3180 HP each
Continuous Service rating		2880 HP each

at 775/750 revs per minute
and fuel consumption,
161/159 grammes per HP hour

The ice strengthened controllable pitch propellers are designed to convert the continuous rated 7500 tm torque of each propeller to the highest possible effective thrust both for the vessel transit mode at 16.5 knots and for low speed icebreaking at 55 tons. The effective thrust is less than the raw propeller thrust due to ship-propeller interaction by 5 per cent at low speed to abt. 20 per cent at transit speed. Tests in a ship model basin were made to obtain the design parameters for the propellers to fit the ship and the objectives.

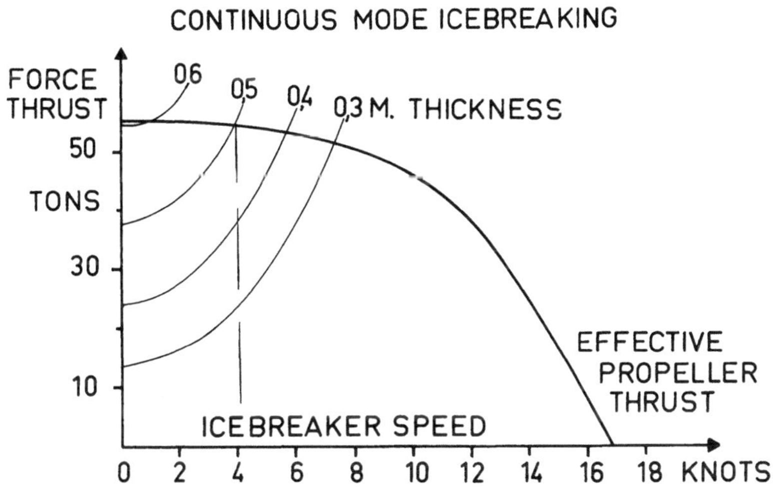

Fig. 4. The propulsion plant in the new icebreaker yields an effective propeller thrust in excess of propulsion resistance as shown by the curve versus speed. The curves marked with ice thickness are the results from table 1. The example shows that 0.5 m thick ice can be broken at 4 knots in continuous mode.

Ice Milling Propellers

The new icebreaker has two variable pitch propellers to convert the output torque of the direct drive diesels into useful thrust. The propellers are of the make B & W Alpha VB 1080 and are outward turning at 240 revolutions per minute at continuous service rating. The diameter is 3050 mm giving a blade tip velocity of abt. 40 m/sec and a passage frequency of 16 blades per second. The torque absorbed corresponds to 5 tons peripheral force at the blade tip. The propellers are specially ice strengthened to operate in ice according to classification criteria.

One of the criteria says that the direction of rotation must be reversible. This is not necessary to reverse the thrust of the variable pitch propeller, but to be able to clear the propeller in case it gets blocked by ice. This might have been necessary in earlier days with large, slow turning propellers, but it is argued that rotation reversal is not necessary for cases like the present having well submerged stern propellers of small diameter and corresponding high number of revolutions. They are fitted underneath of well shaped afterbody and protected in the backing mode by twin rudders located behind each propeller. There is much evidence from ice transiting ships, not being icebreakers and not subject to the rotation reversion criterion, that such propellers do not get clogged by ice. Ref. (4) mentions specific investigations without giving any details.

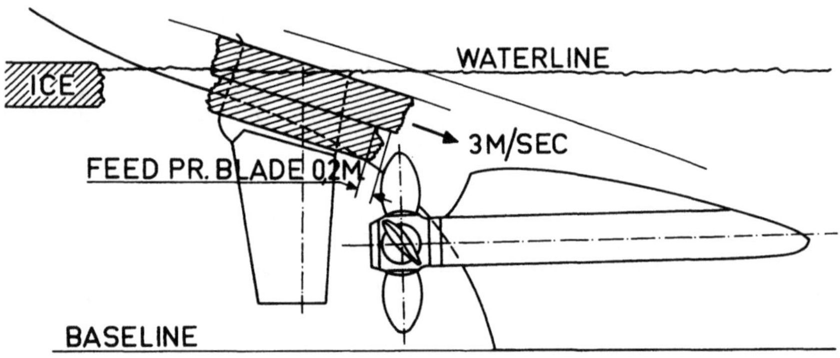

Fig. 5. Ice floes may enter the propeller race from abaft being sucked down by propellers in backing mode. The floes will be guided off centre by the protecting rudders and be exposed to the milling effect of the propeller cutting the ice at 16 blades per second. The feed velocity will be abt. 20 cm per blade passage at 6 knots backing speed. The blade force is comparable to 5 tons at corresponding blade strength.

Fig. 6. Ice floes may enter the propeller race from forward, passing under the bottom and being guided by the propeller bosses at low feed velocity compared to the propeller blade velocity of abt. 80 knots. The propeller will act as a milling machine and not get stuck by ice in this mode.

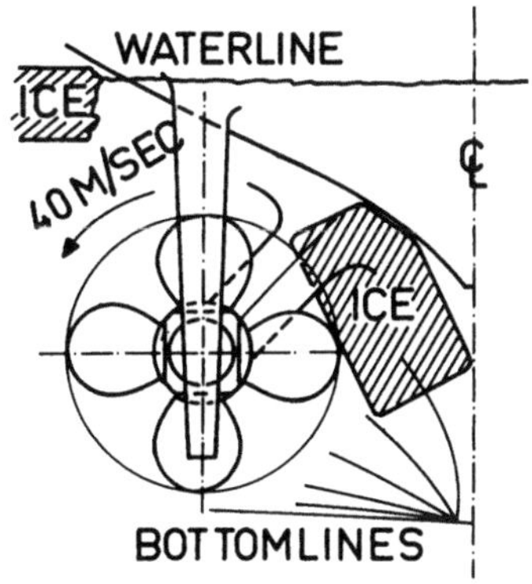

Trials

Special trials have been specified for the now icebreaker after completion. One test covers a longterm breaking of heavy ice according to the capability for the continuous mode. The purpose is to verify the specified capability to break ice at full power in the continuous mode without experiencing vibration caused damage to any part of the ship or its machinery nor to its installations and instrumentation. Such damage would end the icebreaking operation. This test series present itself as a perfect opportunity for icebreaker research to augment the programme by certain additional measurements of the icebreaker behaviour and the environment. The data obtained should be used to verify not only the overall results predicted in this paper from research findings but also for as many individual part contributions as possible, say in the iceforce for the continuous mode. To do this, ice measurements will be necessary and should preferably be made from shipborne sensors not to delay the testing period. Proper instrumentation and planning is needed to make the endeavour successful. Sufficient funding must be raised to finance the work and equipment involved. The efforts will prove beneficial for future icebreaker design and for the understanding of physics of ice and icebreaking which is the background for any new enterprise in icehaunted areas.

Conclusion

The objective of the paper was to use recent basic results from ice and icebreaking research to evaluate the qualities of the new Danish icebreakers now under construction. Bow friction is important in icebreaking. Analytically derived constraints on bow form were applied and showed that the bow was well shaped. It was found that a simplified expression for the constraint presented in the literature would have been sufficient for the evaluation. Icebreaking resistance has been determined in the study and expressed in tabular and graphical form in the paper. It was found that the force required to pure breaking of the ice amounted to only 10 per cent of the total force at 4 knots. The two other components due to ice floe submergence and to ice floe acceleration each accounted for one half of the remainder in the continuous mode icebreaking. For the ramming mode icebreaking it was found that a simple assumption for the deceleration curve in ramming will result in identically the same expression derived from a sophisticated analysis.

A criterion valid by convention for icebreaking ship propellers were questioned and simple arguments given for its withdrawal. Further research on the subject will be useful. A special delivery trial programme is planned. The paper recommends additional instrumentation to use the trial to collect data for verification of the predictions for the benefit of future icebreaker designs.

References

1 Scarton, Henry A: On the Role of Bow Friction in Icebreaking. Journ. of Ship Research, Vol. 19 No. 1, March 1975

2 Kloppenburg, M. Schwartz J.: Neue Wege der Eisbrech Technik. Jahrbuch d. S.T.G. 69 Band 1975.

3 Lewis, Jack W. Edwards, Roderick, Y. Jr.: Methods for Predicting Icebreaking and Ice Resistance Characteristics of Icebreakers. Transaction of SNAME, Vol. 79, 1971.

4 Airaksinen, Kari. Marttila, Markku: On Propulsion of Icebreakers. International Shipbuilding Progress. Vol. No. 1976

Creep and Relaxation in Floating Platforms – an Analysis of Case Histories

D.M.MASTERSON AND A.G.STRANDBERG
FENCO Consultants Ltd.
Calgary, Alberta, Canada

INTRODUCTION

The use of the natural ice sheet in Arctic regions to support stationary, heavy loads of 500 tons or more for periods of greater than 30 days requires analysis of the flexural creep behaviour of ice. Panarctic Oils Ltd. of Calgary, Canada has, since the spring of 1974, been thickening the natural ice cover by flooding and freezing in thin layers and placing conventional land drilling rigs on these constructed platforms. By doing this offshore exploratory drilling for hydrocarbons has proceeded over a great area at relatively little cost. (Baudais, Et. Al. 1974, Baudais, Et. Al. 1976, Masterson and Kivisild, 1978). One offshore gas well was recently completed from a floating ice platform and a pipeline brought to shore. (Watts and Masterson, 1979, Palmer, Et. Al., 1979). Design and monitoring of these platforms have been carried out by FENCO CONSULTANTS LTD. of Calgary, Canada.

The platforms are designed so that maximum allowable elastic flexural stresses are not exceeded and also so that the creep deflections during drilling do not exceed the available freeboard of the ice platform. Stress calculations are conducted using theories applicable to elastic, homogeneous and isotropic plates on elastic foundations. Both closed form and finite element solutions are used (Masterson and Strandberg, 1978). These same methods have been used to predict creep deflections with considerable success and in this paper a method using the elastic formula and the exponential creep formula will be discussed. The formulation presents a rational method which can be easily extrapolated to predict long term deflections for different thickness and load cases.

METHODS OF ANALYSIS

1. Elastic Equations - Effective Modulus Method

The maximum deflection of an infinite, elastic plate on an elastic foundation caused by a lateral load is given by:

$$W = \frac{P}{8k\ell^2} \qquad \text{(Timoshenko, 1940)} \qquad 1)$$

where:

$$\ell = \sqrt[4]{\frac{E\,h^3}{12(1-\mu^2)k}} \qquad 2)$$

If the load deflection history for a platform is known, then the above equations may be employed to calculate the effective or reduced modulus as time progresses. In fact, combining 1) and 2):

$$E_{(t)} = \frac{3(1-\mu^2)}{16kh^3}\left(\frac{P}{W(t)}\right)^2 \quad 3)$$

The effective modulus method has proved very effective for calculating time dependent deflections of ice platforms. It does not, however, take into account stress levels in the ice and is a purely empirical approach. It has also been found that extrapolation very far beyond the range of case histories often results in considerable error in the predicted deflection. A less empirical approach to estimating time dependent deflections using the elastic plate equations is given by Frederking and Gold (1975).

2. Finite Element Time Dependent Solution

The Mark CDC finite element program has been used to calculate instantaneous elastic stresses in ice platforms for complicated loading configurations, boundary conditions and tapered sections. It has also been used for calculation of long term deflections (Masterson and Strandberg, 1978).

Viscoelasticity of the material is modelled using a supplied subroutine which contains expressions for thermal strain, plastic strain, creep strain and Kelvin model strains. The creep law provided is the well known exponential formula:

$$\frac{d\varepsilon}{dt} = \dot{\varepsilon} = K\sigma^n \quad 4)$$

User supplied creep laws can also be used as input.

This procedure was found to work very well once elastic deflections were separated from true creep deflections. With its use, a continuous curve of deflection versus time is obtained. However, for the complicated geometries and loading cases encountered, computer costs are high and can exceed $2,000.00 per case analysed since the solution process is iterative. Since final deflections, and not deflection time curves, are usually required, a simpler and cheaper method of analysis can be used.

The elastic plate equations and the exponential creep law can be used in an iterative process to derive deflection time curves. This has been tried with reasonable success.

At time t_n, the total strain is given by:

$$\varepsilon(t_n) = \varepsilon_e + K(\sigma\,(t_{n-1}))^n \quad 5)$$

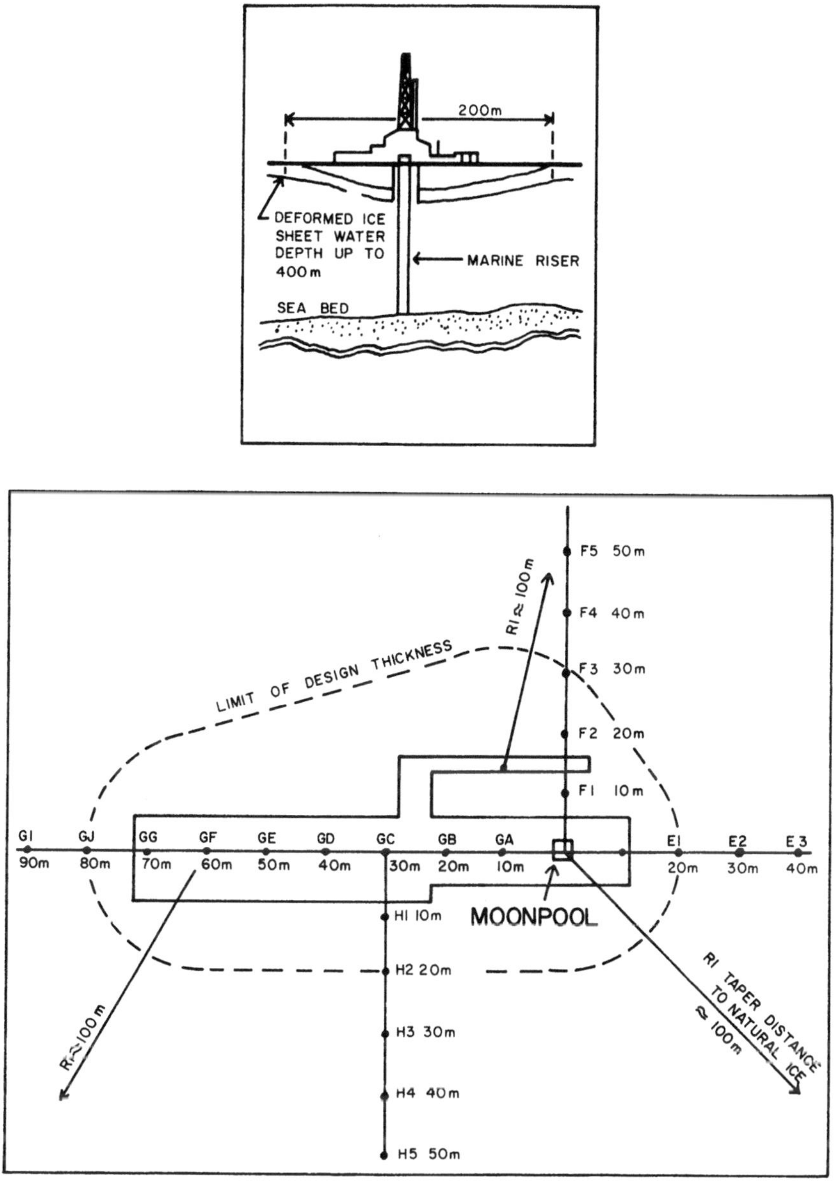

FIGURE 1 RIG 1 ICE PLATFORM FOR OFFSHORE DRILLING

where σ is the initial elastic stress or the one from the previous time increment.

A strain profile at time t will now be nonlinear. It can be linearized by making the area under the strain diagram equal to an idealized linear one (Figure 2).

$$\bar{\varepsilon}(t_n)\,\frac{h}{2} = \int_{-h/2}^{h/2} \varepsilon(t_n)dx$$

$$= \int_{-h/2}^{h/2} \left(\varepsilon_e + k(\sigma\,(t_{n-1}))^n\right) dx \qquad 6)$$

Once $\bar{\varepsilon}(t_n)$ is calculated, a new elastic modulus is calculated from:

$$E(t_n) = \frac{\sigma(t_{n-1})}{\bar{\varepsilon}(t_n)} \qquad 7)$$

A new value of stress is then calculated using the elastic stress formulae and $E(t_n)$. The process is repeated until full time has passed.

Since bending moments and stress vary with distance from the load, some sort of average stress must be used. Obviously this is where the strength of the finite element procedure is appreciated. This procedure was used for one ice platform with reasonable success but was found to be tedious in execution, to involve several assumptions and to yield results which were of little more use than those obtained from the reduced modulus method. Recent measurements with wire resistance strain gauges showing that strain profiles do in fact remain linear with time and showing that the neutral axis is at mid depth of the section would lend credence to this method (Masterson, Et. Al., 1978). The strain data also provide information on the exponent n and the value K.

3. Use of Exponential Creep Law in Simplified Formulation

By combining equation 4) and a simplified version of the elastic stress equation, an expression for time dependent deflection can be obtained (Masterson and Kivisild, 1978).

$$W = K_1 \left(\frac{P}{h^2}\right)^n t \qquad 8)$$

If a set of data for one platform is in hand (i.e. load, ice thickness and deflection) and it is desired to obtain the deflection for another,

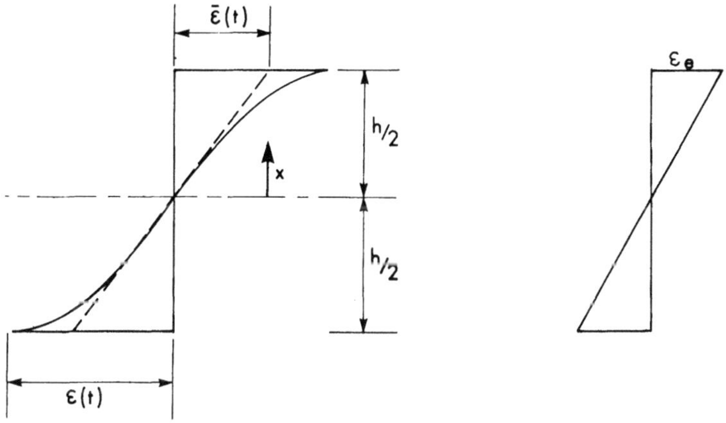

FIGURE 2 STRAIN DIAGRAM

an equation can be written as a ratio.

$$\frac{W(t)}{W_0(t)} = \left(\frac{P}{P_0}\right)^n \left(\frac{h_0}{h}\right)^{2n} \qquad 9)$$

It is this equation which will be used to analyse case histories and values of K_1 and n will be derived.

CASE HISTORIES ANALYSED

The case histories analysed are listed in Table 1. Cases 1 through 7 all have loads of 507 tonnes, 8 through 10 have loads of 679 tons and 11 has a load of 1136 tons. Total duration of load varied and, to conform to the majority of cases, 50 day deflections were used in the analysis. Since case 5 was only loaded 37 days, the deflection time curve had to be extrapolated to 50 days.

Ice thickness is the average thickness measured at the stations on the platforms within the design area of constant thickness. As can be seen from the column of comments that natural ice upon which flooding took place was multi-year except for two cases. Platforms founded on multi-year ice are stiffer than those founded on first-year ice and, all else being equal, the latter deflect 25 percent more with time. This fact will be taken into account in later calculations.

EVALUATION OF N AND K_1 - CONSTANT LOAD CASES

Since cases 1 through 7 were all for a constant load, a regression analysis was performed on them. Equation 8) can be rewritten in the form:

$$W = K_2\left(\frac{1}{h^2}\right)^n \qquad 10)$$

where:

$$K_2 = K_1 P^n t$$

The regression analysis yielded a K_2 value of 25.78 and an n value of 1.49. The correlation coefficient was 0.977 and n will be considered to be 1.5. From 10) $K_1 = 4.52 \times 10^{-5}$.

PREDICTION OF DEFLECTIONS FOR DIFFERENT LOADS

The previously derived value of n was then used in equation 9) to calculate 50 day deflections for cases 8 through 11 where the loads were different. The results are presented in Table 2. The calculated deflections for cases 9 and 11 were multiplied by 1.25 since these platforms were founded on first-year

TABLE 1

CASES OF RIG LOADINGS ON ICE PLATFORMS USED FOR ANALYSIS

CASE NO.	LOAD (TONNES)	TOTAL DURATION OF LOAD (DAYS)	ICE THICKNESS (m)	DEFLECTION AFTER 50 DAYS (m)	COMMENTS
1	507	60	4.48	0.304	Platforms for cases 1 through 8 and no. 10 were founded on multi-year ice
2	507	60	4.60	0.265	
3	507	55	4.97	0.229	
4	507	50	4.97	0.229	
5	507	37	4.85	0.244	37 day deflection was 0.204 m
6	507	50	5.33	0.168	
7	507	50	4.43	0.300	
8	679	102	5.37	0.290	
9	679	108	5.49	0.293	Founded on first-year ice
10	679	105	6.36	0.175	
11	1136	-	7.10	0.244	Load taken as average of 30 and 90 day load

TABLE 2

COMPARISON OF ACTUAL AND PREDICTED DEFLECTIONS FOR VARYING LOAD

50 DAY

CASE NO.	ACTUAL DEFLECTION (m)	CALCULATED DEFLECTION (m)	PERCENT ERROR
8	0.290	0.258	- 11.0
9	0.293	0.245 X 1.25 = 0.306	+ 4.5
10	0.175	0.155	- 11.4
11	0.244	0.241 X 1.25 = 0.301	+ 23.5

$$\frac{W}{W_0} = \left(\frac{h_0}{h}\right)^{3.0} \left(\frac{P}{P_0}\right)^{1.5}$$

$h_0 = 5.0$ m

$P_0 = 507$ t

$W_0 = 25.78\ (5.0)^{-3}$
$= 0.206$ m

ice. Agreement is quite good for cases 8, 9 and 10 and acceptable for 11. At this point, it would be well advised to point out that of all parameters involved in the prediction of deflections and stress, load is usually the one most difficult to determine accurately. Thickness is measured by means of holes drilled in the ice at the end of construction. Strength and stiffness are measured periodically during construction and drilling and are known. Deflections are measured during the total load period with great care using at least two different methods.

However, loads are difficult to measure. Rig pieces are large and supplies for the drilling program come and go. Thus the base load is only known approximately in some cases and the load-time function is definitely not a constant. Deflection-time curves for the platforms (Baudais, Et. Al., 1976 and Watts and Masterson, 1979) are not smooth and show time steps, indicating varying loads. A 15 percent error in estimating the total effective load over 50 days for any of the cases 8 through 11 would result in a 23 percent error in the calculated deflections. Since it is quite conceivable that one or both estimated loads could be in error by this much when using equation 9, the errors in long term deflection calculation are understandable and do not necessarily reflect on the validity of the theory.

Thus the simplified method is useful and generally applicable over a wide range of loads and thicknesses. Caution in its use is advised since it applies to only one load duration, 50 days in this case since the majority of the loads were on the ice for this time. Equation 8 indicates deflection should increase linearly with time. This, however, is not the case (Baudais, Et. Al., 1976, Watts and Masterson, 1979). This is why, if detailed deflection-time information is required, more complicated procedures must be resorted to as discussed in the section on methods of analysis.

As more data with varying loads and durations become available, it will be possible to derive equations of a more general nature.

SUMMARY AND CONCLUSIONS

Long term lateral loading of ice sheets and floating ice platforms requires an ability to predict both maximum stress and creep deflections. The use of ice sheets to support conventional land drilling up for offshore hydrocarbon exploration in the Canadian Arctic Islands has caused much development in prediction methods to take place and has provided a wealth of previously unavailable data regarding the flexural creep of ice sheets subjected to lateral loads greater than 500 tonnes.

The reduced or effective modulus method has been used successfully in the past for predicting creep deflection but extrapolation often leads to considerable errors in prediction.

Finite element time dependent solutions provide detailed accurate information and complete deflection-time curves but are costly and time-consuming to execute. Analytical solutions using the exponential creep law and linearized strain profiles varying with time provide a simplification to the finite element solution but are still tedious in application and involve several assumptions.

Use of the exponential creep law and a simplified elastic stress equation provides a formula which will predict creep deflections after a fixed time with good accuracy. Case histories of ice platforms which had been subjected to a particular constant load for which several combinations of thickness and deflection were available were subjected to a regression analysis. The creep exponent n was derived and found to be 1.5, with a correlation factor of .977. These were then used to predict deflections after 50 days for other platforms subjected to different loads. Maximum error was 23.5 percent and the mean error was 1.4 percent.

This was considered good agreement since rig loads are difficult to determine accurately. Base weights are often only approximately known and the coming and going of supplies means that the load time function is not a constant.

The function derived can only be used for single load durations. A large portion of the loads on ice platforms have been for 50 days and this dictated the time interval analysed. It assumes deflection increases linearly with time, which is not true. Detailed information on deflection-time functions will have to be derived from the more complicated procedures discussed earlier.

As more information on platforms subjected to various loads for different times becomes available, more general formulations can be derived.

ACKNOWLEDGEMENT

The author would like to thank Panarctic Oils Ltd. of Calgary, Canada for granting permission to publish this article.

REFERENCES

Baudais, D. J., Masterson, D. M., and Watts, J. S.,
"A System for Offshore Drilling in the Arctic Islands",
25th Annual Technical Meeting at the Petroleum Society of CIM, Calgary,
May 7 - 10, 1974, Paper No. 374029

Baudais, D. J., Watts, J. S., and Masterson, D. M.,
"A System for Offshore Drilling in the Arctic Islands",
Offshore Technology Conference, Paper No. OTC 2622,
Houston, Texas, May, 1976

Frederking, F. M. W., and Gold, L. W.,
"The Bearing Capacity of Ice Covers Under Static Loads",
Canadian Journal of Civil Engineering, Vol. 3, No. 2,
June, 1976, P. 519 - 538

Masterson, D. M., Anderson, K. G., and Strandberg, A. G.,
"Strain Measurements in Ice Platforms Using 3 m Long Resistance Wire Gauges",
Workshop on the Bearing Capacity of Ice Covers, Snow and Ice Sub-Committee,
Associate Committee on Geotechnical Research, National Research Council
of Canada, Winnipeg, October 16 - 17, 1978

Masterson, D. M., and Kivisild, H. R.,
"Foating Ice Platform: Offshore Oil Exploration",
ASCE Annual Convention, Chicago, Illinois, October 16 - 20, 1978

Masterson, D. M., and Strandberg, A. G.,
"Long Term Loading Analysis of Floating Ice Platforms Using Finite Element
Techniques",
Workshop on the Bearing Capacity of Ice Covers, Snow and Ice Sub-Committee,
Associate Committee on Geotechnical Research, National Research Council
of Canada, Winnipeg, October 16 - 17, 1978

Palmer, A. C., Baudais, D. J., and Masterson, D. M.,
"Design and Installation of an Offshore Flowline for the Canadian Arctic
Islands",
Offshore Technology Conference, Paper No. OTC 3446, Houston, Texas,
April 30 - May 3, 1979

Timoshenko, S.,
"Theory of Plates and Shells",
McGraw-Hill, 1940

Watts, J. S., and Masterson, D. M.,
"Completion of an Offshore Well in the Canadian Arctic Islands",
Offshore Technology Conference, Paper No. OTC 3445, Houston, Texas,
April 30 - May 3, 1979

LIST OF SYMBOLS

E	=	modulus of elasticity
$E(t)$	=	effective modulus at time t
$E(tn)$	=	effective modulus at time tn
h	=	thickness
h_0	=	thickness, case 0
K	=	creep constant
K_1	=	constant
K_2	=	K_1P^nt
k	=	creep constant
ℓ	=	stiffness length
n	=	creep exponent
P	=	load
P_0	=	load, Case 0
t	=	time
tn	=	time at interval n
$tn-1$	=	time at interval n-1
W	=	deflection
$W(t)$	=	deflection at time t
$W_0(t)$	=	deflection at time t, Case 0
x	=	distance
ε	=	strain
$\dot{\varepsilon}$	=	strain rate
εe	=	initial elastic strain
$\varepsilon(tn)$	=	strain at time tn
$\bar{\varepsilon}(t_n)$	=	idealized linear strain at time t_n
μ	=	Poisson's ratio
σ	=	stress

Mechanical Properties of Polycrystalline Ice

Malcolm Mellor
Cold Regions Research and Engineering Laboratory
Hanover, N.H., U.S.A.

Introduction

Ice normally exists at very high homologous temperatures, mostly above 0.9 and commonly above 0.95, and it readily displays a wide range of mechanical properties, including elasticity, nonlinear viscosity, creep rupture and brittle fracture. It is an important material in earth science and engineering, as well as being a valuable model material for exploration of general problems involving the mechanics of solids.

Studies have tended to concentrate on restricted ranges of material behaviour, depending on the requirements of the investigator, and overall impressions of material properties tend to be fragmented and sometimes confusing. However, it is now becoming possible to draw together the results of disparate studies to form a coherent picture.

The following account attempts to give a simple impression of the general mechanical properties of non-saline polycrystalline ice. It is a personal perspective that necessarily includes a certain amount of speculation and personal interpretation, but there should be no difficulty in distinguishing between speculation and well established facts.

Some general characteristics of ice

Although the present concern is with "non-saline" polycrystalline ice, it might be noted that even the purest forms of natural ice contain impurities. These can be solid particles, soluble substances, or gases. Microscopic particles, ionic impurities, and gases can be incorporated into interstitial vacancies, or into the lattice, of ice monocrystals, influencing the nucleation and motion of dislocations. Concentrations of impurities occur within grain boundaries and grain boundary intersections, gas segregating into discrete bubbles and soluble impurities forming liquid films at temperatures above the eutectic point. These liquid films change their volume and concentration as temperature varies. During freezing , entrapment or rejection of soluble or gaseous impurities varies with freezing rate, slow freezing tending to remove more gas or solute.

The basic ice unit for present purposes is the single crystal of ordinary Ice Ih. An ideal crystal has oxygen atoms of the water molecules arranged tetrahedrally at a separation of 2.76 Å, giving rise to hexagonal symmetry about the *c*-axis. There are two hydrogen atoms adjacent to each oxygen atom, but only one hydrogen atom lies between each pair of oxygen atoms, and each molecule is oriented so as to direct its hydrogen atoms towards two of the four neighboring oxygen atoms. Molecules lie close to a set of parallel planes, which are the basal planes of the crystal. A unit cell in the crystal structure contains four molecules, and its dimensions (about $4.5 \times 4.5 \times 7.4$ Å) correspond to a density of 0.917 Mg/m^3 at 0 °C, conferring the important property of positive buoyancy in water. The basal (0001) plane is the sole demonstrated glide plane of the lattice, and shear applied parallel to this plane gives a strain rate about two orders of magnitude higher than that resulting from shear normal to the basal plane. There is no significant preference in glide direction.

Anisotropy of the monocrystal has great influence on the mechanical behaviour of polycrystalline ice. In "isotropic" polycrystalline ice the axes of constituent crystals are randomly oriented, and when such ice is stressed the compliance of individual grains varies, depending partly on how the basal planes are oriented relative to the stress field. Under protracted deviatoric loading, recrystallization can take place so as to bring basal planes into closer coincidence with resolved shears, and consequently the crystal orientation ceases to be random. Preferred orientations of *c*-axes, or fabrics, reflect shear strains and a tendency towards strain softening of the bulk ice.

In all subsequent discussion of polycrystalline ice it will be assumed that any test specimen or control volume is large compared to the size of constituent grains and crystals. In practice, this means that the minimum linear dimension of a specimen is at least an order of magnitude greater than the grain size.

Deformation and failure under uniaxial stress

Much of the information on mechanical properties of ice is derived from laboratory tests that apply uniaxial stress in compression or tension. The basic tests are: (a) creep tests under approximately constant stress, and (b) "strength" tests at approximately constant strain rate. In practice, it is usually load and displacement rate respectively that are held constant. If these tests are carried to sufficiently large strains (say $> 3\%$), they generally display the responses illustrated in figure 1.

The definition of "failure" and "strength" is to some extent arbitrary, but for present purposes they will be defined by the maximum ratio of stress to

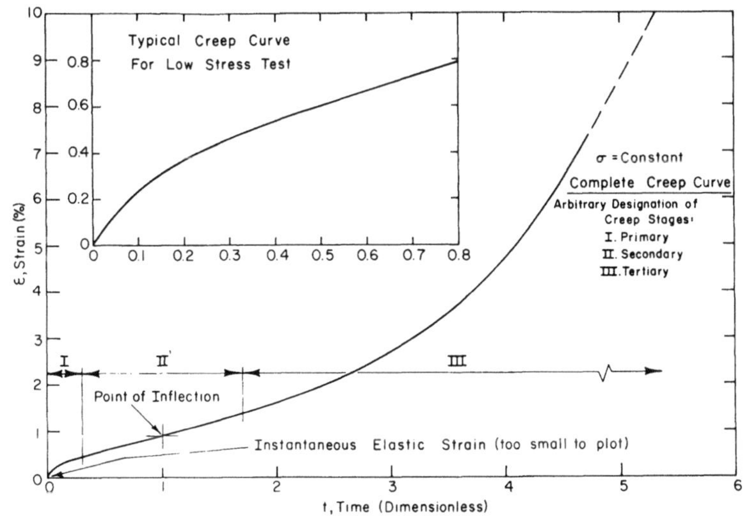

Figure 1a. Complete creep curve for a constant stress test on ice that is initially isotropic.

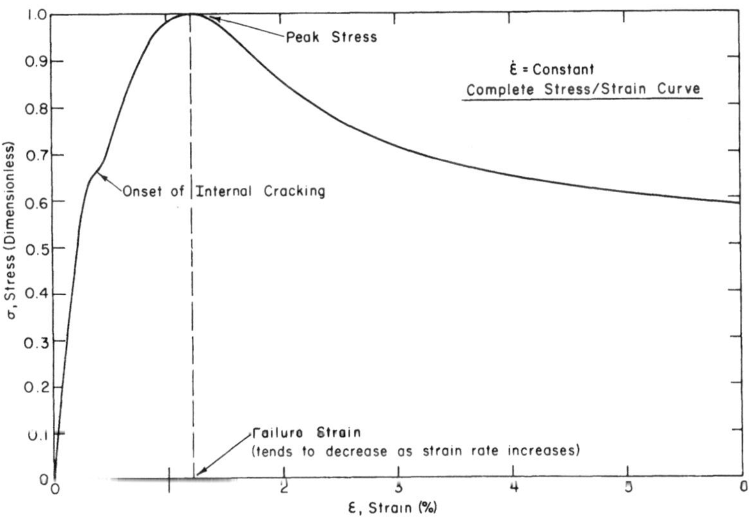

Figure 1b. Complete stress/strain curve for a constant strain-rate test at moderate strain rates where the response is significantly ductile but some internal cracking occurs.

strain rate. In the complete creep curve of figure 1a, the inflection point gives the minimum strain rate for the applied stress, and also the failure strain. In the complete stress/strain curve of figure lb, the peak gives the maximum stress (i.e. peak strength) and the failure strain for the applied strain rate. In both cases the material can retain appreciable strength and deformation resistance at strains well in excess of the "failure strain."

Creep under constant uniaxial stress

Consider abrupt application of constant uniaxial stress to a specimen that is homogeneous and initially isotropic. Temperature is held constant, and it is assumed that the straining process is isothermal. The conventional representation of the resulting behaviour is a plot, on linear scales, of strain as a function of time (fig. 1a). In general, the complete strain/time curve displays: (i) instantaneous elastic strain, (ii) decelerating "primary creep," (iii) a transitional "secondary creep" during which deceleration goes to zero and then reverses sign, (iv) accelerating "tertiary creep," with undetermined limit for acceleration.

Instantaneous elastic strain is controlled by the "true" Young's modulus, although this test is not well adapted for measuring E. In the idealized test, and its mathematical description, stress is applied abruptly at zero time, giving infinite stress rate and infinite strain rate. With older test techniques, load was likely to be applied either gradually or incrementally over a period of 1 to 10 seconds, and for typical stress ranges this appeared to be quite satisfactory. With modern servo-controlled electro-hydraulic testing machines, the step function control can apply the required stress very quickly, perhaps in about 10 milliseconds. We have just discovered at CRREL* that a relatively high stress applied "instantaneously" can damage the specimen prematurely, and stress is now deliberately "ramped on" over a period of seconds to minimize initial internal cracking. In typical creep tests, the instantaneous elastic strain is relatively small—of the order of 0.01% with a stress of 10 bars, and proportionately less at lower stresses.

Decelerating *primary creep* ($\ddot{\varepsilon}$ negative) is a viscoelastic response made up of a recoverable "delayed elastic" strain plus an irreversible viscous strain. In simple rheological models these components are typically represented by a Kelvin–Voigt unit in series with a dashpot (see fig. 2). In a relaxation test, where stress is removed abruptly during the course of a creep test, there is abrupt recovery of the "instantaneous" elastic strain, and gradual recovery of the "delayed elastic" strain.

Secondary creep with $\ddot{\varepsilon} \approx 0$ can cover long time periods when stress or temperature are low, and it has often been referred to improperly as "steady

---

* From tests by D. Cole and W. St. Lawrence, in which acoustic emissions were recorded.

Model	Maxwell	Kelvin–Voigt	3–Element	Burgers
Diagram	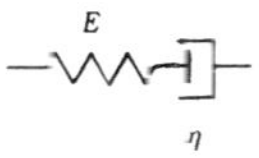	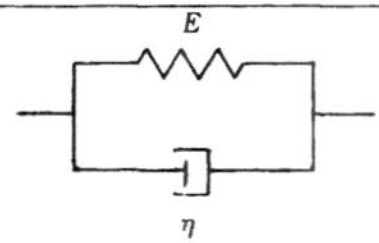	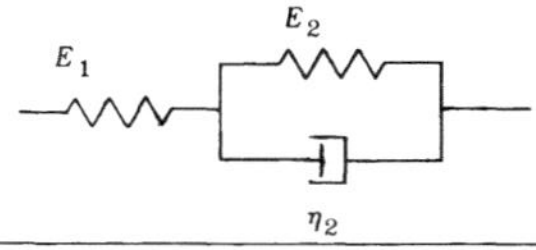	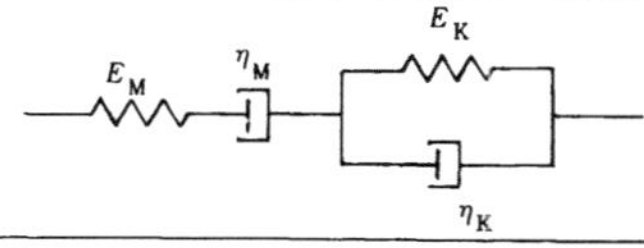
General equation	$\dot{\epsilon}=\frac{1}{E}\dot{\sigma}+\frac{1}{\eta}\sigma$	$\sigma=E\epsilon+\eta\dot{\epsilon}$	$\dot{\sigma}+\left(\frac{E_1+E_2}{\eta_2}\right)\sigma=E_1\dot{\epsilon}+\left(\frac{E_1E_2}{\eta_2}\right)\epsilon$	$\ddot{\sigma}+\left(\frac{E_M}{\eta_M}+\frac{E_M}{\eta_K}+\frac{E_K}{\eta_K}\right)\dot{\sigma}+\frac{E_ME_K}{\eta_M\eta_K}\sigma=E_M\ddot{\epsilon}+\frac{E_ME_K}{\eta_K}\dot{\epsilon}$
Constant stress sol'n (σ_0 applied at $t=0$)	$\epsilon=\sigma_0\left(\frac{1}{E}+\frac{t}{\eta}\right)$	$\epsilon=\frac{\sigma_0}{E}\left[1-\exp\left(-\frac{E}{\eta}t\right)\right]$	$\epsilon=\sigma_0\left\{\frac{1}{E_1}+\frac{1}{E_2}\left[1-\exp\left(-\frac{E_2}{\eta_2}t\right)\right]\right\}$	$\epsilon=\sigma_0\left\{\frac{1}{E_M}+\frac{t}{\eta_M}+\frac{1}{E_K}\left[1-\exp\left(-\frac{E_K}{\eta_K}t\right)\right]\right\}$
Constant strain rate sol'n ($\dot{\epsilon}=k_1$ applied at $t\geq 0$, $\sigma=0$)	$\sigma=k_1\eta\left[1-\exp\left(-\frac{E}{\eta}\right)\right]$	$\sigma=k_1(\eta+Et)$	$\sigma=k_1\left\{\eta_2\left(\frac{E_1}{E_1+E_2}\right)^2\left[1-\exp\left(-\frac{E_1+E_2}{\eta_2}t\right)\right]+\frac{E_1E_2}{E_1+E_2}t\right\}$	$\sigma=k_1\left\{\frac{E_M}{r_1-r_2}(e^{r_1t}-e^{r_2t})+\frac{\eta_M}{r_1-r_2}(r_2e^{r_1t}-r_1e^{r_2t})+\eta_M\right\}$ where $r_1,r_2=-\frac{1}{2}\left(\frac{E_M}{\eta_M}+\frac{E_M}{\eta_K}+\frac{E_K}{\eta_K}\right)\pm\frac{1}{2}\left[\left(\frac{E_M}{\eta_M}+\frac{E_M}{\eta_K}+\frac{E_K}{\eta_K}\right)^2-4\left(\frac{E_ME_K}{\eta_M\eta_K}\right)\right]^{1/2}$
Constant stress rate sol'n ($\dot{\sigma}=k_2$ initiated at $t=0$, $\epsilon=0$)	$\epsilon=k_2\left(\frac{t}{E}+\frac{t^2}{2\eta}\right)$	$\epsilon=k_2\left\{\frac{t}{E}-\frac{\eta}{E^2}\left[1-\exp\left(-\frac{E}{\eta}t\right)\right]\right\}$	$\epsilon=k_2\left\{\left(\frac{1}{E_1}+\frac{1}{E_2}\right)t-\frac{\eta_2}{E_2^2}\left[1-\exp\left(-\frac{E_2}{\eta_2}t\right)\right]\right\}$	$\epsilon=k_2\left\{\left(\frac{1}{E_M}+\frac{1}{E_K}\right)t+\frac{1}{2\eta_M}t^2-\frac{\eta_K}{E_K^2}\left(1-\frac{\eta_K}{\eta_M}\right)\left[1-\exp\left(-\frac{E_K}{\eta_K}t\right)\right]\right\}$
Complex modulus $Y(i\omega)=Y_1+iY_2$: Y_1	$\frac{\omega^2\eta^2E}{E^2+\omega^2\eta^2}$	E	–	–
Y_2	$\frac{\omega\eta E^2}{E^2+\omega^2\eta^2}$	$\omega\eta$	–	–
Complex compliance $J(i\omega)=J_1+iJ_2$: J_1	$\frac{1}{E}$	$\frac{E}{E^2+\omega^2\eta^2}$	$\frac{1}{E_1}+\frac{E_2}{E_2^2+\omega^2\eta_2^2}$	$\frac{1}{E_M}+\frac{E_K}{E_K^2+\omega^2\eta_K^2}$
J_2	$\frac{1}{\omega\eta}$	$\frac{\omega\eta}{E^2+\omega^2\eta^2}$	$\frac{1}{\eta_2}-\frac{E_2^2}{\omega\eta_2(E_2^2+\omega^2\eta_2^2)}$	$\frac{1}{\omega\eta_M}+\frac{1}{\omega\eta_K}-\frac{E_K^2}{\omega\eta_K(E_K^2+\omega^2\eta_K^2)}$

Figure 2. Constitutive equations derived from standard rheological models with linear elastic and viscous elements.

state creep." In general, secondary creep is a minimum rate transition between the decelerating (strain hardening) stage and the accelerating (strain softening) stage, and true secondary creep rate is given by the tangent through the point of inflection. The strain at the inflection point has not yet been determined systematically, but it appears to be slightly less than 1% for typical stresses. The corresponding time is the "time to failure," although engineers have tended to use this term for the time to onset of clearly perceptible tertiary acceleration.

Tertiary creep is characterized by an initial acceleration ($\ddot{\varepsilon}$ positive) that is identified with recrystallization and development of preferred crystal fabric. Tertiary creep at large strains ($\sim$ 10%) has not been well studied, but a reasonable supposition is that acceleration gradually ceases ($\ddot{\varepsilon} \rightarrow 0$), so that strain rate tends asymptotically to a limit ($\dot{\varepsilon} \rightarrow$ const.). The alternative is unlimited acceleration ($\dot{\varepsilon} \rightarrow \infty$), as assumed by Nevel (1976) in an unusual rheological model that employs a dashpot in parallel with a spring of *negative modulus* to simulate acceleration.

There are some indications that the distinction between secondary and tertiary creep may become less significant as stress decreases. Below about 1 bar, it may be that the decelerating processes of primary creep eventually equilibrate with the accelerating processes of tertiary creep, giving the appearance of indefinitely protracted secondary creep. Where such a situation prevails, or in cases where interest is confined to the period before onset of tertiary creep, the rheological response at a given stress level can be described by the Burgers model, i.e. a series combination of the Kelvin–Voigt and Maxwell rheological models (see fig. 2).

Shortcomings in experimental technique and data interpretation have sometimes clouded the picture. There have been unreliable measurements of absolute strain, and the various stages of creep have been identified arbitrarily. Many tests have terminated at small strains while still in decelerating creep, but the terminal strain rate has been erroneously accepted as minimum secondary creep rate. There has been widespread use of extrapolation procedures for estimating minimum strain rate, most notably use of the so-called Andrade Law, but extrapolations from small-strain test data have now been shown to be unreliable.

One safeguard against misinterpretation of data is provided by alternative data plots. One useful procedure is to plot strain rate against strain or against time, using either linear or logarithmic scales. A plot such as that illustrated in figure 3 shows immediately whether a minimum strain rate has been reached.

The uncertainty about tertiary creep, indicated by the broken lines in fig. 1 and 3, arises partly from the practical difficulty of running uniaxial tests to large strains without undue distortion of the test geometry. In future work it may be necessary to re-shape specimens during the course of a test, or to design loading platens that do not exert lateral restraint, positive or negative.

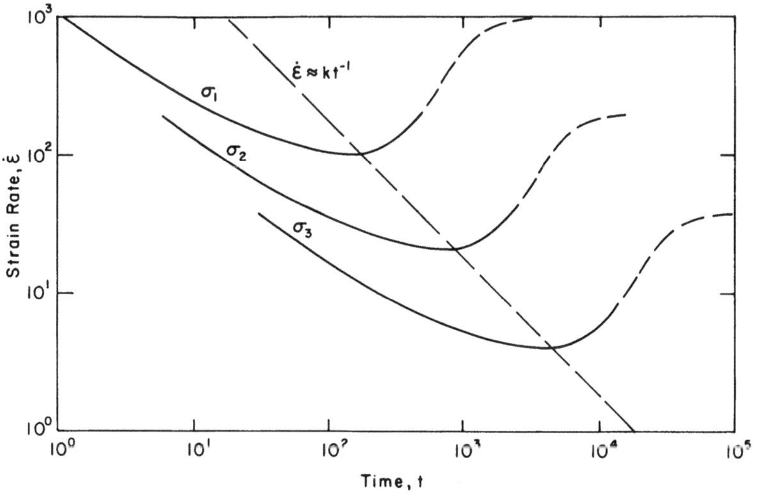

Figure 3a. Alternative presentation of creep curves, using a log-log plot of strain rate against time.

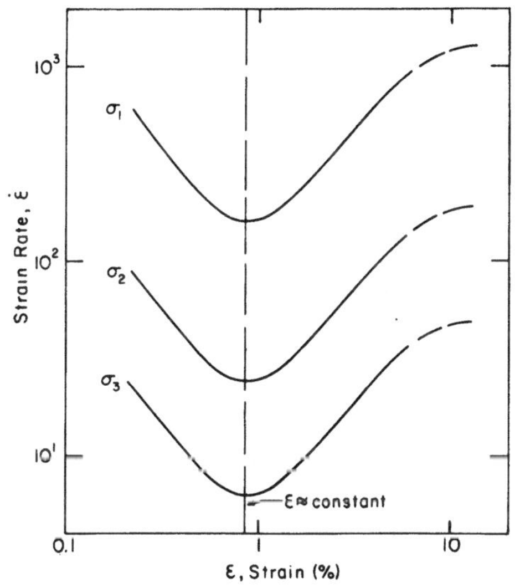

Figure 3b. Alternative presentation of creep curves, using a log-log plot of strain rate against strain.

The foregoing discussion is based on *smoothed* data curves, but there are hints in some experimental results that creep deformation may sometimes involve tiny jerks.

Deformation at constant strain rate

Consider the application of constant axial strain rate to a laterally unconfined specimen that is homogeneous, initially isotropic, and isothermal. Although ambient temperature is held constant, the process is not strictly isothermal in large specimens subjected to high strain rates, but even when the process is adiabatic the average temperature rise in the ice mass is negligibly small because of the low flow stress (by constrast, high-rate metal forming can produce large temperature increases).

The conventional presentation of data is a plot of stress as a function of strain (fig. 1b). In general, the complete stress/strain curve from a careful test in a stiff (non-accelerating) machine displays: (1) a non-linear increase of stress with strain, the slope at the origin, $(d\sigma/d\varepsilon)_{\varepsilon=0}$, representing the true Young's modulus ("initial tangent modulus"), (2) a peak where σ is a maximum and $d\sigma/d\varepsilon$ is zero, (3) a non-linear decrease of stress with strain, with $d^2\sigma/d\varepsilon^2$ changing sign and $d\sigma/d\varepsilon$ gradually decreasing in magnitude to a small value at some finite value of σ.

In the *first stage*, where $d\sigma/d\varepsilon$ is generally positive, it is usual for $d^2\sigma/d\varepsilon^2$ to *appear* negative or zero throughout at typical plotting scales, provided that the ice is of low porosity (i.e. low compressibility). However, the true behaviour can be more complicated. First of all, there are some data suggesting initial strain hardening, with $d^2\sigma/d\varepsilon^2$ positive near the origin. This is perhaps what would be expected, but some experimental indications of positive curvature near the origin have to be discounted because strain is measured between platens, and the closure of platen/specimen interfaces is known to give spurious indications of strain hardening. The other complication is that stress/strain curves for a certain range of strain rates display a discontinuity at about 0.3% to 0.4% strain (see fig. 1b). This may represent an abrupt decrease of slope, a step increase of strain, or a "blip" where stress drops slightly. It seems to be associated with the onset of internal cracking, which involves the release of elastic strain energy and acoustic noise when multiple cracks form, predominantly along grain boundaries that are: (a) parallel to the finite principal stress in compression, or (b) normal to the finite principal stress in tension. This preliminary failure occurs at strains of about 0.3% to 0.4%. Overlooking these complications and considering the general shape of the ascending stress/strain curve, we note that at high strain rates or low temperatures the response is predominantly elastic, so that $d^2\sigma/d\varepsilon^2 \rightarrow 0$ and the average value of $d\sigma/d\varepsilon$ ("secant modulus") is high. By contrast, at low strain rates or high temperatures there is pronounced curvature ($d^2\sigma/d\varepsilon^2$ negative) and the secant modulus is much lower than the true Young's modulus. The constant strain rate solution for the Burgers model in figure 2 gives a more systematic description of this behaviour.

The *peak* of the stress/strain curve gives the maximum stress (σ_{max}) that can be developed at a given strain rate, irrespective of whether the "failure" is by brittle fracture (strain prior to failure predominantly elastic), by creep rupture (strain prior to failure predominantly dissipative and irrecoverable), or by some mixed mechanism involving both flow and fracture. The strain at which the peak is reached (ε_f), termed the "failure strain," has not been studied systematically, but it appears to be of the order of 1% for ductile yielding in compression, decreasing towards 0.1% for tensile fracture and compressive brittle fracture at high strain rates. Knowing the failure strain and the applied strain rate, the "time to failure" is also known.

The *post-peak* deformation and stress decay is easily observed during ductile yielding. When fracture processes dominate, more sophisticated techniques and equipment are needed to trace out a complete stress/strain curve, using a stiff system or rapid servo control to avoid uncontrolled acceleration from release of elastic strain energy in the testing machine. The asymptotic limit of stress for large strains (σ_∞) has not been investigated systematically, but it appears that in general it is finite, and such that the stress drop ($\sigma_{max}/\sigma_\infty$) decreases as strain rate decreases. For very low strain rates, it may be that ($\sigma_{max}/\sigma_\infty$) $\approx$ 1, i.e. there is no post-peak stress decay.

In the past, there has been a tendency to confuse constant strain rate with constant stress rate when conducting strength tests. Minimum loading rate for strength tests has been specified in terms of stress rate, and some old testing machines used a pacer to permit manual control of force rate. For a purely elastic material and a test terminating in brittle fracture the practical distinction may be unimportant, but in general the two conditions are very different.

Correspondence between test data for constant stress and constant strain-rate

When samples of the same material are subjected to both constant stress and constant strain rate tests, the same material properties ought to be revealed by each test. None of the common rheological models simulate the complete behaviour of ice, so that we cannot readily demonstrate the correspondence by comparing appropriate solutions of a common differential equation, as in figure 2. However, one might speculate as follows.

In principle, Young's modulus E is given on the ε–t creep curve by the applied stress σ and the intercept on the ε axis at $t = 0$. On the σ–ε stress/strain curve it is given by $d\sigma/d\varepsilon$ at $\sigma = 0$, $\varepsilon = 0$. About this there is little argument.

The minimum strain rate $\dot{\varepsilon}_{min}$ for a given stress σ is given on the ε t creep curve by $\dot{\varepsilon}$ at the inflection point. The maximum stress for a given strain rate is given on the σ–ε stress/strain curve by the peak stress σ_{max}. Thus each test gives a pair of σ–$\dot{\varepsilon}$ values for which $\sigma/\dot{\varepsilon}$ is a maximum. If both tests give the same information, the plot of σ against $\dot{\varepsilon}$ for $(\sigma/\dot{\varepsilon})_{max}$ will be the same for both tests. Present indications are that this is the case, at least where the yielding is predominantly ductile. A further speculation is that the failure strains for the two tests (ε at the inflection point of the ε–t curve, and beneath the maximum of the σ–ε curve) might be the same. Limited data suggest that this could be approximately true for certain ranges of stress and strain rate. Work is in progress to check these suppositions experimentally.

As regards the behaviour at large strains, we might look for equality in the limiting ratio of stress to strain rate. On the complete creep curve, the final rate of tertiary creep $\dot{\varepsilon}_\infty$ has a special relation to the applied stress σ. On the complete stress/strain curve, the residual stress σ_∞ also bears a special relation to the imposed strain rate $\dot{\varepsilon}$. The expectation is that a plot of $\dot{\varepsilon}_\infty$ against σ for a set of ε–t curves would coincide with a plot of $\dot{\varepsilon}$ against σ_∞ for a set of σ–ε curves. Furthermore, a correspondence is to be expected between $\dot{\varepsilon}_\infty/\dot{\varepsilon}_{min}$ as a function of σ in creep, and between $\sigma_{max}/\sigma_\infty$ as a function of $\dot{\varepsilon}$ under constant strain rate (noting that, in general, $\dot{\varepsilon}$ will be a strong non-linear function of σ). These speculations about the behaviour at large strains are unsupported by direct experimental evidence.

Simple flexure

Simple flexure of a long beam produces uniaxial stresses in tension and compression, with gradients of stress and strain rate normal to the neutral axis. Beam bending is sometimes used for large scale field tests, but fundamentally it is unsuitable for measuring ice properties. The test is an *indirect* one, in that a constitutive relation has to be assumed in order to analyze the results. The usual assumption of linear elasticity, with equal moduli for tension and compression, can only be justified for very high loading rates or very low temperatures. Furthermore, because of the stress gradient the extreme stress is applied only to a thin surface layer rather than to a large multigrain volume. These general criticisms can probably be relaxed when high frequency flexural vibrations of small amplitude are used to measure Young's modulus and viscous damping.

Diametral compression of an annulus is theoretically capable of measuring uniaxial tensile strength, but the problems just mentioned apply with even greater force, especially in the case of the extreme stress gradient at the critical point.

Sinusoidal oscillations, stress waves, and cyclic loading

Quasi-static creep tests and strength tests have to be conducted very carefully if they are to yield accurate values for the elastic moduli of ice. An alternative approach is to use dynamic methods involving stress waves or pulses of small amplitude and high frequency.

Young's modulus E can be found directly from longitudinal vibration of a column or from flexural vibration of a beam by measuring the resonant frequency. It can also be found from the velocity of a pulse traversing a column, or from the velocities of elastic dilatational waves and shear waves in large ice masses. The latter methods also give Poisson's ratio ν. Values of E found by such methods are approximately 80 to 90 kilobars, while ν is about 0.3 to 0.36. There does not seem to be much frequency dependence in a range from less than 10^2 Hz to more than 10^3 Hz, and the temperature dependence is quite small—an increase of about 0.1% per degree as temperature decreases.

When a suitable rheological model is assumed (see fig. 2), vibrational methods give a measure of both elastic and viscous properties. For sinusoidal oscillation at angular frequency ω, with stress and strain amplitudes of σ_0 and ε_0, the stress/strain relationships can be expressed as $\sigma = Y(i\omega)\varepsilon_0 e^{i\omega t}$ and $\varepsilon = J(i\omega)e^{i\omega t}$, where $Y(i\omega)$ is the complex modulus and $J(i\omega)$ is the complex compliance. In the latter the real parts, the storage modulus Y_1 and the storage compliance J_1, give elastic properties and stored strain energy. The imaginary parts, the loss modulus Y_2 and the loss compliance J_2, relate to dissipation of energy by viscous damping, or internal friction, which itself is characterized by a loss factor given in terms of the phase lag angle δ between stress and strain, i.e. $\tan\delta = Y_2/Y_1 = J_2/J_1$. The loss factor tends to decrease with increasing frequency (between 10^2 and 10^3 Hz), and it decreases sharply with decreasing temperature (down to -30 °C).

Large amplitude cyclic loadings and fatigue failure have not been studied systematically for ice, although they are of considerable practical significance. Some unpublished test results* show the stress/strain curve for repetitive loading to a given stress level as a hysteresis loop, with its lower tip progressing along the strain axis at a decreasing rate. New work in this area is planned.

Relationship between stress and strain-rate

A primary concern of glaciologists has been the relationship between minimum creep rate $\dot{\varepsilon}_{min}$ and applied stress σ. Although this is often referred to as "*the* flow law of ice," it is, in fact, far from being a complete constitutive relation, and there are other stress/strain-rate relations that are important to engineers. The σ–$\dot{\varepsilon}$ relation for large strains clearly is significant, as is the σ–$\dot{\varepsilon}$ relation for a specified instant of time t. Furthermore, it is desirable to extend relationships for purely ductile yielding into the range of stresses and strain rates where fracture occurs.

The relationship between $\dot{\varepsilon}_{min}$ and σ for creep at constant temperature is usually presented graphically as a log-log plot. For the range of conditions that give relatively rapid creep, say 2 to 20 bars compressive stress and temperatures above -20 °C, the data on a log-log plot can be represented by a straight line with a slope between 3 and 4, and consequently the best known relationship is one expressed in the form $\dot{\varepsilon}_{min} = A\sigma^n$, where the exponent n is between 3 and 4, and A is a temperature-dependent constant (whose dimensions can be stabilized by writing the whole equation in dimensionless form).

* By Hawkes and Mellor at CRREL.

When the range of stress is extended, the relationship becomes a little more complicated (fig. 4). At high stresses, the log-log plot takes on curvature in the sense that n increases with increasing stress. At low stresses, the curvature appears to be in the sense that n decreases with decreasing stress, although early indications of $n = 1$ (Newtonian viscosity) at stresses below 1 bar turned out to be false (tests terminated at low strain before $\dot{\varepsilon}$ had decelerated to $\dot{\varepsilon}_{min}$). Since the simple power relation does not fully describe this behaviour, various alternative expressions have been proposed, including simple polynomials and the hyperbolic sine. The latter, $\dot{\varepsilon}_{min} = \dot{\varepsilon}_0 \sinh(\sigma/\sigma_0)$, has the attraction of meeting reasonable boundary conditions and being derivable from reaction-rate theory. However, it might be noted that the simple reaction-rate theory is developed for "steady-state" creep, and for secondary ice creep it is strictly justifiable only for relation of strain rate and stress at the inflection points of complete creep curves (it may also be justifiable for tertiary creep at large strains).

There should be a corresponding relationship between applied stress σ and final tertiary creep rate $\dot{\varepsilon}_\infty$, but there is very little information on creep at strains of the order of 10%. Current studies at CRREL indicate that creep is still decelerating at strains near 10%. Taking the tertiary creep data that do exist, and plotting final strain rates on the same logarithmic graphs where corresponding minimum strain rates have been plotted, the tertiary points obviously lie above the points for $\dot{\varepsilon}_{min}$. The ratio $\dot{\varepsilon}_\infty/\dot{\varepsilon}_{min}$ has values typically in the range 3 to 8, and data from some sources suggest that the ratio increases as stress increases. The latter effect, which was mentioned earlier in suggesting that $\dot{\varepsilon}_\infty$ and $\dot{\varepsilon}_{min}$ might not be much different at low stress, could be real, but existing data cannot be relied upon if the final strain in a given test is itself a function of stress.

In many engineering problems the main concern is with creep transients at varying points along the creep curve. In particular, there are boundary value problems in which each element has been stressed for the same length of time. An appropriate stress/strain-rate relation is the one obtained by taking a constant-time cut through a set of creep curves or through a set of stress/time curves from constant strain rate tests. It has been suggested that this procedure might give a simple linear relation, i.e. a Newtonian viscous type of relation, but data available to this writer contradict such a notion ($\dot{\varepsilon}$ increases with higher powers of σ). What may happen is that n could tend to unity for low stresses.

The uniaxial strength of ice

The strength of a material is usually defined in terms of the maximum stress that can be reached when the material is loaded at either constant strain rate or constant stress rate. For ice, this stress is strongly rate-dependent, and so it must be specified as a function of strain rate. Essentially the same limiting relationship between stress and strain rate, i.e. $(\sigma/\dot{\varepsilon})_{max}$, ought to be

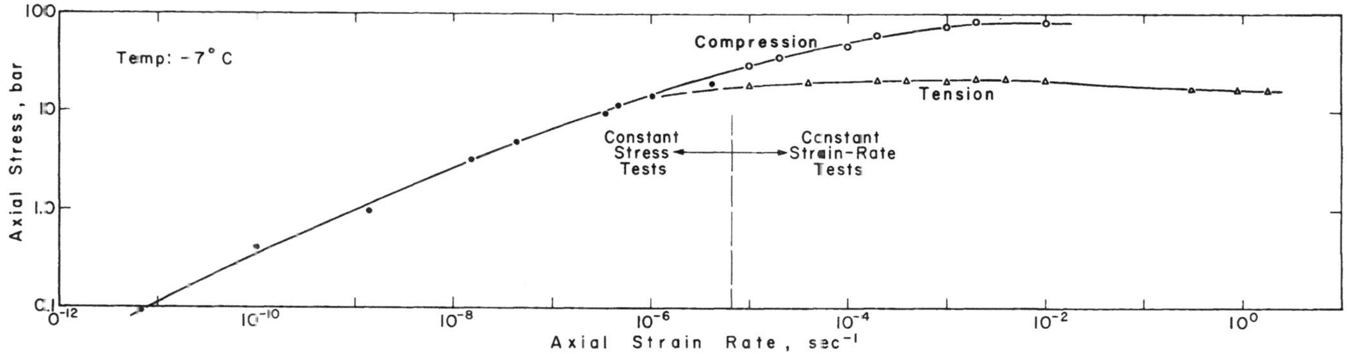

Figure 4. Stress/strain-rate data for uniaxial tests on ice at -7°C.

given by the minimum strain rate in a constant stress test. Thus figure 4, which has already been given as an example of the dependence of creep rate on stress, can also be interpreted as an example of the variation of strength with strain rate.

Several investigators have shown compressive strength increasing with strain rate in the ductile range of behaviour, but then reaching a peak and dropping to a lower constant value above a strain rate characterized as the "ductile/brittle transition." Such an effect can arise from contact stresses when the end preparation of specimens is imperfect, and there are some indications that the same effect can be produced by a compliant loading system which releases strain energy as the specimen yields. There is a temptation to identify this apparent lower strength in brittle fracture with the incipient failure that occurs at about 65% of peak stress in low-rate tests (see fig. 1b).

While there is ample evidence that the uniaxial tensile strength becomes more or less constant at high strain rates (fig. 4), it seems unlikely that compressive strength has reached its limit at the comparatively low strain rates used for the experiments mentioned above. One might expect compressive strength to keep increasing until it is at least an order of magnitude higher than tensile strength, perhaps reaching very high values when the specimen is subjected to stress pulses traveling at sonic velocity.

Although the main interest centers on the "ultimate strength," or peak stress, there is also some concern with the "residual strength" σ_∞, as represented by the post-peak asymptote of the complete stress/strain curve (fig. 1b). For failure that involves both flow and internal fracture, the residual strength seems to be at a stress level somewhat below that at which internal cracking initiates. For very low strain rates, where yielding is purely by flow, the residual strength σ_∞ appears to be not much below the ultimate strength σ_{max}, as mentioned earlier. If our speculations about correspondence between constant stress tests and constant strain rate tests are correct, the residual strength as a function of strain rate can also be determined from tertiary creep data, since these give a high-strain limit for $(\sigma/\dot{\varepsilon})_\infty$.

In theoretical fracture mechanics, the ratio of ultimate uniaxial compressive strength σ_c to ultimate uniaxial tensile strength σ_T is of considerable significance. For the *brittle* fracture of solids, classical Griffith theory predicts a ratio of $\sigma_c/\sigma_T = 8$, while other derivatives of the same theory call for even higher ratios. By contrast, theories of *plastic* yielding, as exemplified by the Tresca and Von Mises yield criteria, require a ratio of $\sigma_c/\sigma_T = 1$. For ice, the latter is approximately true for slow creep and ductile yielding, while the trend is for σ_c/σ_T to increase with increasing strain rate when internal cracking begins to control the failure (see fig. 4). However, under typical testing conditions σ_c/σ_T is likely to be less than 5 for ice, and perhaps less than 3. This is one reason why certain tests borrowed from other fields of technology do not work for ice—the conditions for tensile failure in a multiaxial stress state may be complicated by the low value of σ_c/σ_T.

Effect of temperature on creep rate and strength

The effect of temperature on the *creep* of ice has usually been described in terms of activation energy Q, i.e. it has been assumed that creep is a thermally activated process such that $\dot{\varepsilon}$ varies with absolute temperature T according to the Arrhenius equation $\dot{\varepsilon} = A \exp(-Q/RT)$, where R is the gas constant. This assumption appears to be justified for ice monocrystals and also for polycrystalline ice of typical purity at temperatures below -10 °C; approximate values of Q are about 70 kJ/mol for monocrystals and perhaps slightly less for polycrystalline ice below -10 °C. Above -10 °C, a plot of ln $\dot{\varepsilon}$ against $1/T$ for typical polycrystalline ice is strongly non-linear (fig. 5) in the sense that point values of Q defined by the tangent to the curve increase with increasing temperature. There are several modifications of the Arrhenius equation that account for the temperature dependence of Q, but they seem dubious. In fact, although the simple Arrhenius equation is based on valid physical principles, it has no special merit as a mathematical description of temperature dependence, since the significant properties of the function appear only at temperatures far above 273K, where ice has ceased to exist (fig. 6).

The Arrhenius equation is likely to remain in use until a different formulation is found, and it certainly has some merits. However, it is important to recognize that use of this equation with values of Q around 70 kJ/mol will seriously underestimate the temperature sensitivity of ice creep at temperatures near 0 °C. For the rather numerous problems where the temperature range is from about 0 °C to -5 °C, an apparent value of Q can be taken from experimental data, or a simple exponential of the form $\dot{\varepsilon} = A \exp(-k\theta)$ can be adopted (taking θ as the number of degrees departure from a 0 °C datum).

The effect of temperature on the *strength* of ice is rather poorly defined, at least for the higher strain rates where internal cracking is important. The *tensile* strength at moderately high strain rates is quite insensitive to temperature (fig. 7), just as it is insensitive to strain rate (fig. 4). In fact, the variation of σ_T with temperature is not much greater than the variation of Young's modulus E with temperature (fig. 7). At typical testing rates, the *compressive* strength σ_c is much more strongly dependent on temperature (fig. 7), and it might be noted in passing that σ_c is still varying with strain rate in this range (fig. 4).

At very low strain rates, where yielding is predominantly ductile, the temperature dependence of strength ought to be predictable from the relationship between strain rate and temperature. In other words, if creep tests give a stress/strain-rate/temperature relationship of the form $\dot{\varepsilon} = A\sigma^n f(\theta)$ for the minimum strain rate, then the "strength" ought to vary with temperature according to a relation of the form $\sigma = [\dot{\varepsilon}/Af(\theta)]^{1/n}$. In figure 7 the data of figure 5 have been used to construct a temperature relation for ductile yield strength, assuming a value of $n = 3.3$.

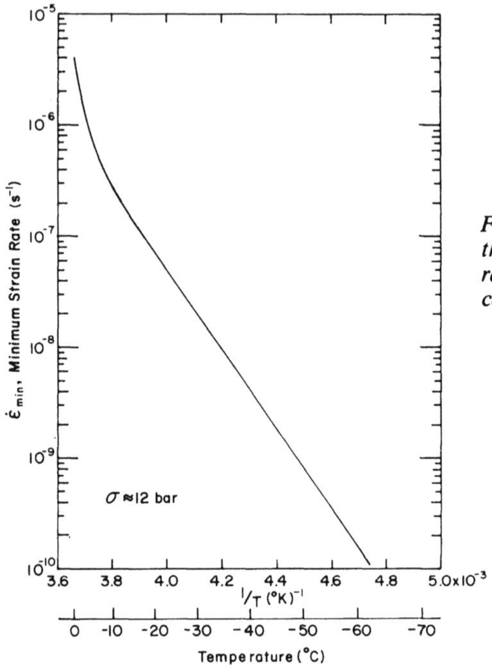

Figure 5. Experimental curve for the logarithm of minimum strain rate plotted against the reciprocal of absolute temperature.

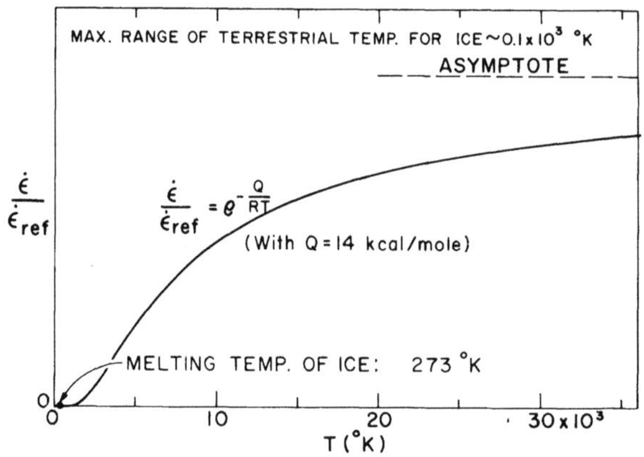

Figure 6. Illustration of the Arrhenius function.

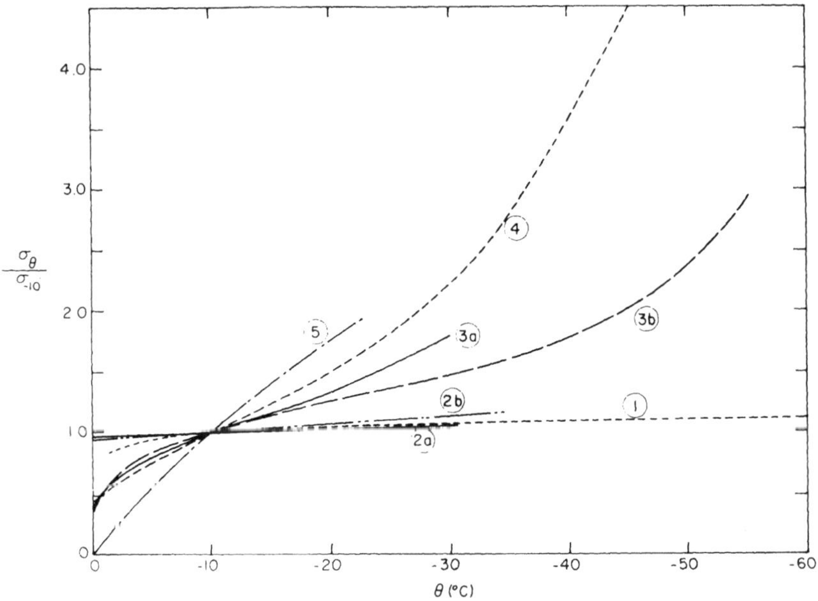

Figure 7. Compilation of temperature relationship. All the stress values are normalized with respect to the value for -10°C. 1) Variation of Young's modulus with temperature. 2a) Uniaxial tensile strength according to data by D. Carter (1970). 2b) Uniaxial tensile strength according to data by F.D. Haynes (1978). 3a) Uniaxial compressive strength according to data by D. Carter (1970). 3b) Uniaxial compressive strength according to data by F.D. Haynes (1978). 4) Ductile yield stress deduced by creep data by M. Mellor and R. Testa (1969). 5) Pressure for phase transition from Ice Ih to water under isothermal hydrostatic compression.

The final item of interest in figure 7 is the temperature characteristic for the hydrostatic pressure at which phase change occurs, from Ice Ih to water. This is the phase boundary for slow isothermal compression, and since the transition from solid ice to liquid water represents a virtual disappearance of shear resistance, it can be regarded as a type of failure.

To sum up, there are indications that the temperature-sensitivity of strength varies with strain rate. At very high rates, and especially in pure tension, there is likely to be little variation of strength with temperature. At low strain rates, where yielding is ductile, there is a strong temperature effect, both in compression and tension.

Effects of grain size on creep rate and strength

The effects of grain size on creep rate and strength have not been studied systematically to any great extent. Part of the difficulty in experimental work is that grains smaller than about 0.5 mm tend to grow spontaneously, while grains bigger than 10 mm are very large relative to typical specimen sizes.

Above the 1 mm grain size, several investigators have found creep rate increasing in approximate proportion to the square of grain size, but the only study extending to smaller grain sizes showed creep increasing rapidly again as grain size *decreased* from 1 mm to 0.5 mm. These puzzling results need further study, and some new work on the subject is currently in progress at CRREL.

As far as strength, or yield stress, is concerned, a decrease with increasing grain size is to be expected if ice behaves in a similar way to metals, rocks, and other multi-grain solids. A first guess is that strength might be inversely proportional to the square root of grain size. However, systematic experimental data are not yet available, and the peculiar creep behaviour referred to above would, if validated, be reflected by a corresponding effect on strength.

Effects of anisotropy on creep rate and strength

For relatively pure polycrystalline ice, two types of "initial" anisotropy can be distinguished: (i) textural anisotropy, developed by the shape and assembly of grains, and (ii) fabric anisotropy, arising from preferred orientation of crystallographic axes in the constituent crystals. In contrast to "initial" anisotropy, i.e. anisotropy that exists when a specimen is first considered, there is *induced* anisotropy that develops during the course of loading or straining.

Systematic studies of the general mechanical effects of anisotropy are few, and the probable trends have to be pieced together from fragmentary evidence.

In *creep*, textural anisotropy might be most significant at the highest temperatures, where grain boundary slip is important. Initial fabric anisotropy is clearly significant if the stress field is such that shears can be resolved in some consistent orientation relative to preferred crystallographic directions, since shear along basal planes takes place much more easily than shear normal to basal planes. Induced fabric anisotropy develops naturally when deviatoric strains become large (1 to 10%), and the strain softening effect reflected by tertiary creep is generally attributed to recrystallization, or development of induced fabric anisotropy.

In situations where *brittle fracture* or *internal cracking* occur, it is to be expected that textural anisotropy will be important. For example, under uniaxial tension or compression one might expect columnar-grained ice to be strongest when the finite principal stress is parallel to the grain axes, and the limited experimental data tend to confirm this expectation. Fabric anisotropy might also be expected to influence fracture strength, but there are some problems in separating the effects of textural anisotropy and fabric anisotropy because there is often an implicit relation between the two in natural ice. For example, in lake ice that has vertical columnar grains, the *c*-axes of crystals are likely to be horizontal. In this case, uniaxial compression parallel to the grain axes and normal to the *c*-axes would form cracks within the crystals that tend to be normal to the *c*-axes, i.e. cleaving the crystals parallel to basal planes.

Effect of porosity on deformability and strength

The effect of porosity on the mechanical properties of ice is best indicated by the results of studies on well sintered high density snow and on bubbly ice (for snow and ice, bulk density ϱ and porosity n are related by $\varrho = 0.917(1-n)$ Mg/m^3.

Because snow is highly compressible, it is difficult to interpret the results of creep tests, and to relate them to the creep behaviour of almost incompressible ice. Figure 8 combines results of constant stress creep tests and constant strain rate strength tests in an attempt to show how the limiting stress/strain-rate relation for $(\sigma/\dot{\varepsilon})_{max}$ is affected by porosity (or density). A major complication arises from the fact that compressible "snow" tends to equilibrate to compressive stress by relatively rapid quasi-plastic collapse, as indicated by

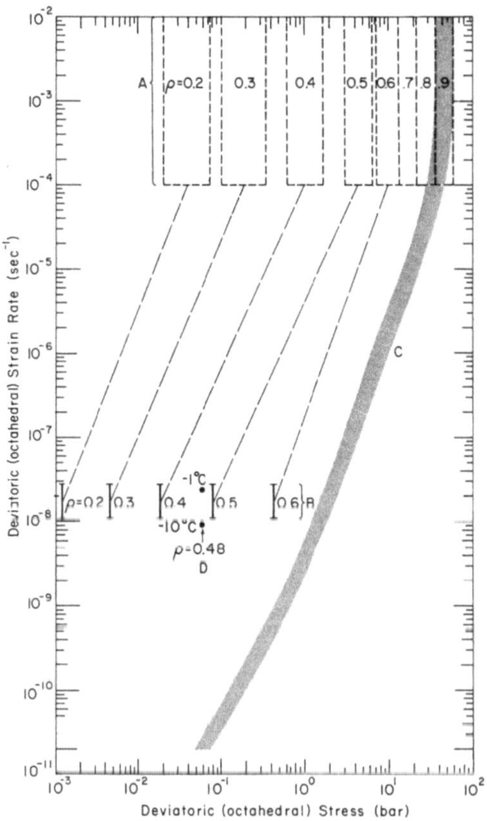

Figure 8. Deviatoric stress/strain-rate relationships for snow and ice over a wide range of bulk density. Trends deduced indirectly from a variety of data sources.

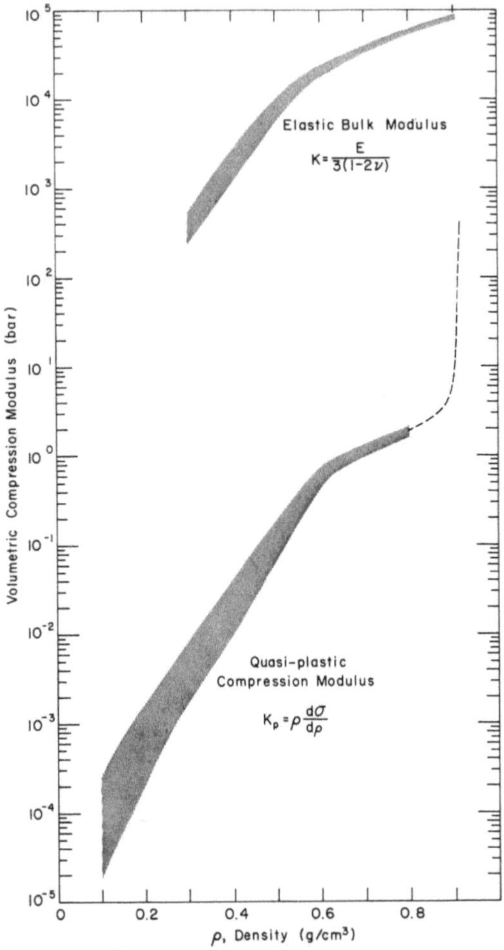

Figure 9. Volumetric compression moduli plotted against bulk density for snow and ice.

the modulus k_p in figure 9. The variation of elastic bulk modulus with density is also shown.

The variation of strength with porosity, or density, is easier to describe. Figure 10a gives data for uniaxial compressive strength and uniaxial tensile strength at strain rates in the range 10^{-4} to 10^{-2} s^{-1}. Figure 10b gives independent results for the direct shear strength at small normal pressures.

Failure strain

If strain rate, temperature and ice type vary, the uniaxial strength of ice can vary by orders of magnitude, and it becomes very difficult to make a clear statement about the strength of ice. By contrast, the deviatoric failure strain ε_f appears to vary by less than an order of magnitude over the whole

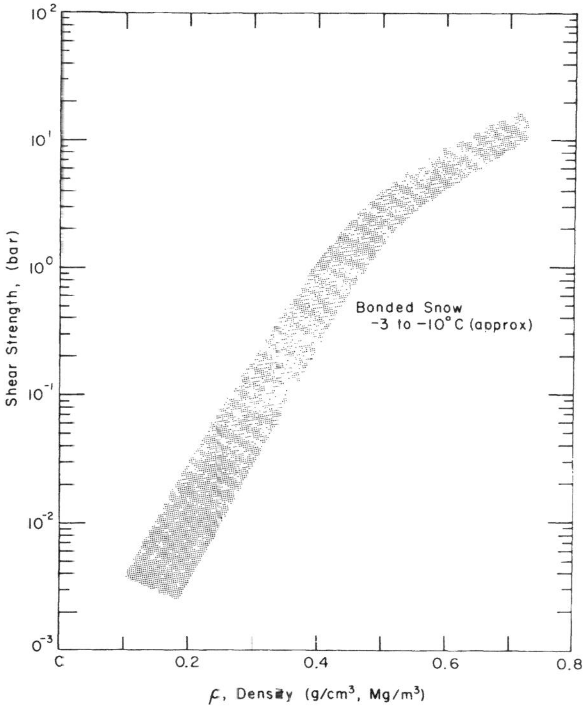

Figure 10a. Uniaxial strength of snow and ice as a function of density, for relatively high strain rates and temperatures approximately -10°C (±5°C).

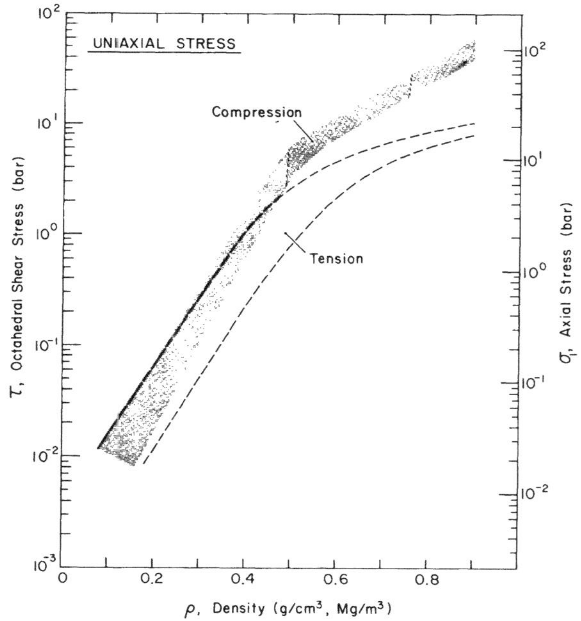

Figure 10b. Direct shear strength of snow and ice as a function of density at temperatures from -3 to -10°C.

range of conditions for ductile yield and brittle fracture. The range is from about 0.3% for brittle fracture at high strain rates, to about 2% for very slow ductile yielding.

Time to failure

Among engineers the concept of "time to failure" has found wide acceptance. The usual approach is to relate stress level to the time taken for ice to begin losing strength. The detailed definitions have been somewhat arbitrary. For example, in a creep test the time to failure has been taken as the time needed for perceptible acceleration to appear in the creep curve.

For a consistent definition, time to failure t_f should be taken as the time to the inflection point in a creep test, or as the time to the peak stress in a constant strain rate test. Thus, in creep, $t_f \approx \varepsilon_f/\dot{\varepsilon}_{min}$ (see fig. 1a), and since ε_f is approximately constant for ductile yielding, t_f is inversely proportional to $\dot{\varepsilon}_{min}$ (see fig. 3a and 3b). Under constant strain rate $\dot{\varepsilon}_0$ (fig. 1b), $t_f = \varepsilon_f/\dot{\varepsilon}_0$, and if ε_f is approximately constant t_f is again inversely proportional to strain rate. If the relationship between stress and strain rate is approximated by a simple power function $\dot{\varepsilon} = A\sigma^n$, the relationship between stress level σ and time to failure t_f would thus be expected to be of approximate form $t_f = k\sigma^{-n}$.

Deformation and yielding in multiaxial stress states

There has been surprisingly little systematic attention given to the deformation and yielding of ice in biaxial and triaxial stress states. One result is that formal failure criteria are lacking, even though such criteria are of prime importance for analysis of numerous practical problems.

In the following notes we consider: (i) the compressibility of ice under hydrostatic pressure and rapid isotropic compression, (ii) the effect of compressive bulk stress on deviatoric stress/strain-rate relations for creep and ductile yield, (iii) the failure envelope in three-dimensional stress space and the formulation of failure criteria.

Compressibility of ice

Under moderate pressures, say up to 100 bars or so at temperatures significantly below 0 °C, ice of zero porosity has low compressibility, to the extent that it can be assumed to be incompressible in many problems. The elastic bulk modulus of "solid" ice is about 90 kilobars. When the ice has finite porosity, usually in the form of discrete air bubbles, the elastic bulk modulus may not be much affected for rapid loading and small stress, but the compressibility under slow loading increases significantly. If the ice is under pressure for a sufficiently long time, pore volume tends to adjust in accordance with the gas laws.

When ice is subjected to slow isothermal compression up to high pressures, phase changes occur in accordance with the P-V-T phase diagram, high den-

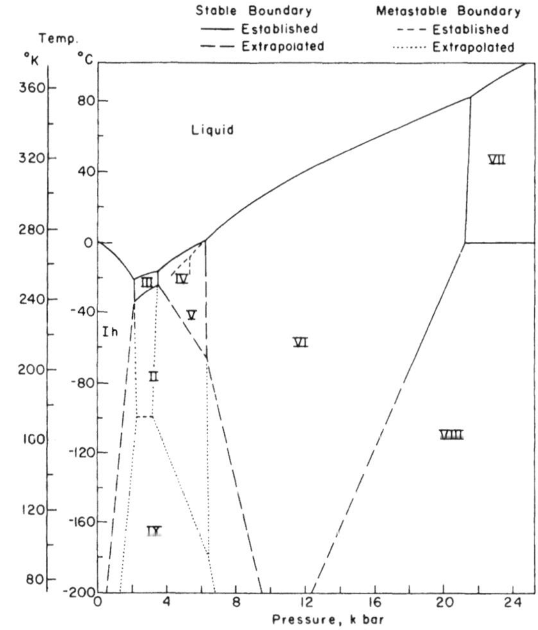

Figure 11a. Phase diagram for ice.

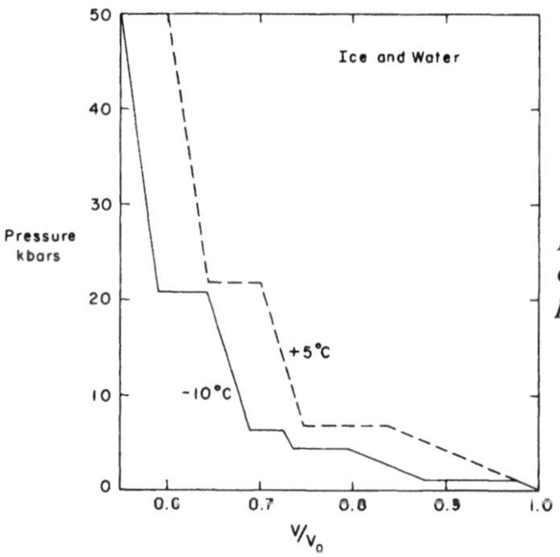

Figure 11b. Volumetric stress/strain characteristics for isothermal compression of ice at −10°C and water at +5°C.

sity polymorphs may be formed (fig. 11a), and large volumetric strains can occur (fig. 11b).

When high compressive stresses are induced adiabatically by shock loading, details of the volumetric stress/strain curve, or Hugoniot, are more difficult to ascertain. However, from the limited available data it seems that the overall compressibility (fig. 12) is not much different from the isothermal compressibility (fig. 11b), even though slightly higher pressures occur for a given strain during shock loading.

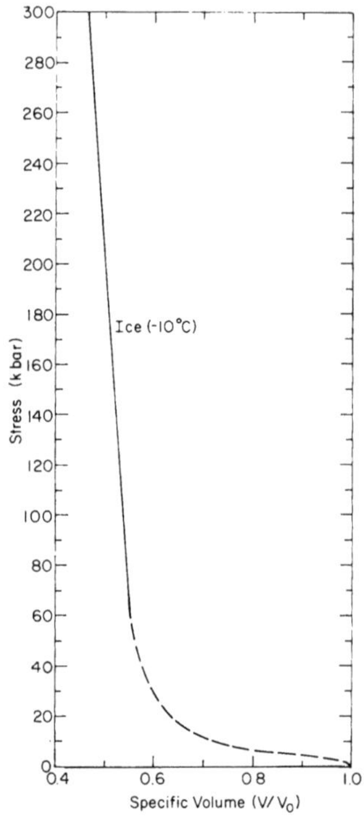

Figure 12. Stress/strain characteristic for high-rate adiabatic compression of ice at -10°C.

Effect of pressure on deviatoric stress/strain-rate relationship and on shear strength

Glaciologists concerned solely with creep at very low strain rates have long assumed, on the basis of limited experimental evidence, that the shearing of ice is virtually unaffected by hydrostatic pressure. The only caveat here is that the effect of pressure on the melting point should be accounted for. With this interpretation, ductile yielding can be considered to conform to a failure criterion similar to the Tresca or the von Mises criteria. In other words it is assumed that the flow of ice can be described by a relationship between the deviatoric components of stress and strain rate, i.e., a relationship involving only the second invariants of the stress and strain rate tensors. The first invariant is not considered to have any effect, and there is no convincing evidence of any influence from the third invariant.

From a formal point of view this interpretation is open to criticism, in that it does not represent the behaviour for constant temperature. If constant temperature is specified, then the indications are that creep rate will increase with increasing hydrostatic pressure. For example, with a creep rate of the order of 10^{-8} s^{-1} and a temperature near -10 °C, the creep rate increases by about 20% as pressure increases from 0 to 0.4 kbar. Looked at another way, this means that the deviatoric yield stress *decreases* with increasing pressure.

When ice is deformed at relatively high strain rates, so that internal microcracking and dilatation occur, one would expect compressive bulk stress to increase the shear resistance and the shear strength. Limited experimental data tend to support this expectation. For example, triaxial tests made by Stephen Jones show that at strain rates of the order of 10^{-4} s^{-1} a pressure of 0.3 kbar approximately doubles the shear strength. This implies that an appropriate failure criterion might be similar to a Griffith or Mohr–Coulomb type of criterion. However, the same tests indicate that the effect of pressure peaks at 0.3 to 0.4 kbar at −11.5 °C, so that higher pressures begin to *reduce* the shear strength.

Criteria for yielding and failure in multiaxial stress states

In glaciology, ice is usually considered to follow a non-linear viscous constitutive law, and if a ductile yield criterion is considered at all it is one that approximates a Tresca or von Mises, i.e. maximum shear stress constant or the second invariant of stress constant. Thus, for isotropic ice, the failure envelope in three-dimensional stress space would be either a hexagonal tube or a cylindrical tube disposed symmetrically about the hydrostat line (shown in two dimensions in fig. 13).

By contrast, in those ice engineering problems where the concern is with rapid loading and quasi-brittle fracture, ice is often considered to follow a linear elastic constitutive equation, and the failure criterion is often implicitly assumed to be a maximum principal stress criterion. A less primitive approach is to assume a criterion which predicts an increase of shear strength with increasing bulk stress. The most obvious is the Mohr–Coulomb criterion, which in its simplest form predicts linear increase of shear strength with normal stress, or alternatively proportionality between the greatest and smallest principal stresses at failure (fig. 13). There is also a modification of the Mohr–Coulomb criterion that assumes a parabolic failure envelope. The simple Griffith criterion, which requires that $\sigma_c/\sigma_T = 8$, is not directly applicable, but modifications for the range $3 \leqslant \sigma_c/\sigma_T \leqslant 8$ can be derived by generalizing the ellipticity of the "Inglis crack" in the basic theory.

Although any of the foregoing criteria may be useful in certain restricted ranges of behaviour, none of them really represents the failure characteristics of ice.

In attempting to formulate a general failure criterion, the following characteristics have to be taken into account.

1. For ductile yielding at low strain rates, $\sigma_c \approx \sigma_T$ and bulk stress has little effect on shear resistance, although the trend is for decrease of strength with increasing pressure.

2. As strain rates increase and internal cracking develops, σ_c/σ_T increases, as do the absolute values of σ_c and σ_T.

3. At moderately high strain rates, the shear strength increases with increasing confining pressure up to a certain limit, but then starts to decrease

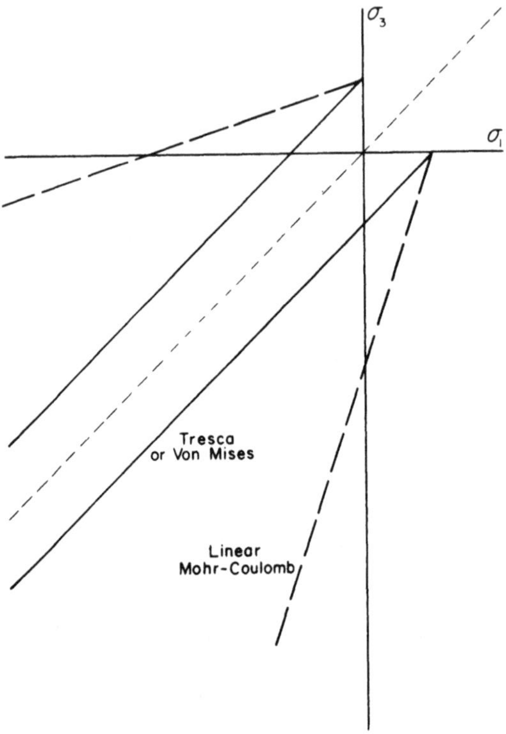

Figure 13. Two-dimensional representations of the Tresca and von Mises criteria, and the linear Mohr–Coulomb criterion.

with further increase of pressure.

4. Under quasi-static loading and temperatures above -20 °C or so, hydrostatic pressure alone can destroy the shear resistance because of phase change to water.

The last of these is a unique and highly significant characteristic that seems to have been overlooked by theoreticians. Whereas most failure criteria predict finite shear strength under extreme pressure, ice eventually loses its shear strength under high pressure.

Taking into account all of this evidence, we can begin to sketch out the form of the failure criterion, at least qualitatively.

Figure 14 shows a simple envelope that conforms to the known facts about ductile yielding (noting that nothing is known about the tension-tension quadrant). At temperatures near -10 °C, the phase change pressure is two orders of magnitude greater than the uniaxial yield stress, so that convergence of the envelope with the hydrostat is very gradual. By the same token $\sigma_c/\sigma_T \approx 1$, but if the converging trend is continuous through the tension-compression quadrants σ_c/σ_T should be slightly less than unity.

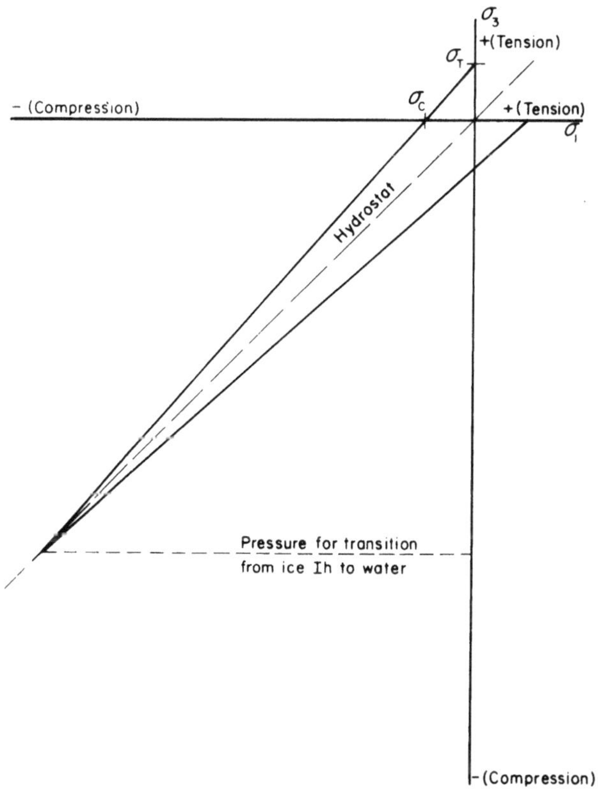

Figure 14. Two-dimensional representation of a simple yield criterion for slow flows of ice at temperatures above −20 °C.

Figure 15 shows what a complete failure envelope might look like at temperatures near −10 °C and strain rates of the order of 10^{-5} to 10^{-3} s^{-1}. $\sigma_c/\sigma_T >$ 1, and convergence of the envelope from the hydrostat continues until the crack suppression effects of pressure begin to be outweighed by "pressure melting" effects. Thereafter the envelope converges with the hydrostat, meeting it at the pressure for phase change from Ice Ih to water. As strain rate increases, neither σ_T nor the phase change pressure are likely to be much affected, but σ_c will tend to increase in magnitude. As strain rate decreases, σ_c/σ_T decreases, σ_T will eventually decrease, and the shape of the envelope may tend to the simplified form illustrated in figure 14.

It is interesting to speculate on the shape of the envelope as temperature approaches 0 °C, since the phase change pressure then moves up towards the origin. The failure envelope has to develop an equant shape, and with small

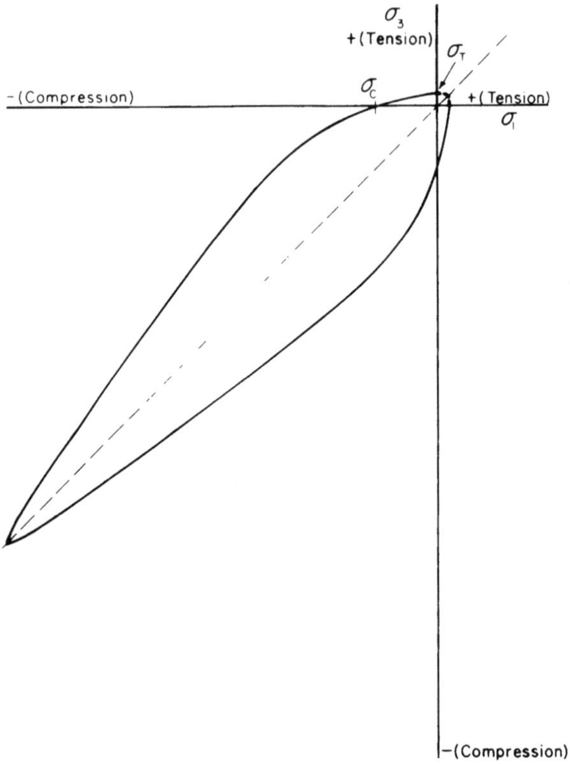

Figure 15. Two-dimensional representation of a yield criterion that embodies the characteristics of ice subjected to moderately high strain rates at temperatures in the vicinity of -10°C.

strain rates and isothermal conditions the strength should tend to zero as the phase transformation pressure tends to zero. At higher strain rates the strength will presumably depend on thermodynamics rate limitations.

All of the foregoing comments apply to failure criteria based on stress, but as was mentioned earlier there is a possibility that it might be preferable to work in terms of strain.

Conclusion

The rather complicated mechanical behaviour of ice is now becoming easier to understand, and the time is ripe for new initiatives in experimental and theoretical investigations.

The recently improved understanding of failure strain, at which $\sigma/\dot{\varepsilon}$ is a maximum, is of great significance. The stress/strain-relation for $(\sigma/\dot{\varepsilon})_{max}$

can be refined by eliminating erroneous data, and the transition to accelerating creep can be defined more systematically. The relatively small variation of failure strain compared to failure stress raises the intriguing possibility that it might be advantageous to formulate failure criteria in terms of strain rather than stress.

It is becoming clear that steady-state creep, if it exists at all, is only likely to occur at large deviatoric strains, where induced anisotropy is fully developed. The old idea of identifying steady-state creep with the minimum strain rate of a creep test is, in general, completely inappropriate. Very little is known about stress/strain-rate relationships for large devatoric strains, and more experimental work in this area is essential for better understanding of ice engineering and glacier flow.

Not much attention has been given to flow at small strains, or to the important matter of stress/strain-rate relationships for constant time. Work is needed to define the appropriate constitutive equations.

The effects of cyclic loading are not well understood, and virtually nothing is known about fatigue failure.

There has been very little systematic consideration of failure criteria for multiaxial stress states, but there are plenty of clues to the general form of the relation. The unusual, or even unique, feature is that the shear resistance of ice can be destroyed purely by hydrostatic pressure at typical temperatures.

Acknowledgment

Without wishing to spread the blame for any errors in this account, the writer gratefully acknowledges the guiding influence of extended discussions and joint writings with his colleagues of the ICSI/NSF Working Group on Ice Mechanics: Roger Le B. Hooke, Stephen J. Jones, Robert Torrance Martin, Mark F. Meier, and Johannes Weertman.

Mechanical Model of Creep of Polycrystalline Ice with Cracking Activity

B. MICHEL

Université Laval

Québec, Canada

Abstract

Recently the author has presented a two-dimensional mechanical model for the deformation and creep of polycrystalline ice that was based on the elastic and plastic deformation within each crystal of the aggregate.

The plastic properties of ice were studied with the principle of movement and multiplication of dislocations within each crystal.

In this paper we will show that the same basic physical model can be used to represent creep of ice during the formation of micro-cracks within crystals. The equations of creep during micro-cracking are developed and shown to fit well with experimental data. This model explains most of the physical processes that occur at relatively high rate of loading of ice in the ductile range, and at the limit, in the brittle one.

Introduction

A few investigators have observed the physical processes in the deformation under load of polycrystalline ice (Steineman 1958; Rigsby 1960; Gold 1960; Wakahama 1962).

They found that various mechanisms would intervene at one time or another in the creep process. In addition to elastic deformation and plastic deformation by movement of dislocations they observed grain boundary slip, cavity formation at the grain boundaries, formation of low-angle boundaries, polygonization and recrystallization. Recrystallization was noticed mainly when the ice deformation became large. It either took the form of grain boundary migration, i.e. some grains growing at the expense of others, or the formation of completely new crystals.

Gold (1960, 1970) was the first to observe extensively the formation of micro-cracks in ice. These cracks, which formed within a crystal of the aggregate, are either parallel or perpendicular to the basal plane of the grain. He found that the cracks would start to appear only at a certain stress level in compression. They would then increase, during transitory creep, but come to a stop if the stress did not exceed certain stress limit. At still higher stresses, the process continued and creep accelerated leading to complete fracture of the ice. Gold also observed cracks forming at grain boundaries that were related to the process of grain boundary slip.

Because of the complexity of ice deformation in the ductile range, it seems appropriate to make a distinction between the physical processes that are involved and classify the types of creep curves.

Michel (1978a) symbolizes the various types of deformation that are involved in the creep processes: ε = pure elastic deformation; ν = pure plastic deformation; α = deformation caused by an increase in the number of mobile dislocations in the ice and corresponding to a decreased viscosity of the ice (it is in itself a strain softening process); κ = deformation accompanied by microcracking activity in the ice, which also decreases the apparent viscosity of the ice (it is more softening that the α type creep process); and δ = deformation with decreased viscosity caused by syntectonic recrystallization (Lliboutry 1964).

In this paper we will develop the basic equation of κ type creep.

Basic model

Let us consider a two-dimensional simplified model of polycrystalline ice as shown in Fig. 1 where there are j rows of crystals, each containing i crystals. The whole block is subjected to a shear stress τ and a hydrostatic stress σ. Each crystal in this group is acted upon by shear stress τ_{ij} in one direction and τ_{ji} in the other, which are producing the total shear deformations γ_{ij} and γ_{ji}.

Because of the extreme difficulty in moving dislocations and producing plastic flow in directions other than that of the basal plane we consider, as for the case of monocrystals, that each crystal slips only along its basal plane, which has an orientation α_{ij} with the j shear direction.

It is obvious that in such a non-uniform model, some grain boundaries will be subjected to intense stressing at the beginning of loading and local boundary slip will occur to accomodate the total deformation of the ice aggregate. Even some small boundary cracks might appear. The worst oriented crystals will be subjected to a moment of rotation and will initially rotate in order to reduce the angle of their basal plane to that favorizing an easier flow direction. We will thus consider the rotation as the first mechanism to intervene in the ductile behavior of ice; reducing the angle of some crystals in key positions to the critical angle of equilibrium α_m where the orthogonal stresses on each crystal become equal, i.e. $\tau_{ij} = \tau_{ji}$.

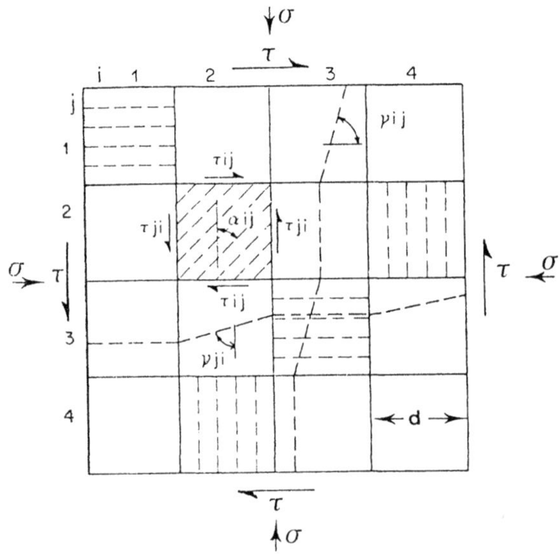

Fig. 1. Basic model of polycrystalline ice containing four crystals in the horizontal rows i and the same in the vertical rows j. It is acted upon by a shear stress τ and a hydrostatic compressive stress σ. Dashed lines indicate the orientation of basal planes in each crystal and also the line of shear deformation.

Once a first rearrangement of crystals has taken place by preliminary rotation, experiments show that preferential slip bands form, where crystals in a row deform considerably under plastic flow, while the neighbouring rows deform little and only in an elastic manner. The deformation pattern is shown with dashed lines on Fig. 1. The major hypothesis of this mechanical model is that the bulk of each crystal, outside a thin boundary zone, if deformed in a manner that is compatible with that of its neighbours in the same slipping band.

Thus, on any horizontal or vertical row:

$$\begin{aligned} \gamma_{ij} &= \gamma_j = \text{constant} \\ &= \\ \gamma_{ji} & \gamma_i = \text{constant} \end{aligned} \qquad (1)$$

where the γ_i are the deformation at 90 to those of the main considered shear plane γ_j.

The condition described by (1) obviously produces shear stresses that are different from one crystal to another, on the same slip band.

We have to consider that the bulk of each crystal deforms like a monocrystal where there is an addition of elastic and plastic deformation as in a Maxwell rheological model. Taking into account the values of the plastic deformations in function of the angles α_{ij} for each grain, but only in the slip direction, we get:

$$\dot{\gamma}_j = \dot{\gamma}_{j\varepsilon} + \dot{\gamma}_{j\nu} \tag{2}$$

where the subscripts ε and ν refer respectively to the elastic and viscous component of the deformation. Thus:

$$\dot{\gamma}_{j\varepsilon} = \frac{1}{G_{ij}} \frac{d\tau_{ij}}{dt} \tag{3}$$

where G_{ij} is the shear modulus in the ij direction. The viscous component is given by an expression, also based on the movement and multiplication of mobile dislocations, resolved on the basal plane of each crystal:

$$\dot{\gamma}_{j\nu} = C\ (\tau_{ij} \cos 2\alpha_{ij})^n \sin \alpha_{ij} \tag{4}$$

in which the $\cos 2\alpha_{ij}$ resolves the stress along the basal plane, $\sin \alpha_{ij}$ resolves the strain along the main shear direction and n is the Glen's exponent.

For the multiplication of mobile dislocations Michel (1978a) proposed the following expression:

$$C = C_0(1 + \alpha\gamma^m) \tag{5}$$

where $$C_0 = bn_0v_0 \exp\ [Q_c(1/273R - 1/\theta^*R)]$$

in which b is the Burger's vector, v_0 = velocity of reference of mobile dislocations at 0°C, n_0 = total number of mobile dislocations, α = coefficient of multiplication of mobile dislocations and m = characteristic power of growth of mobile dislocations, Q_c is the activation energy for creep, R the uniaxial gas constant = 8.315 J/mol and θ^* the absolute temperature in K.

The shear modulus of ice depends little on crystal orientation and its variability is not significant in viscous flow so that $G_{ij} \simeq G$. In any one row, if we assume identical dislocation characteristics, we then have:

$$\dot{\gamma}_{j\nu} = \dot{\gamma}_{ij\nu} = C(\tau_{ij})^n\ CR(\alpha_{ij}) \tag{6}$$

The creep function $CR(\alpha_{ij})$ compares the creep rate of each crystal for a given shear stress as a function of its orientation α_{ij}:

$$CR(\alpha_{ij}) = (\cos 2\alpha_{ij})^n \sin \alpha_{ij} \tag{7}$$

This function is shown on Fig. 2. It is theoretically equal to 0 for $\alpha_{ij} = 0$ and $\alpha_{ij} = \pi/4$. On the same figure is also shown the curve $CR(\alpha_m)$ where the angles are classed irrespective of their actual values, but as a function of the increasing values of $CR(\alpha_{ij})$. It gives an overall picture of the resistance to viscous flow of a number of grains in the polycrystalline aggregate.

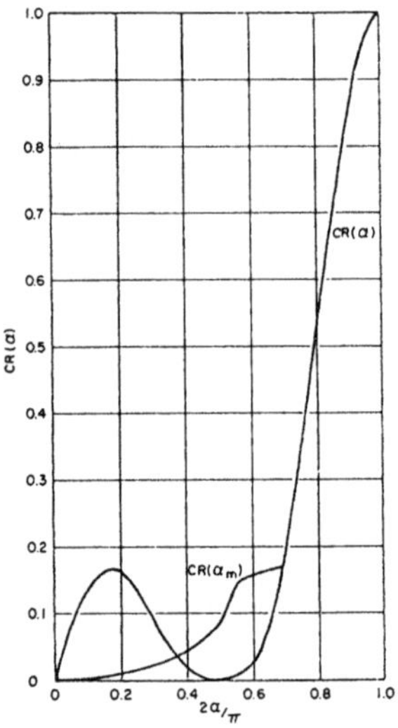

Fig. 2. Creep function $CR(\alpha)$ in relation to the angle of the corresponding crystal α. Also shown is the classified value $CR(\alpha_m)$ for increasing values of the function.

Because many crystals are blocking viscous flow in one direction it can be expected and it is in fact observed that slip bands will form in preferred directions whereas other bands will be rigid. This is why, in this proposed model, we may expect that in a randomly oriented ice sample half of the slip bands will form in one direction and the other half in the other. Certain rows, alternating with slip bands, will deform only in an elastic manner. For these rows j + 1, the strain rate is:

$$\dot{\gamma}_{j+1} = \frac{1}{G}\frac{d\tau}{dt} \tag{8}$$

Once a slip band is mobilized to flow in one direction, the deformation is controlled by that of the worst oriented crystal, which will have pre-rotated to a critical angle α_m. From (3), (4) and (7):

$$\dot{\gamma}_j = \frac{1}{G}\frac{d\tau_{j\alpha m}}{dt} + C(\tau_{j\alpha m})^n CR(\alpha_m) \qquad (9)$$

As only half of the rows are subject to plastic flow, as shown in Fig. 1, and the other half deform only elastically under the average stress τ, then the total deformation of the sample $\dot{\gamma}$ is given by the addition of (8) and (9):

$$\dot{\gamma} = \frac{1}{\ell}\sum_1^{\ell}\dot{\gamma}_j = \frac{1}{2G}\frac{d\tau}{dt} + \frac{1}{2G}\frac{d\tau_{\alpha m}}{dt} + \tfrac{1}{2}C(\tau_{\alpha m})^n CR(\alpha_m) \quad (10)$$

where the critical angle of pre-rotation α_m is the same for each slip band, $\gamma_{j\alpha m} = \gamma_{\alpha m}$.

Equation 10 is the general law of creep of polycrystalline ice. It can also be expressed as:

$$\dot{\gamma} = \frac{1}{G}\frac{d\tau}{dt} + \frac{1}{2G}\frac{d(\tau_{\alpha m} - \tau)}{dt} + \tfrac{1}{2}C(\tau_{\alpha m})^n CR(\alpha_m) \qquad (11)$$

The first term represents pure elastic deformation, the second term delayed elasticity and the third term pure viscous flow. The rheological representation of this mechanical model is shown in Fig. 3. It is made of a series of springs representing the pure elastic bands with a network of parallel elements each combining a spring with a non-linear dashpot and representing a crystal in the slip bands.

The stresses in this model are not the same on each crystal. The stress distribution during creep can be obtained in the following manner.

If we admit that the stress distribution during any stage of creep is very close to that of permanent creep, we then have:

$$\dot{\gamma}_{ij\nu} = \dot{\gamma}_{j\nu} = 2\dot{\gamma}_\nu = C(\tau_{ij})^n CR(\alpha_{ij}) \qquad (12)$$

Isolating the τ_{ij}, we have:

$$\tau_{ij} = (2\dot{\gamma}_\nu/C)^{1/n}/CR(\alpha_{ij})^{1/n} \qquad (13)$$

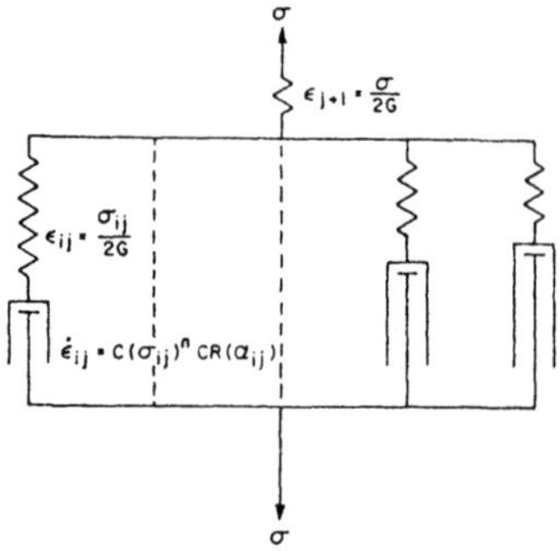

Fig. 3. Sketch of rheological model of creep of ice under a tensible stress σ.

If the distribution of α is uniform from α_m to $\alpha = \pi/2$ and if the $CR(\alpha_{ij})$ is equal to $CR(\alpha_m)$ from $\alpha = 0$ to $\alpha = \alpha_m$, the average stress on a slip band is given by:

$$\tau_0 = \frac{1}{\ell}\sum_1^{\ell} \tau_{ij} = \frac{(2\dot{\gamma}_\nu/C)^{1/n}}{\Omega'} \qquad (14)$$

where

$$\frac{1}{\Omega'} = \frac{2}{\pi}\int_{\alpha_m}^{\pi/2} \frac{d\alpha}{CR(\alpha)^{1/m}} + \frac{\alpha_m}{CR(\alpha_m)^{1/n}} \qquad (15)$$

because in the same row:

$$\tau_{90} = (2\dot{\gamma}_\nu/C)^{1/n} \qquad (16)$$

where τ_{90} is the crystal that slips more on a slip band, being in the slip direction.

The value of Ω' can also be expressed, using (14) and (16) by:

$$\Omega' = \tau_{90}/\tau_0 \qquad (17)$$

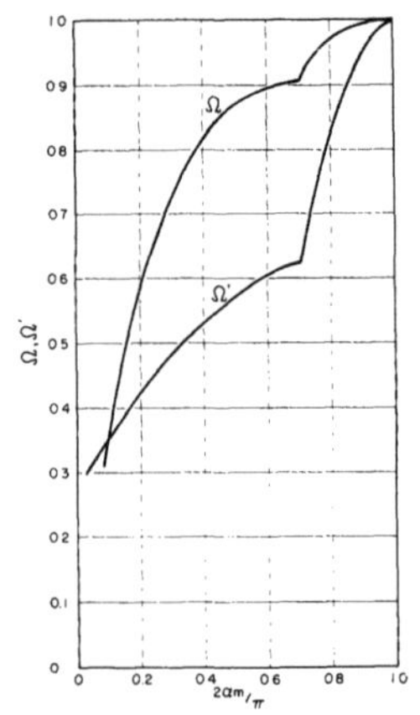

Fig. 4. Values of the functions Ω and Ω' in function of the worst angle of rotation of the head crystal α_m.

The function Ω' has been integrated numerically and its value is given in Fig. 4. Also given in this figure is the value of Ω that relates the stress on the worst oriented crystal to the average stress:

$$\Omega = \tau_0/\tau_{\alpha m} = CR(\alpha_m)^{1/n}/\Omega' \tag{18}$$

Cracking activity in polycrystalline ice

The basic mechanical model of ice creep that we have just discussed shows high stress concentration on the worst placed crystals. There is, however, a limiting stress field that can support a crystal without brittle fracture. The mechanisms of crack formation have been studied by Gold (1960, 1970) which shows that cracks start to form at a compressive stress of around 5×10^5 Pa for S2 ice at -10°C. For higher loads, the rate at which the cracks form depends upon stress, strain and temperatures. His experimental results are shown on the full lines of Fig. 5. He found that if the stress was smaller than 10×10^5 Pa, the rate of cracking tended towards zero with continued deformation. For higher loads, the cracking activity continued at a constant or increasing rate. He also found that initially, the cracks were uniformly distributed within each specimen, but with increasing load they tended to concentrate in zones parallel to the planes of maximum shear and in many cases, these zones became well defined "fault zones".

This mechanical behavior can be well represented with the previous basic model. The worst placed crystals with high stress concentration will first rotate, and then fail. The maximum stress will then be transmitted to the next worst placed crystal of a slip band (lowest α), and the process of micro-fracture will continue either to a stop or complete fracture of the ice by cataclasis.

Under a normal compressive stress, we know from Michel (1978b) that a crystal will fail under brittle conditions along the basal plane, when the stress in the basal plane τ^* is:

$$\tau^* = \tau_i + kd^{-\frac{1}{2}} \tag{19}$$

where τ_i is the cohesive strength of ice that depends only on temperature for polycrystalline ice, k is a numerical coefficient depending only on physical characteristics of the ice and d is the average equivalent dimension of a grain.

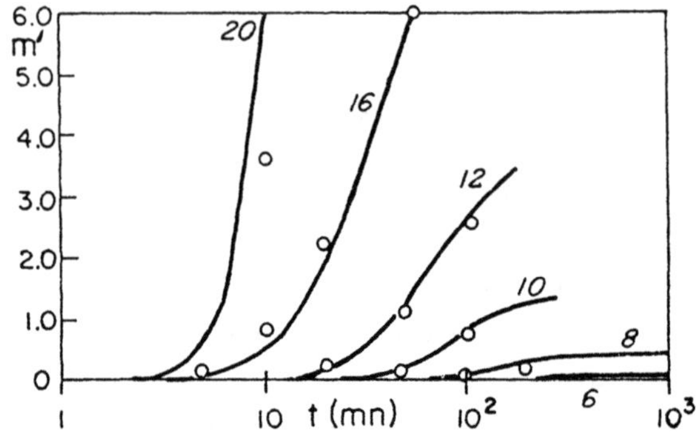

Fig. 5. Time dependance of cracks per cm^2 (m') for various compressive stresses (Kg/m^2) at $\theta = -9.5°C$. (Gold, 1960). Circles represent values computed from Eq. (43)

We have also found experimentally that there are non-basal shear fracture mechanisms (mainly cleavage) when compressive stress makes an angle of more than 75° with the basal plane. The resistance of the crystal in shear τ_c, was then found to be, from a study on monocrystals, as shown on Fig. 6.

$$\tau_c \simeq 3\tau^* \tag{20}$$

In the case of normal tensile stresses, the appearance of a first crack leads to sudden fracture of the whole sample so there is no further creep of ice. This study is thus limited only to the case of normal compressive stresses.

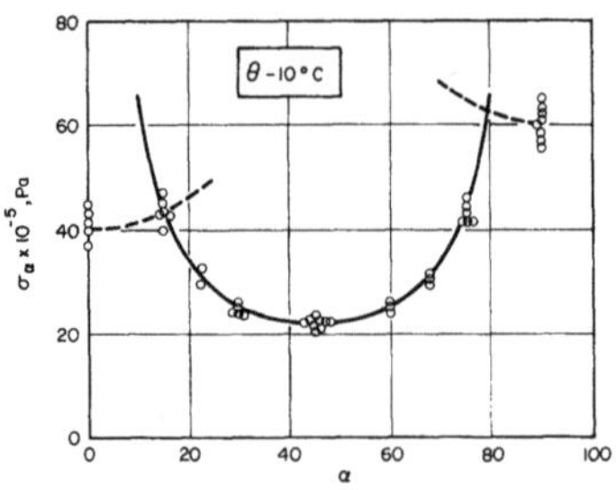

Fig. 6. Variation of crushing strength of S_1 ice as a function of angle α of basal plane (Carter and Michel 1971).

Now, if we call m the crack density defined as the ratio of the number of cracks to the total number of grains in a given area; m will be directly related, in polycrystalline ice with a uniform grain distribution, to the crystal orientation of the worst stressed crystal which is still unbroken, defined as α_c:

$$m = 4 \quad (\alpha_m - \alpha_c)/\pi \tag{21}$$

where α_m is the orientation of the most stressed crystal, before cracking activity starts. When $\alpha_c = \alpha_m$, $m = 0$. The crack density m is related to m' the number of cracks per unit area, by:

$$m' = m\,\zeta \tag{22}$$

where ζ is the number of grains per unit area.

When some crystals are broken in a slip band, the real tangential stresses on remaining crystals τ in increased to:

$$\tau = \tau_o/(1 - m) \tag{23}$$

where τ_o is the average stress for the bulk of the ice.

The controlling mechanism of fracture for all crystals when $0 > \alpha_c > 15^o$ is non-basal fracture. The tangential stress on the crystal of orientation α_c is then equal to τ_c (Eq. 20).

If we relate τ_c to the average shear stress τ_o with equations (17) and (23), we get:

$$\tau_o = \tau_c\ \Omega\ (1 - m) \tag{24}$$

With τ_c in (20) this gives:

$$\tau_o/\tau^* = 3\Omega\ (1 - m) \tag{25}$$

This function is shown on Fig. 7. It is an increasing function which shows that the number of cracks increases with load, but without failure of the ice. In order to simplify this function is has been computed and set on log log paper on Fig. 8. It can be seen that this function is closely represented, for the range of its use by:

$$CR\ (\alpha) = 2.5 \times 10^{-3} \left(\frac{\tau_o}{\tau^*}\right)^5 \tag{26}$$

For $\alpha_c > 15^o$ the fracture will occur in the basal plane of the crystals. The stress on the maximum stressed crystal is then given at fracture by:

$$\tau_c = \tau^*/\cos 2\alpha_c \tag{27}$$

Relating τ_c to the average stress with (17) and (24), we get:

$$\tau_o/\tau^* = \frac{\Omega(1 - m)}{\cos 2\alpha_c} \tag{28}$$

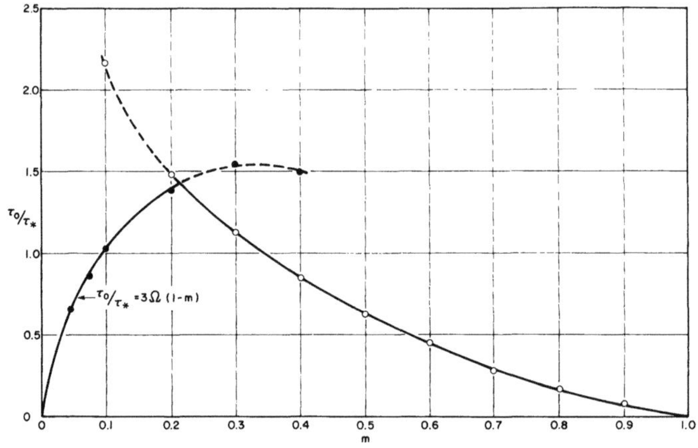

Fig. 7. Variation of average stress τ_0/τ^* during cracking activity in function of crack density m.

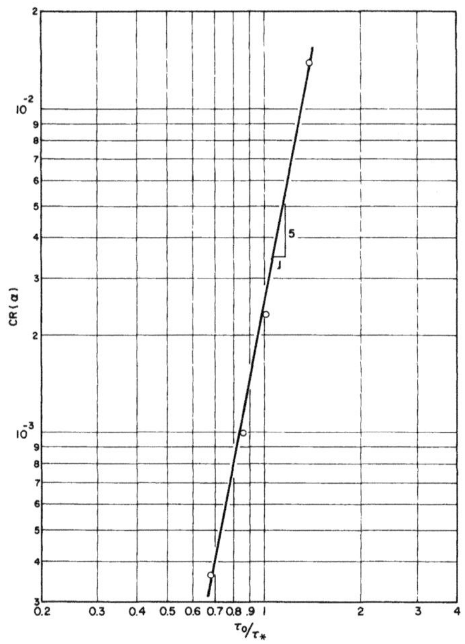

Fig. 8. Creep function CR(α) in function of shear stress for low levels of cracking activity, m < 0.1.

This relation is also shown in the second part of Fig. 7. It is a decreasing function that shows that once the cracking activity has reached a certain level, the ice is unstable. An increase in cracking activity leads to a decrease in resistance.

Thus, for a value of τ_c/τ^* of the order of 3, the ice becomes unstable for a value of m of around 20%. It can also be observed that the maximum stress that the ice can resist under ductile conditions is higher than the stress it will resist under pure brittle fracture, where it is τ^*. This assumes indeed, with equation (17), complete plastification of the ice, which is never completely the case so that the stress on the worst placed crystal may not be as high. For all stresses lower than this maximum value τ_y max, the cracking activity will increase to a certain level and stop at a value given by equation (24). For conditions leading to a crack density over the critical value, the cracking activity will accelerate until complete failure by cataclasis of the sample.

Equations of κ creep

When cracking activity starts, the stress on the more stressed crystal in a creeping row becomes, $\tau_{\alpha m} = \tau_c$, so that the basic creep equation of the mechanical model, Eq. 11, becomes:

$$\dot{\gamma} = \frac{1}{G}\frac{d\tau}{dt} + \frac{1}{2G}\frac{d(\tau_c - \tau)}{dt} + \tfrac{1}{2}\, C\, (\tau_c)^n\, CR\, (\alpha_c) \qquad (29)$$

Constant stress

For tests under a constant stress τ_o, we then have:

$$\frac{d\tau}{dt} = 0$$

$$\dot{\gamma} = \tfrac{1}{2}\, C\, (\tau_c)^n\, CR\, (\alpha_c)$$

With the value of C in Eq. (5) and of CR (α_c) in Eq. (26) we get:

$$\dot{\gamma} = 1.25 \times 10^{-3}\, Co\, (1 + \alpha\, \gamma^m) \left(\frac{\tau_o}{\tau^*}\right)^5 \qquad (30)$$

It is interesting to see that κ creep is never strain hardening. This is because the stress on the worst placed crystal does not increase to augment the delayed elastic deformation but stays a constant as the leading crystal change from one to the next as soon as it cracks, keeping a constant stress level. Furthermore, κ creep under a constant stress is slightly strain softening in the first part of the deformation as the number of mobile dislocations is increased. When $\gamma = \gamma_0$, so that all the dislocations are in movement, we attain permanent creep with cracking activity, at a strain rate $\dot{\gamma}_0$:

$$\dot{\gamma}_0 = 1.25 \times 10^{-3}\ Co\ (1 + \alpha\ \gamma_o^m) \left(\frac{\tau_o}{\tau^*}\right)^5 \tag{31}$$

The stress exponent in permanent creep with cracking activity, which is 5, is much higher than the Glen's exponent which is usually n = 3 for creep without crack formation.

With reduced variables:

$$\bar{\alpha} = \alpha\ \gamma_o^{\ m}$$

$$\bar{\gamma} = \gamma/\gamma_o$$

$$\bar{t} = \frac{1.25 \times 10^{-3}\ Co\ (\tau_o)^5}{\gamma_o\ (\tau^*)^5} \cdot t$$

we obtain the dimensionless equation of $\alpha\kappa$ creep and permanent κ creep:

$$\frac{d\bar{\gamma}}{d\bar{t}} = (1 + \bar{\alpha}\ \bar{\gamma}^m) \tag{32}$$

when $\bar{\gamma} = 1$, $\frac{d\bar{\gamma}}{d\bar{t}} = 1$ for the permanent stage of creep.

Obviously when the crack density attains a certain critical value, the cracking activity starts to accelerate and lead to failure of the sample by cataclasis. This condition is not analyzed here.

Eq. (32) for κ creep is identical to that of $\alpha\gamma$ type creep (Michel 1978 a). It differs only by the time scale. For α type creep the time scale is given by (Michel 1978 a):

$$\bar{t}_\alpha = \frac{Co\ (\Omega')^n\ \tau_o^{-n}}{2\gamma_o} \tag{33}$$

For κ type creep the time scale is given by:

$$\bar{t} = \kappa\ \bar{t}_\alpha \tag{34}$$

with

$$\kappa = \frac{2 \times 10^{-3}\ (\tau_o)^{(5-n)}}{(\tau^*)^5} \tag{35}$$

where κ is always larger than one, showing that κ creep is always more ductile than α creep.

Constant strain rate

At constant strain rate $\dot{\gamma}_o$, basic equation (29) becomes:

$$\dot{\gamma}_o = \frac{1}{2G}\frac{d\tau}{dt} + \tfrac{1}{2}\ C\ (\tau_c)^n\ CR\ (\alpha_c) \tag{36}$$

With values of C and CR (α_c) in eqs 5 and 26:

$$\dot{\gamma}_o = \frac{1}{2G}\frac{d\tau}{dt} + 1.25 \times 10^{-3} \text{ Co } (1 + \alpha \gamma^m) \left(\frac{\tau}{\tau^*}\right)^5 \qquad (37)$$

with $t = \gamma/\dot{\gamma}_o$.

With reduced variables:

$$\bar{\gamma} = \gamma/\gamma_o \quad \text{with} \quad \gamma_o = \tau_o/2G$$

$$\bar{\alpha} = \alpha \gamma_o^m \qquad \tau_o = \tau^* \left(\frac{\dot{\gamma}_o}{1.25 \times 10^{-3} \text{ Co}}\right)^{1/5}$$

$$\tau = \tau/\tau_o$$

The dimensionless equation of κ creep under a constant strain rate is:

$$\frac{d\bar{\tau}}{d\gamma} = 1 - (1 + \bar{\alpha}\,\bar{\gamma}^m)\,\bar{\tau}^5 \qquad (38)$$

The highest value of $\tau = \tau_y$ at yield point, corresponds to the end of the cracking activity. Otherwise if cracking would continue with decreasing stresses this would lead to failure of the sample by cataclasis.

Here again the equation for κ creep is very similar to that of α creep except that the Glen's exponent increases from 3 to 5 and the time scale is increased by a factor of κ.

Experimental verification

Cracking activity

The more stressed crystals in a row of creeping crystals will deform mostly in an elastic manner as their basal plane is normal to the main shear direction and they do not allow much plastic deformation in that direction. Each time a crack is formed the stress giving elastic deformation is released to load the remaining crystals of the row. Thus, in this elastic process it can be expected that a number of cracks will depend only on the total deformation of the row. This can be expressed by a relation of the form:

$$m' = A\,(\gamma - \gamma_m) \qquad (39)$$

where m' is the crack number per unit area γ_m the deformation at the beginning of the cracking activity and A a constant.

Under a constant load the basic equation of creep is given from Eq. (30):

$$\gamma = 1.25 \times 10^{-3} \text{ Co} \left(\frac{\tau_o}{\tau}\right)^5 \int (1 + \alpha \gamma^m) \, dt \qquad (40)$$

Assuming an average value of γ_a in the integral during the cracking activity we get:

$$\gamma = 1.25 \times 10^{-3} \text{ Co } (1 + \alpha \gamma_a{}^m) \left(\frac{\tau_o}{\tau^*}\right)^5 t + \text{cte} \qquad (41)$$

Eq. (41) combines with Eq. (39) to give an expression of the form:

$$m' = K_o \tau_o{}^5 t + K_1 \qquad (42)$$

where K_o and K_1 are constants.

On Fig. 5 is shown, in full lines, the curves giving the actual number of cracks m' in function of stress σ_o and t as measured by Gold (1972). The circles on the figure represent Eq. (42) with best fit values for the constants but using σ_o instead of τ_o. The equation is:

$$m' = 0.03 \left(\frac{\sigma_o}{12 \times 10^5}\right)^5 t - 0.5 \qquad (43)$$

The measured curves represent not only steady creep but also accelerated creep for the two higher stresses. Even then, it is interesting to see that the basic model of cracking activity looks rather good to represent such a complex phenomenon.

Creep under constant strain rates

We have been using the experimental results of Hawkes and Mellor (1972) to check the validity of the present theory. On Fig. 9 is shown, with points, the results of one uniaxial compressive test on snow ice at various strain rates. There is a rather sharp break in the curves when cracking activity begins.

Three types of this creep curve have to be represented with the equations:

1 - The strain-hardening first part of the curves which is $\alpha\varepsilon$ creep with multiplication of dislocations.

2 - The second part of the curves which is κ creep with cracking activity up to the yield point.

3 - The last part of the lower curves which is strain-softening creep with increased number of dislocations down to permanent creep.

The first part of the curve has been fitted with the equation for $\varepsilon\alpha$ and $\alpha\nu$ creep under a constant strain rate given by Michel (1979). This equation is:

$$\frac{d\bar{\tau}}{d\bar{\gamma}} = 1 - (1 + \bar{\alpha}\,\bar{\gamma}^{m})\,\bar{\tau}^{3} \qquad (44)$$

where $\bar{\tau} = \tau/\tau_o$

$\bar{\alpha} = \alpha\,\gamma_{\varepsilon t}^{m}$

$\bar{\gamma} = \gamma/\gamma_{\varepsilon t} = \dot{\gamma}_0 t/\gamma_{\varepsilon t}$

in which τ_o would be the equivalent yield point if the same process would continue and $\gamma_{\varepsilon t}$ is the total elastic plus delayed elastic deformation.

The third part of the lower curves is also represented by the same equation, but where new values of τ_o and $\gamma_{\varepsilon t}$ have to be adjusted to take into account the displacement of the creep curve caused by the cracking activity in between.

Finally the central part with cracking activity has been adjusted with Eq. (38) where the parameters τ_o and γ_o are obtained from the data points.

The adjusted parameters for the curve with a strain rate of $4.3 \times 10^{-5}\ s^{-1}$, are:

For the first part with $\alpha\varepsilon$ creep:

$$\sigma_o = 157 \text{ bars}$$

$$\varepsilon_{\varepsilon t} = 0.0031$$

$$\alpha = 50,\ m = 0.8,\ n = 3$$

For the second part with κ creep:

$$\sigma_o = 93 \text{ bars}$$

$$\varepsilon_o = 0.0077$$

$$\alpha = 50,\ m = 0.8,\ \kappa = 2.0$$

For the third part with $\alpha\nu$ creep:

$$\sigma_0 = 150 \text{ bars}$$

$$\varepsilon_0 = 0.0090$$

$$\alpha = 50,\ m = 0.8$$

The measured curve and computed one are shown in Fig. 9. It can be seen that the theory gives a very good fit for these types of tests.

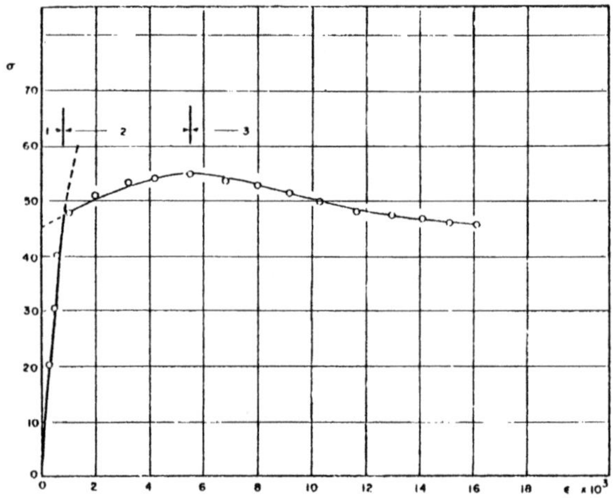

Fig. 9. Creep curve in function of time in snow ice under a constant strain rate of $4.3 \times 10^{-5}\ s^{-1}$. The measurement (in circles) are from Hawkes and Mellor (1972). The full line represents the equation for $\alpha\varepsilon$ creep in zone 1, κ creep in zone 2, and $\alpha\nu$ creep in zone 3.

Conclusions

We can say, in conclusion, that the mechanical model that we had previously developed for the creep of polycrystalline ice without crack formation could be extended to the case of crack formation.

The model then shows that cracking activity is a very softening process and that the exponent in the equivalent Glen's equation is increased from 3 to 5.

The theory can be shown to prove the direct observations by Gold (1962) of the number of cracks in polycrystalline ice, and check well with creep data, although the numerical parameters differ from one test to the other in function of the physical properties of the crystals of one sample compared with another (size of grains, initial number of dislocations, growth of dislocations etc...).

The theory also shows that the yield point during ductile deformation can be as much as 40% higher that for brittle fracture. That had already been found by many experimenters, but never completely explained.

References

CARTER, O., MICHEL, B. 1971. "Lois et mécanismes de l'apparante rupture fragile de la glace de rivière et de lac" Report S-22, Civil Engineering, Université Laval, Quebec.

GOLD, L.W. 1960. "The cracking activity in ice during creep" Canadian Journal of Physics 38, pp. 1137-1148.

GOLD, L.W. 1970. "Process of failure in ice" Canadian Geotechnical Journal, 7, pp. 405-413.

GOLD, L.W. 1972. "The failure process in columnar-grained ice" Technical Paper 369, Division of Building Research, National Research Council of Canada, Ottawa, Ontario NRCC 12637.

HAWKES, I. and MELLOR, M. 1972. "Deformation and fracture of ice under uniaxial stress" Journal of Glaciology, Vol. 11, No. 61, p. 103-131.

LLIBOUTRY L. 1964. "Traité de glaciologie" Tome 1. Masson et Co. Paris.

MICHEL, B. 1978 a. "A mechanical model of creep of polycrystalline ice" Can. Geotech. J., Vol. 15, p. 155-170.

MICHEL, B. 1978 b. "The strength of polycrystalline ice" Can. J. Civ. Eng., Vol. 5, p. 285-300.

MICHEL, B. 1979. "Ice Mechanics" Les Presses de l'Université Laval. 500 p.

RIGSBY, G.P. 1960. "Crystal orientation in glacier and in experimentally deformed ice" Journal of Glaciology, 3(27), pp. 589-606.

STEINEMAN, S. 1958. "Experimentelle Untersuchungen zur Plastizität von Eis" Beiträge zur Geologie der Schweiz, Geotechnische Serie, Hydrologie, No. 10, 72 p.

WAKAHAMA, G. 1962. "On the plastic deformation of ice" Low Temperature Science, Series A, 20, pp. 57-100.

The Application of Fracture Mechanics to Ice Problems

K J MILLER
University of Sheffield
England

Abstract

Linear elastic fracture mechanics concerns the application of crack tip characterizing parameters to the study of instability phenomena of cracked bodies. These parameters are G, the strain energy release rate, K, the stress intensification factor, and CTOD, the crack tip opening displacement. Successful applications are associated with brittle fracture studies and fatigue crack propagation in metals when plasticity is limited. For non-linear elastic behaviour of crack tip material the J contour integral is an improved parameter.

None of these parameters however quantify the propagation of cracks, rather they are concerned with the initiation situation; cf. the relation between yield point and plastic flow behaviour of materials. Thus G, K, CTOD and J do not quantify stable crack growth although they have been used as characterizing parameters for crack growth in ductile materials. The difficulty in the quantification of crack initiation and growth are the roles played by plasticity, time dependent deformation and differing fracture processes and recent developments have centred on C* and a crack growth parameter G^{Δ}.

All of these parameters will be briefly reviewed and their relevance to the fracture of ice outlined both from a theoretical and an experimental viewpoint.

Introduction

The engineering design of safe structures in the aeronautical, chemical, and nuclear industries today requires a theoretical and experimental knowledge of the mechanics of fracture. Calculations are made concerning, for example, allowable stress limits for detectable defects and fatigue lifetime calculations for growing but sometimes unobservable defects. Indeed all structures are today assessed on the basis that they inherently contain defects. Such defects may only be parts of a millimetre in length prior to fracture.

Some of the earth's largest natural defects occur in ice; crevasses. This paper focuses attention to the application of engineering mechanics to the study of fracture of ice and was stimulated from the author's unintentional close study of numerous crevasses and a desire to have information concerning their depth and distribution. [1].

Linear Elastic Fracture Mechanics (LEFM)

If materials behave in a purely elastic manner Griffith [2] showed that unstable crack propagation takes place when an increment of crack growth occurs which results in more stored energy being released than is absorbed by the creation of the new crack surfaces. However in real materials considerable plastic and/or time-dependent deformation can take place which presents great difficulties in evaluating the energy release term. Nevertheless Griffith's equation

$$\sigma_{crit} = \sqrt{\frac{2E\gamma}{\pi a}} \qquad (1)$$

laid the basis of fracture mechanics.

Here σ_{crit} is the critical stress for unstable fracture, E is Young's modulus, γ is surface energy and a is crack length. Griffith showed that the product of σ_{crit} and $\sqrt{a}$ was constant for a given material, and so it is possible to analyse large scale structures by performing tests on smaller models in a laboratory. He also showed that stresses σ_x, see Fig 1, had no effect on $\sigma_{crit}\sqrt{a}$.

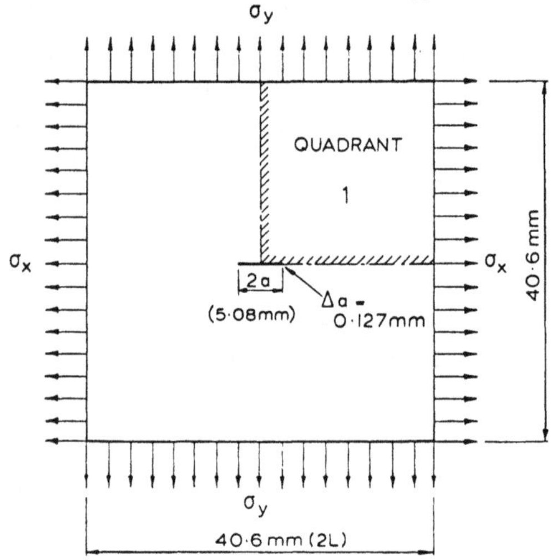

Fig 1. Configuration used for elastic-plastic analysis of crack growth in biaxially stressed plates

Irwin [3] and Orowan [4] suggested that (i) γ_p should replace γ in equation (1) since γ_p, the energy absorbed by plastic deformation, can be three orders of magnitude greater than γ in most materials and (ii) provided the plastic zone is small in comparison to the crack length and body dimensions the energy released could still be calculated from elastic analyses with sufficient accuracy. Irwin and Kies [5] then proposed that if the fracture process was similar, albeit in different specimens and with different loading conditions, fracture occurred when the strain energy release rate attained a critical value, G_c, a material property; here

$$G = \frac{\partial U}{\partial a} = \frac{F^2}{2} \cdot \frac{\partial C}{\partial a} \qquad (2)$$

where U is the strain energy in an elastic body, F = force and C = compliance. Whilst this approach has much appeal the evaluation of C in real-life situations is frequently difficult.

An alternative approach developed by Irwin [6] was to consider the spatial stress distribution at the tip of a crack. Solutions show that

$$\sigma = \frac{K}{\sqrt{2\pi r}} \cdot f(\theta) + \ldots, \text{ series} \qquad (3)$$

where r and θ are polar coordinates with the origin at the crack tip. Thus the stress intensity factor K helps describe the mechanical environment of the crack tip and is equivalent to G, i.e.

$$GE = K^2 \text{ for plane stress conditions}$$

$$GE = K^2 (1-\nu^2) \text{ for plane strain conditions} \qquad (4)$$

where ν is Poisson's ratio.

There are three modes of extending a crack [7], see Fig 2, but here we shall confine ourselves to the tensile opening mode, designated I, and the critical value of the mode I stress intensity factor K_{Ic}. The value of K is dependent on loading conditions and specimen shape and these effects are embodied in the Y term of the K calibration i.e.

$$K = Y \sigma\sqrt{\pi a} \qquad (5)$$

For the specimen shown in Fig 3a

$$K = \frac{P\sqrt{a}}{BW}\left[29.6 - 185.5\left(\frac{a}{W}\right) + 655.7\left(\frac{a}{W}\right)^2 - 1017\left(\frac{a}{W}\right)^3 + 638.9\left(\frac{a}{W}\right)^4\right] \qquad (6)$$

here P = applied force and B = specimen thickness.

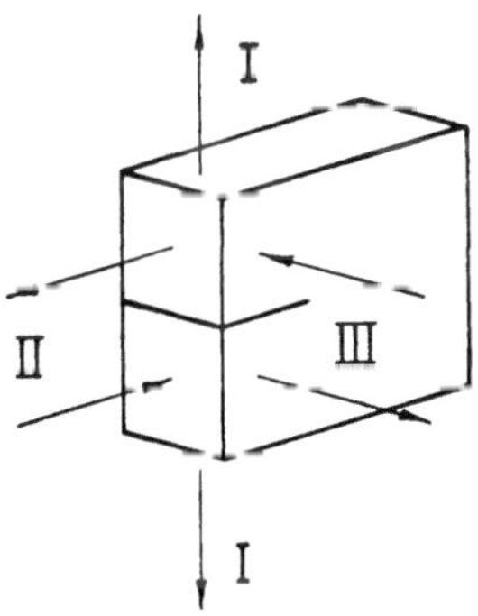

Fig 2. Three modes of crack extension

The K calibration of the specimen for crack arrest studies illustrated in Fig 3b is given in Ref(8). Unlike G the parameter K is relatively easy to determine and now may be obtained from compendia, eg. reference[9].

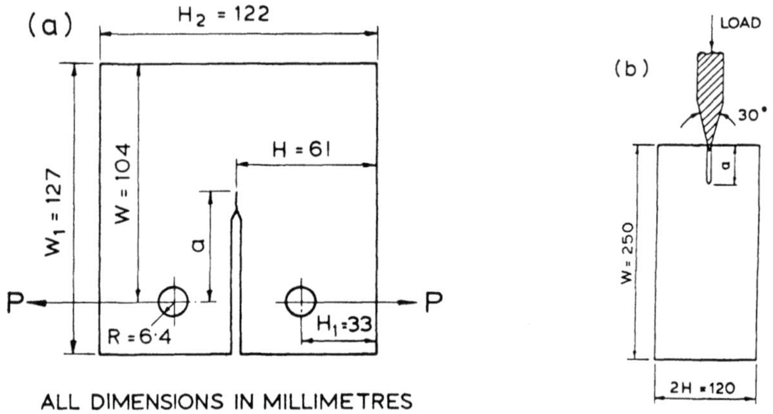

Fig 3. Fracture toughness specimens. (a) Compact Tension Specimen (CTS) and (b) Wedge opening specimen for crack arrest studies.

Elastic-Plastic Fracture Mechanics (EPFM)

The difference between LEFM and elastic-plastic fracture mechanics is difficult to define because even brittle-like materials, such as glass, exhibit some degree of plasticity at the crack tip prior to fracture. Indeed it was Liu (10) who pointed out that it is not the elastic stresses and strains outside the plastic zone which cause fracture but rather the mechanical state within the plastic zone. As in LEFM, a single parameter characterization for fracture from EPFM analyses was next attempted. This was required because LEFM was too conservative since it could not quantify the energy absorbed at crack tips that required an increase in stress before criticality occurred. The first attempt by Dugdale [11] and Barenblatt [12] considered a small plastic strip zone developing ahead of the crack tip thus allowing the crack tip to blunt by an extent later termed the crack tip opening displacement CTOD (or δ) by Wells (13) who proposed that fracture occurred at a critical value, δ_c.

Burdekin and Stone (14) then evaluated CTOD as

$$\delta = \frac{8\sigma_y}{\pi}\frac{a}{E}\log\sec\frac{\pi\sigma}{2\sigma_y} \tag{7}$$

and so after expanding the log.sec.term, it was found that

$$\delta = \frac{\pi\sigma^2 a}{E\sigma_y}\left[1 + \frac{\pi^2}{24}\left(\frac{\sigma}{\sigma_y}\right)^2 + \cdots\right] \tag{8}$$

It follows that $G = \alpha\delta\sigma_y$, with $\alpha = 1$ to 2.
Whilst the CTOD approach is a valuable attempt to realistically quantify events at the crack tip it is difficult to apply in design situations for ductile materials. This is why the K approach is still often preferred. Nevertheless the K approach requires modification to permit non-linear deformation at the crack tip. Thus the J contour integral was used by Rice (15) where

$$J = \int_\Gamma \left(w dy - T_i \frac{du_i}{dx} \right) ds \tag{9}$$

Here J is an anticlockwise contour from the lower crack face around the crack tip to the upper face, whilst s is the path along the contour and w is the strain energy density of the form $\int_0^e \sigma_{ij} d\varepsilon_{ij}$. Finally $T_i du_i$ are work terms when components of surface tractions T_i (on the contour path) move through displacements du_i. The benefits of the J approach are (1) the result is independent of the path when the crack faces are stress free and so the path can be chosen remote from the crack tip thereby avoiding the complexity of crack tip plasticity, and (2) non-linear behaviour of crack tip material can be accounted for albeit that the stress and strain functions derived by Hutchinson (16) and Rice and Rosengren (17) are based on deformation rather than incremental plasticity considerations. It should be noted that the contour integral had previously been developed separately by Eshelby (18) and Cherepanov (19). The approach is conceptually the same as that of Griffith and equivalence is maintained by G, J and K^2/E.

A further development was to consider the growth of cracks due to creep. Some researchers adopted a K approach whilst others claimed better correlations with the nett section stress across the ligament or the use of a reference stress. Understandably the first approach is suitable for brittle materials whilst the second approach favours ductile materials. Recently some experimenters (20,21,22) have attempted to introduce the non-linear behaviour characterized by J, to account for creep crack growth. The creep equivalent of J has been called C*.

Thus if J equals $-\frac{1}{B}\frac{dU}{da}$ where B is the plate thickness (crack width) and U is potential energy then

$$C^* = \frac{1}{B}\frac{d\dot{U}}{da} \qquad (10)$$

This approach is now being actively pursued but will probably suffer the same limitations as the J approach, and will be geometry dependent. It should be noted that none of the parameters yet mentioned consider growth, only initiation and none take true plasticity into consideration.

A crack growth parameter

All the previous parameters do not consider real material events at the instant of crack growth and what occurs immediately after the start of growth. With reference to Fig 4a if the crack growth step

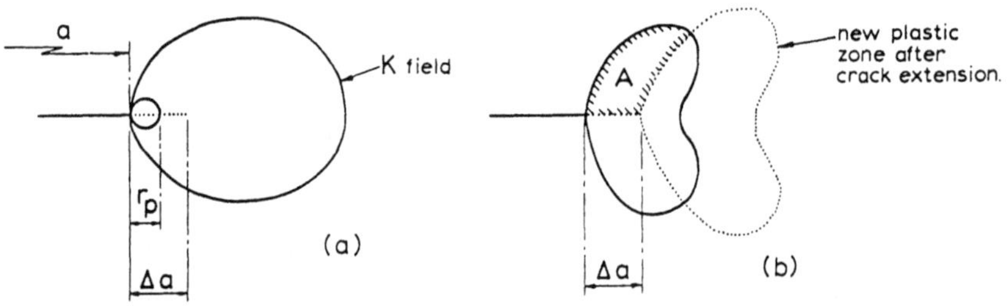

Fig 4. Crack tip zones

Δa is much greater than the plastic zone size, r_p, which is an alternative way of saying the latter term is exceedingly small, then the growth step is adequately characterized by the elastic stress field and LEFM applies. If as in Fig 4b Δa is smaller than r_p then LEFM does not apply and growth must be determined from EPFM parameters. In this case immediately crack growth commences, elastic recovery takes place in zone A that tends to close the crack which thus requires more force to open it for propagation to continue. It is this energy redistribution within the fracture process zone Δa that provides the complication. However an anomaly can now be explained. In a conventional elastic energy balance approach $\partial w/\partial a$ (=G) is defined as the limit of $\delta \bar{W}/\delta a$ as δa tends to zero. The energy balance is therefore calculated over a distance δa which is always infinitely smaller than r_p, however small the latter. Thus it is not surprising that the fracture separation energy cannot be accounted for from a continuum mechanics energy balance (23). Thus there is a requirement for a process fracture zone Δa which requires ΔW, the work absorbed during separation of the crack surfaces over the distance Δa. The growth parameter G^{Δ} is defined as $\Delta W/\Delta a$ and if G^{Δ} exceeds the cohesive strength of the material then extension will occur even if $\partial W/\partial a$ is zero. To appreciate the significance of these events an elastic-plastic finite element analysis was performed on the plate shown in Fig 1. (24). Different states of biaxiality were introduced since K and possibly J cannot account for any biaxial effect at

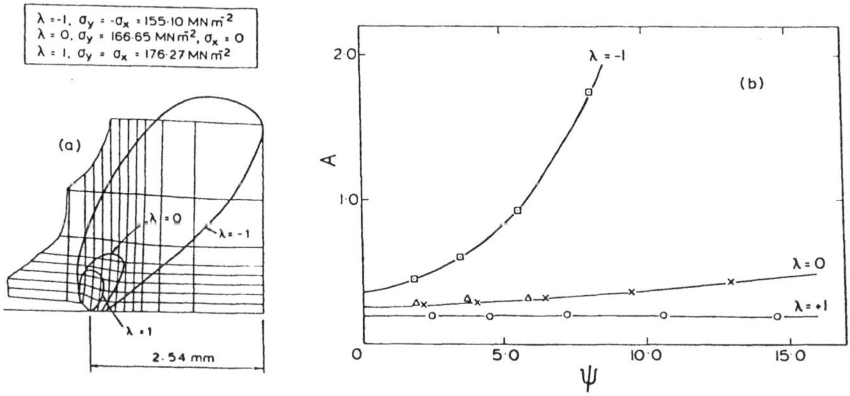

Fig 5. Plastic Zone sizes

the singularity. The effect of biaxiality is shown in Fig 5a. For $\lambda = 0$ (uniaxial loading; $\sigma_x = 0$) an intermediate plastic zone size is created when compared with $\lambda = +1$ (equi-biaxial tension) and $\lambda = -1$ (shear loading); the latter producing a very large plastic zone. In these cases K is nominally constant. In Fig 5b, A is a crack-tip plastic zone size factor which is usually related to the applied load and yield stress by

$$r_p = A\left(\frac{K_1}{\sigma_y}\right)^2 \qquad (11)$$

The value of A is assumed to be 0.175 (25) but it is seen to be a function of load, ψ, and biaxiality except for the case of the smallest almost negligible plastic zone of the equi-biaxial tension mode. Here $\psi = G/G_o$ where G = Griffith's energy release rate and G_o is the value of G at incipient yielding (26).

Fig 6. shows what happens when plasticity occurs in relation to purely elastic behaviour at the crack tip. It is obvious that as the crack begins to propagate there is a partial self-healing effect

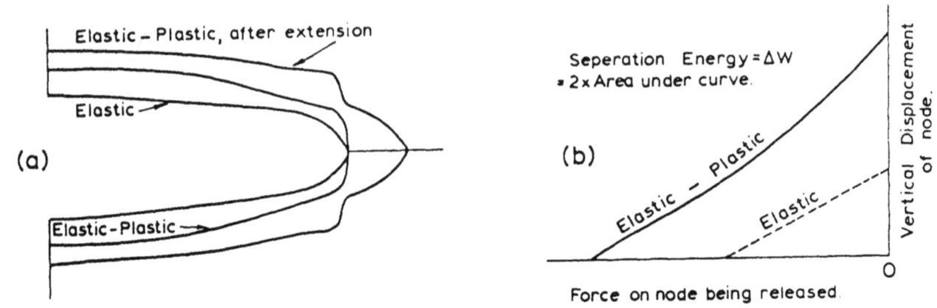

Fig 6. Elastic and Elastic-Plastic Effects of Crack Tip behaviour

that helps close the crack and so a larger force is required to maintain a constant critical CTOD. Meanwhile the plastic zone acts not only as an energy sink limiting the amount of energy required for crack separation but also blunts the crack and causes a stress redistribution in the crack tip zone. It follows that the greater the plastic zone size the lower G^{Δ}. Fig 6b indicates that under elastic-plastic conditions there is a lack of linearity in the nodal force vs node displacement characteristic from which ΔW and hence G^{Δ} $(=\Delta W/\Delta a)$ is determined in a finite element analysis.

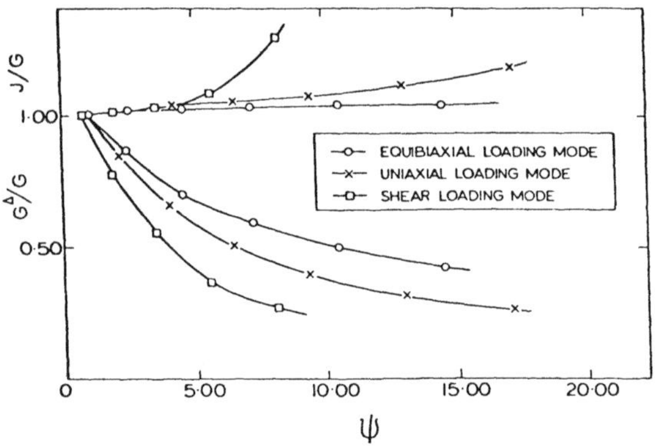

Fig 7. A comparison of J and G^{Δ} parameters for different degrees of stress biaxiality.

Fig 7 shows the increasing divergence of parameters J and G^{Δ} as load increases (27). Note the tendency to zero of G^{Δ} for increasing loads. Using the Dugdale-Barenblatt model Kfouri and Rice (28) established the relationship between the G^{Δ} and J uniaxial curves but this relationship must also be a function of stress biaxiality.

Finally Fig 8 shows that if the ratio of process zone size to plastic zone size is (a) constant, then G^{Δ} is constant irrespective of the state of stress biaxiality in the plate of Fig 1 and (b) when r_p is small or Δa large then $G^{\Delta} = G$ (= J) and LEFM applies.

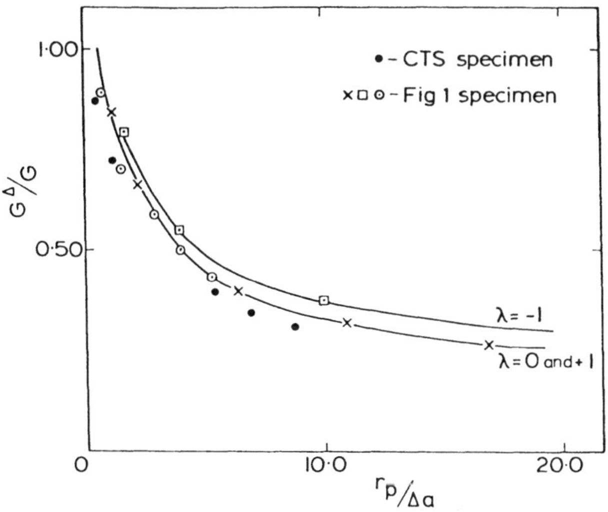

Fig 7. Normalization of G^{Δ} with $r_p/\Delta a$

Using this approach Kfouri and Miller [24] accurately predicted the variation of K_{1c} with temperature for three materials simply by accounting for variations in plastic zone size as temperature increased. It should be noted that since r_p is a function of both the mode I opening stress and the state of biaxiality λ a single parameter characterization for fracture is not possible. Similarly in multi-axial fatigue studies Miller [29] and co-workers have shown that in complex stress situations it is necessary to describe fracture by two parameters, the maximum shear strain (or stress) and the strain normal to the plane of maximum shear strain. The first controls the direction of crack growth and the second affects the speed of growth.

To conclude it should be recorded that G^{Δ} has also been used successfully to explain the reduction of fracture toughness in steels due to hydrogen embrittlement (30) and is now being used to quantify fatigue crack growth.

Fracture toughness of ice

Very little work has been published to date concerning the fracture toughness of ice. Liu and Miller [8] reported some 63 tests on bubble-free fresh-water ice at temperatures in the range -1 to -46°C at loading rates in the range 0.5 to 480 mm/min. Eleven other tests were also reported concerning the effect of speed of freezing of samples, crack arrest and environmental effects. Goodman & Tabor [31] reported 5 fracture toughness calculations from indentation tests and 3 point bend tests. Gold [32] had previously deduced the toughness characteristics of ice under thermal shock conditions. These latter test data are difficult to interpret if fracture toughness is related to crack tip opening events since maximum CTOD does not necessarily occur at the maximum K value due to yield variations with temperature throughout a cycle; an identical problem to thermal shock studies of gas turbine blades. More than 100 tests have recently been conducted on glacier ice samples although these results have not yet been analysed [33].

The application of fracture mechanics theory to the problem of crevasse penetration was first introduced by Smith [34] who assumed that K_{1c} was negligibly small. The stress intensity factor was a superposition of three effects namely the longitudinal stress in the glacier σ_t, the overburden pressure and the wedge effect due to water in the crevasse.
From the general equation,

$$K = 1.12\sigma_t\sqrt{a} + 0.683\rho_w g d^{3/2} - 0.683\rho_i g a^{3/2} \qquad (12)$$

where a is the depth of the crevasse, d the depth of water, and ρ_i, ρ_w the density of ice and water respectively, he simply deduced the depth of dry crevasses to be 36m as well as ascertaining the effects of crevasse spacing and water filling.

The problems of testing ice samples are many. Grips can cause pressure melting which is a particularly difficult problem when testing at low strain rates and high temperatures. Another problem is to obtain suitably shaped ice samples especially from depths. From both these considerations the pressurization of annular discs containing edge slits would appear to be the best compromise. External radial slits have a K calibration strongly dependent on crack length as well as the internal and external radii but internal slits are not so strongly influenced [9] although pressure fluctuations due to seeping into the crack could be a problem.

Fracture toughness of ice results

Fig 9 shows some unusual trends [8], e.g. as temperature decreases K_{1c} increases indicating more energy is absorbed at the crack tip

Fig 9. Fracture Toughness of Ice

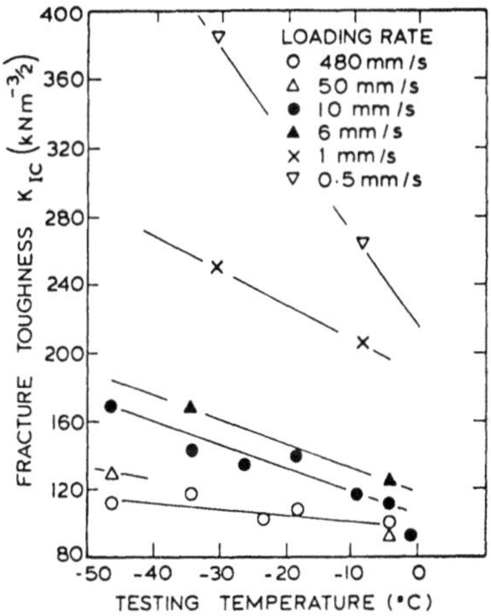

prior to fracture, the reverse happens in metals when a decrease in temperature causes a decrease in plastic zone size since yield stress increases. Toughness is also strongly dependent on crack tip strain rate, characterized by $\dot{K}$, much more than in metals. For ice at these relatively high temperatures creep processes can take place causing stress redistribution at the crack tip and hence a greater force is required to cause fracture. This effect will be enhanced the slower the straining rate.

Another effect is that of recrystallization to produce stress free material at the crack tip. None of these effects however account for the decreasing temperature effect which can only be explained by an increase of cohesive strength. Fletcher (35) has shown that a transition from a crystalline to a quasi-liquid surface exists at some temperature between -2 and -10°C. This thin liquid film will lower the total free energy of the ice surface. Tests carried out in a water + antifreeze mix indicated a 50% reduction in fracture toughness [8].

Conclusions

1. Only in the past few years has fracture mechanics been applied to ice. In the now expected multiplication of studies care should be exercised when selecting a relevant fracture toughness parameter.
2. Studies of ice, particularly those concerned with the effects of temperature and strain rate, will help elucidate the role of stress redistribution, creep, recrystallization and surface energy.

References

[1] Miller, K. J. Traverse of the Staunings Alps, North-East Greenland. Alpine Journal 81 (1976) 143 - 152

[2] Griffith, A. A. The phenomena of rupture and flow in solids. Phil.Trans.Roy.Soc. 1921 A221, 163 - 198

[3] Irwin, G. R. Fracture dynamics. Fracturing of metals A.S.M. Cleveland 1948

[4] Orowan E. Fracture and strength of solids Rep. Phys. Soc. Prog. Phys. 12 (1948/49) 185 - 232

[5] Irwin, G. R. and Kies, J. A. Fracturing and fracture dynamics. Weld. J. Res. Suppl. 17 (1952) 95s - 103s

[6] Irwin, G. R. Analysis of stresses and strains near the end of a crack traversing a plate. J.App.Mech. 24 (1957) 361 - 364

[7] Knott, J. F. Fundamentals of Fracture Mechanics (1973) Butterworths

[8] Liu, H. W. and Miller K. J. Fracture toughness of fresh-water ice. J. Glac. 22 (1979) 135 - 143

[9] Rooke, D. P. and Cartwright, D. J. Compendium of Stress Intensity factors. HMSO 1976

[10] Liu, H. W. Discussion in Fracture toughness testing and its applications. ASTM (1965) S.T.P. 381 p. 23 - 29

[11] Dugdale, D. S. Yielding of steel sheets containing slits J. Mech. Phys. Solids. 8 (1960) 100 - 104

[12] Barenblatt, G. I. The mathematical theory of equilibrium cracks in brittle fracture. Advances in Applied Mechanics 7 (1962) 55 - 111

[13] Wells, A. A. Unstable crack propagation in metals: cleavage and fast fracture. Crack Propagation Symposium, Cranfield 1961

[14] Burdekin, F. M. and Stone, D.E.W. The crack opening displacement approach to fracture mechanics in yielding materials. J. Strain Analysis 1 (1966) 145 - 153

[15] Rice, J. R. A path independent integral and the approximate analysis of strain concentration by notches and cracks. J. Appl. Mechanics. (1968) 379 - 386

[16] Hutchinson, J. W. Singular behaviour at the end of a tensile crack in a hardening material J.Mech.Phys.Solids. 16 (1968) 13 - 31

[17] Rice, J. R. and Rosengren, G. F. Plane strain deformation near a crack tip in a power-law hardening material J. Mech.Phys.Solids 16(1968) 1 - 12

[18] Eshelby, J. D. Solid State Physics 3, Ed Seitz F. & Turnbull D. New York (1956) 79 - 144

[19] Cherepanov, G. P. J.App.Math.Mech. 31 (1967) 503 - 512

[20] Landes, J. D. and Begley J. A. A Fracture Mechanics Approach to creep crack growth. Westinghouse Res.Lab.Rep. (1974) 74-1E7-FESGT-P1

[21] Nikbin, K. M., Webster, G. A., and Turner, C. E. A comparison of methods of correlating creep crack growth. Proc. ICF 4 (1977) 627-634

[22] Ellison, E. G., and Walton, D. Fatigue, creep and cyclic creep crack propagation in a 1% Cr-Mo-V Steel. Instn.mech.engrs. (1973/74). 1,p173.1 - 173.12

[23] Rice, J. R. An examination of the fracture mechanics energy balance from the point of view of continuum mechanics ICF 1 (1966) p309 - 340. 1st Inst.Conf. on Fracture, Sendai, Japan 1965, Japanese Society for Strength & Fracture of Materials 1966

[24] Kfouri, A. P., and Miller, K. J. Crack Separation Energy Rates in Elastic - Plastic Fracture Mechanics. Proc.Instn.mech.Engs. 190 (1976) 571-584. First published as a Cambridge University Report (1974) CUED/C-Mat/TR18

[25] Rice, J. R. and Johnson, M. A. Inelastic behaviour of Solids Ed. M.F. Kanninen et al. McGraw-Hill, New York (1960) 641-672

[26] Miller, K. J. and Kfouri, A. P. A comparison of Elastic-Plastic Fracture Parameters in Biaxial Stress-States ASTM STP 668 (1979) 214-228

[27] Kfouri, A. P. and Miller, K. J. The effect of load biaxiality on the fracture toughness parameters J and G^{Δ}.Proc. ICF 4 (1977) 241-245

[28] Kfouri, A. P. and Rice J. R. Elastic/Plastic Separation Energy Rate for Crack Advance in Finite Growth Steps Proc. ICF4, 1, (1977) 43 - 59

[29] Miller, K. J. Fatigue under complex stress. Metal Science Journal August/September 1977 432 - 438

[30] Howard, I. C. Models of the reduction of fracture toughness due to hydrogen in strong steels. Proc. ICM3 2 (1979) 463 - 474

[31] Goodman, D. J. and Tabor, D. Fracture toughness of ice: a preliminary account of some new experiments. J. Glac. 21 (1978) 651 - 660

[32] Gold, L. W., Crack Formation in Ice Plates by Thermal Shock. Canadian J. Physics 41 (1963) 1712 - 1738

[33] Miller, K. J., Sheffield University 1978 Greenland Expedition Report. To be published.

[34] Smith, R. A., The application of fracture mechanics to the problem of crevasse penetration. J. Glac. 17(1976) 223-228

[35] Fletcher, N. H. The Chemical Physics of Ice Cambridge University Press 1970

Bending and Buckling of a Wedge on an Elastic Foundation

by

D.E. Nevel

US Army Cold Regions Research and Engineering Laboratory

Hanover, NH 03755

Summary

When an ice sheet begins to slide up a sloping structure, the ice cracks radially from the structure creating wedges. Beam theory is used to analyze these wedges under the influence of both horizontal and vertical forces. Buckling and bending of these wedges are considered.

Introduction

In a series of previous papers [1,2,3,4], the author has considered the bending of a wedge on an elastic foundation due to vertical loads. This problem is important in the final failure of a floating ice sheet which initially cracks radially. When a sloping structure or an icebreaker acts against the edge of an ice sheet, the ice will crack into wedges that are loaded by both horizontal and vertical forces. The solution of this problem is presented in this paper.

Formulation

Let us consider an infinite wedge on an elastic foundation as shown in figure 1.

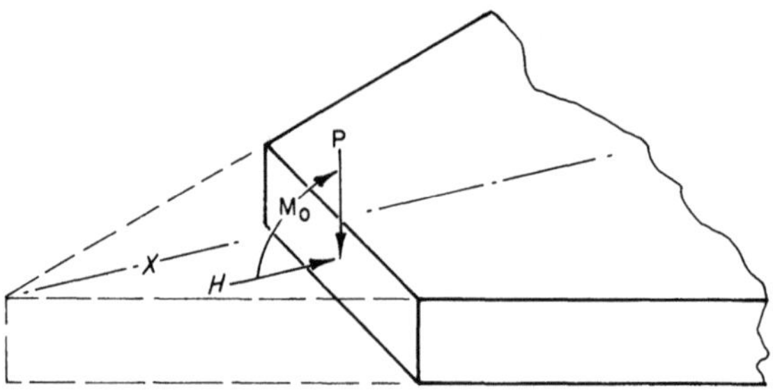

Figure 1. Infinite wedge on an elastic foundation.

By an elastic foundation we mean that the pressure from the foundation is proportional to the deflection of the wedge. We assume that the wedge angle is sufficiently small such that the theory for beam bending is adequate. The differential equation for the bending of a beam with a horizontal compressive force is

$$\frac{d^2}{dx^2}\left[EI\,\frac{d^2w}{dx^2}\right] + H\,\frac{d^2w}{dx^2} + k\,b\,w = q \tag{1}$$

where

w is the deflection of the beam,

x is the coordinate along the beam,

E is Young's modulus,

$I = bh^3/12$ is the moment of inertia,

b is the width of the beam,

h is the thickness of the beam,

H is the horizontal compressive force,

k is the foundation proportionality constant, and

q is the applied vertical load per unit length of the beam.

The bending moment M is determined from the local curvature of the beam

$$M = -\,EI\,\frac{d^2w}{dx^2} \tag{2a}$$

or from the moment equilibrium of the beam

$$M = \int_x^\infty k\,w\,b\,x\,dx - x\int_x^\infty k\,w\,b\,dx + H\,w. \tag{2b}$$

The vertical shear force V is determined from the moment equilibrium of a differential element

$$V = \frac{dM}{dx} - H\,\frac{dw}{dx} \tag{3a}$$

or from the force equilibrium of the wedge

$$V = -\int_x^\infty k\,w\,b\,dx. \tag{3b}$$

The stress σ on the top and bottom surface of the beam are determined respectively by

$$\sigma = -\frac{H}{b\,h} \mp \frac{M\,h}{2\,I}. \tag{4}$$

The width of the wedge is written as $b = b_o\,x$.

Introducing the dimensionless coordinate $\xi = x/\ell$ where $\ell = Eh^3/12k$, the differential equation 1 becomes

$$\xi \frac{d^4 w}{d\xi^4} + 2 \frac{d^3 w}{d\xi^3} + \alpha \frac{d^2 w}{d\xi^2} + \xi w = \frac{q}{b_o k\ell} \tag{5}$$

where the dimensionless horizontal load $\alpha = H/(b_o k\ell^3)$. For the sake of brevity, the differential operator

$$F_\xi \equiv \xi \frac{d^4}{d\xi^4} + 2 \frac{d^3}{d\xi^3} + \alpha \frac{d^2}{d\xi^2} + \xi \tag{6}$$

is introduced so that the differential equation 5 can be written as

$$F_\xi \, w = \frac{q}{b_o k\ell}. \tag{7}$$

Solution

We look for solutions of equation 7 in the form

$$w(\alpha,\xi) = \int_c v \, e^{\xi t} \, dt \tag{8}$$

where v is a function of the complex variable t, and c is a curve in the complex plane. Substituting equation 8 into equation 7 with q = 0, we obtain

$$\int_c v \, F_\xi(e^{\xi t}) \, dt = 0. \tag{9}$$

We define the operator

$$G_t = (1 + t^4) \frac{d}{dt} + (\alpha t^2 + 2t^3) \tag{10}$$

such that

$$F_\xi(e^{\xi t}) = G_t(e^{\xi t}) \tag{11}$$

and then equation 9 becomes

$$\int_c v \, G_t(e^{\xi t}) \, dt = 0. \tag{12}$$

If we try to make a perfect derivative of the integrand in equation 12, we obtain

$$\int_c \left\{ \frac{d}{dt} [v(1+t^4)e^{\xi t}] + e^{\xi t} \, \overline{G}_t(v) \right\} dt = 0 \tag{13}$$

where

$$\overline{G}_t = - (1+t^4) \frac{d}{dt} + (\alpha t^2 - 2t^3) \tag{14}$$

is the adjoint operator of G_t. Setting $\overline{G}_t(v) = 0$ in equation 13 gives

$$\frac{1}{v}\frac{dv}{dt} = \frac{\alpha t^3 - 2t^3}{1 + t^4} \tag{15}$$

which by partial fractions integrates to

$$v = (t - t_1)^{\beta_1} (t - t_2)^{\beta_2} (t - t_3)^{\beta_3} (t - t_4)^{\beta_4} \tag{16}$$

where

$$t_1 = \frac{1+i}{\sqrt{2}},\ t_2 = \frac{-1+i}{\sqrt{2}},\ t_3 = \frac{-1-i}{\sqrt{2}},\ t_4 = \frac{1-i}{\sqrt{2}}$$

and

$$\beta_1 = -\frac{1}{2} + \frac{\alpha}{4}\ t_4,\ \beta_2 = -\frac{1}{2} + \frac{\alpha}{4}\ t_3,\ \beta_3 = -\frac{1}{2} + \frac{\alpha}{4}\ t_2,$$
$$\beta_4 = -\frac{1}{2} + \frac{\alpha}{4}\ t_1.$$

This shows that v has four (4) branch points at t_1, t_2, t_3, and t_4 because a function of the form $(t-t_o)^\beta$ increases by a factor of $e^{i2\pi\beta}$ when going clockwise around t_o once.

The remaining part of equation 13 is a perfect derivative and integrates to

$$v\ (1+t^4)e^{\xi t}\ \Big|_{t_A}^{t_B} = 0 \tag{17}$$

where t_A and t_B are the beginning and end of the path of integration respectively. For $\xi > 0$, equation 17 can be satisfied by choosing a path that encloses at least one branch point and whose real part of t_A and t_B goes to minus infinity. Two such convenient paths are shown in figure 2.

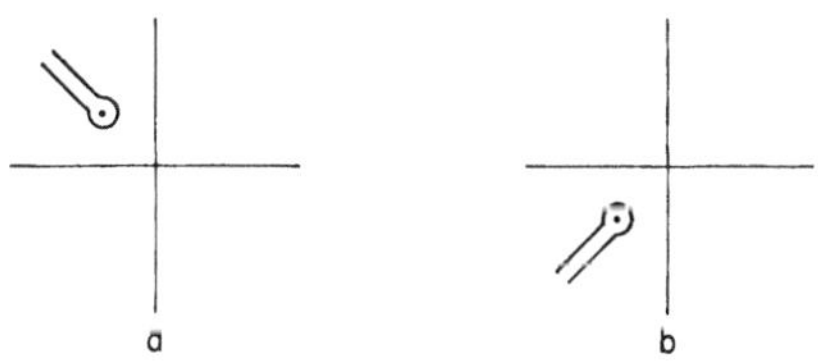

Figure 2. Integration paths.

For any ξ, equation 17 can be satisfied by choosing a closed path ($t_A = t_B$) around more than one branch point provided that the function $v(1+t^4)e^{\xi t}$ returns to its initial value. There are two obvious choices and these are shown in figure 3.

Figure 3. Integration paths.

In addition to the solution as given in equation 8, the derivation of the solution with respect to ξ will be needed. Thus we need

$$w^n = \int_C ve^{\xi t} t^n dt \qquad (18)$$

where n indicates the nth derivative of w with respect to ξ.

The integration of $ve^{\xi t}t^n$ along the path shown in figure 2a is designated by $w_4^n + iw_3^n$ where w_4^n and w_3^n are the real and imaginary parts of the integral respectively. This integration is divided into three parts, which must be continuous between the adjacent parts. The first part is along the radial line $t = r\ e^{-i5\pi/4}$ from $r = \infty$ to $r = 1 + \Delta$ where Δ is a small positive number. The second part is along the circle $t - t_2 = \Delta\ e^{i\theta}$ from $\theta = -5\pi/4$ to $\theta = 3\pi/4$. The third part is along the radial line $t = r\ e^{i3\pi/4}$ from $r = 1 + \Delta$ to $r = \infty$.

The integration along the circle is performed by expanding the analytic part of the integrand into a Taylor's series about t_2. Factoring out the complex constant $(1-e^{2\pi i\beta_2})$, the solution becomes

$$w_4^n + iw_3^n = \int_{1+\Delta}^{\infty} \mathrm{EXP}\left[\frac{-\xi r+\alpha R}{\sqrt{2}} + i\left[\frac{\xi r+\alpha R}{\sqrt{2}} + \frac{(1+3n)\pi}{4}\right]\right]\frac{r^n}{\sqrt{r^4-1}}\,dr$$

$$- \sum_{m=0}^{\infty} \frac{1}{m!} \left[\frac{d^m}{dt^m} \frac{v\, e^{\xi t} t^n}{(t-t_2)^{\beta_2}} \right]_{t=t_2} \frac{[\Delta e^{-i5\pi/4}]^{\beta_2+m+1}}{\beta_2 + m + 1} \tag{19}$$

where $R = (1/4) \log \frac{r+1}{r-1} + (1/2) \arctan \frac{1}{r}$ with the $\arctan \frac{1}{r}$ defined in the first quadrant. When $\alpha < 2\sqrt{2}$, the limit as Δ approaches zero may be taken, but when $\alpha \geqslant 2\sqrt{2}$, Δ must remain finite. The integral part of equation 19 can be numerically integrated from 1 + Δ to r_o where r_o is a large number. The remaining part of the integral form r_o to ∞ can be intergrated if one (1) is neglected when compared to r^4. The real part w_4^n converges except that the first and higher order derivatives diverge when $\xi = 0$. The imaginary part w_3^n converges for all values of ξ.

The solution obtained by integrating along the path as shown in figure 2b is proportional to $w_4^n - iw_3^n$ which is the complex conjugate of the previous solution. Hence the real and imaginary parts are solutions. When $\alpha = 0$, these solutions reduce to

$$[w_4 + iw_3]_{\alpha=0} = y_4 - iy_3/\sqrt{2} \tag{20}$$

where y_4 and y_3 are the solutions previously presented [3] when $\alpha = 0$.

The integral along the path in figure 3a is divided into parts similar to that previously done. Factoring out of the integral the complex constant $(1 - e^{2\pi i\beta_1})e^{i\pi/4}$, the result can be expressed as

$$w_1^n + iw_2^n = \int_0^{1-\Delta} \text{EXP}\left[\frac{\xi r - \alpha R}{\sqrt{2}} + i\,\frac{\xi r + \alpha R}{\sqrt{2}}\right] \frac{(re^{i\pi/4})^n}{\sqrt{1 - r^4}}\, dr$$

$$+ \int_0^{1-\Delta} \text{EXP}\left[-\frac{\xi r - \alpha R}{\sqrt{2}} - i\,\frac{\xi r + \alpha R}{\sqrt{2}}\right] \frac{(re^{-i3\pi/4})^n}{\sqrt{1 - r^4}}\, dr$$

$$+ \sum_{m=0}^{\infty} \frac{1}{m!} \left[\frac{d^m}{dt^m} \frac{v\, e^{\xi t} t^n}{(t-t_1)^{\beta_1}} \right]_{t=t_1} \frac{\Delta(\Delta e^{-i3\pi/4})^{m+\beta_1}}{m + \beta_1 + 1}$$

$$+ \sum_{m=0}^{\infty} \frac{1}{m!} \left[\frac{d^m}{dt^m} \frac{v\, e^{t\xi} t^n}{(t-t_3)^{\beta_3}} \right]_{t=t_3} \frac{\Delta(\Delta e^{i\pi/4})^{m+\beta_3}}{m + \beta_3 + 1} \qquad (21)$$

where R is defined as before. When $-2\sqrt{2} < \alpha < 2\sqrt{2}$, the limit of equation 21 as Δ approaches zero may be taken, but otherwise Δ must remain finite. This solution and its derivatives converge for all values of ξ except when ξ approaches infinity.

The solution obtained by integrating along the path shown in figure 3b is proportional to $w_1^n - i\, w_2^n$. Hence the real and imaginary parts are solutions. For $\alpha = 0$, these solutions reduce to

$$[w_1 + i\, w_2]_{\alpha=0} = y_1 + i\, y_2 \qquad (22)$$

where y_1 and y_2 are solutions previously presented [3] when $\alpha=0$.

Buckling

The buckling load of the wedge is the minimum α which causes instability of the wedge. This buckling load depends upon the boundary conditions of the wedge. We will assume that one end of the wedge extends to infinity while the other end is truncated at $x = x_o$. At this truncated end, we can apply the usually boundary conditions of beam theory. That is a simple support with zero deflection and bending moment, a rigid support with zero deflection and slope, or a free support with a specified vertical force and bending moment. These buckling loads were numerically determined as a function of the dimensionless truncation distance $\xi_o = x_o/\ell$ and are shown in table I. For the rigid support there appears to be a singularity at $\xi_o = 0$. Additional values were computed near $\xi_o = 0$ for this case.

Table I. Buckling Loads

ξ_o	FREE	SIMPLE	RIGID
0	0.7785	2.4713	-
.0001	-	-	2.9
.001	-	-	3.17
.01	-	-	3.57
.1	0.9115	2.9266	4.6240
.2	1.0356	3.2892	5.2844
.3	1.1557	3.6175	5.8233
.4	1.2732	3.9260	6.3000
.5	1.3888	4.2214	6.7404
.6	1.5027	4.5072	7.1548
.7	1.6155	4.7856	7.5500
.8	1.7271	5.0582	7.9302
.9	1.8378	5.3261	8.2984
1.0	1.9477	5.5899	8.6565

D.S. Sohdi[5] has recently considered buckling of a wedge using plate theory rather than beam theory. Polar coordinates were used, and the wedge truncated along the arc of finite radius. The horizontal load was distributed in the radial direction proportional to the cosine of the angle from the centerline of the wedge. The finite element method was used and he found the results agreed with those of this paper when the wedge angle was small.

Bending

As an ice sheet begins to slide up or down on a sloping structure, radial cracks form, creating a number of wedges. Let us assume that these wedges are of infinite length, and that the wedges are truncated at a distance x_o. We assume that the structure exerts on the wedge a vertical force P, a bending moment M_o, and a horizontal compressive force H as shown in figure 1. For this problem the solution is

$$w = D_3\, w_3(\alpha,\xi) + D_4\, w_4(\alpha,\xi). \tag{23}$$

The unknown constants D_3 and D_4 are determined from the equilibrium of the entire wedge using equations 2b and 3b. Equations

2a and 3a can be also used except when $\xi_o = 0$. This exception occurs because the first and higher order derivatives of w_4 diverge when $\xi = 0$. The summation of the vertical forces and the summation of the moments about $\xi = 0$ yield respectively

$$\int_{\xi_o}^{\infty} \xi \, w \, d\xi = \frac{P}{b_o k \ell^2} \tag{24a}$$

$$\int_{\xi_o}^{\infty} \xi^2 \, w \, d\xi + \alpha \, w(\xi_o) = \frac{P \, \xi_o}{b_o k \ell^2} + \frac{M_o}{b_o k \ell^3} \, . \tag{24b}$$

Substituting equation 23 for w and solving for D_3 and D_4, the deflection becomes

$$w = \frac{P}{b_o k \ell^2} \, F(\alpha,\xi_o,\xi) + \frac{M_o}{b_o k \ell^3} \, G(\alpha,\xi_o,\xi) \tag{25}$$

where

$$F(\alpha,\xi_o,\xi) = \left[\int_{\xi_o}^{\infty} \xi^2 w_3 d\xi + \alpha w_3(\xi_o) - \xi_o \int_{\xi_o}^{\infty} \xi w_3 d\xi\right] w_4/\mathrm{DEN}$$

$$- \left[\int_{\xi_o}^{\infty} \xi^2 w_4 d\xi + \alpha w_4(\xi_o) - \xi_o \int_{\xi_o}^{\infty} \xi w_4 d\xi\right] w_3/\mathrm{DEN}$$

$$G(\alpha,\xi_o,\xi) = \left[w_3 \int_{\xi_o}^{\infty} \xi w_4 d\xi - w_4 \int_{\xi_o}^{\infty} \xi w_3 d\xi\right]/\mathrm{DEN}$$

$$\mathrm{DEN} = \int_{\xi_o}^{\infty} \xi w_4 d\xi \left[\int_{\xi_o}^{\infty} \xi^2 w_3 d\xi + \alpha w_3(\xi_o)\right]$$

$$- \int_{\xi_o}^{\infty} \xi w_3 d\xi \left[\int_{\xi_o}^{\infty} \xi^2 w_4 d\xi + \alpha w_4(\xi_o)\right].$$

From this solution the bending moment is determined from equation 2a and the stress along the top and bottom surface of the wedge is determined from equation 4. The stress σ is made dimensionless with the usual factor P/h^2 and is

$$\frac{\sigma h^2 b_o}{6P} = - \frac{Hh}{6P\ell\xi} \pm \left[F''(\alpha,\xi_o,\xi) + \frac{M_o}{P\ell} \, G''(\alpha,\xi_o,\xi)\right] \tag{26}$$

where the primes indicate differentiation with respect to ξ. The term $- Hh/(6P\ell\xi)$ is the horizontal compressive stress which is added to the bending terms. The horizontal force also influences the bending part of the solution through the dimensionless factor α. The applied moment M_o can be easily attained if the horizontal force does not act at the neutral axis of bending. For example if H acts at the top surface, then $M_o = Hh/2$.

The stress as given in equation 26 is a function of the five parameters $Hh/P\ell$, $M_o/P\ell$, $\alpha = H/(b_o k\ell^3)$, $\xi_o = x_o/\ell$, and $\xi = x/\ell$. Although it is easy to evaluate the stress for any combination of these parameters, it is a large task to systematically vary each parameter and present the results in this paper. Therefore only a typical result for the bending analysis will be given. We shall choose a wedge that is truncated at the distance $\xi_o =$ 0.2. The horizontal force H will be applied at the neutral axis $M_o = 0$ or at the top edge $M_o = Hh/2$ of the ice sheet. Then the factor $Hh/P\ell$ occurs in the $M_o/P\ell$ factor as well as in the compressive force term. If the coefficient of friction between the structure and the ice is given as $\tan \phi$ and the slope of the structure as $\tan \theta$, then $H/P = \tan (\theta + \phi)$. The factor h/ℓ varies from 0.03 to 0.1. Choosing $H/P = 2$ and $h/\ell = 0.1$, then the factor $Hh/P\ell = 0.2$. For these two cases the stress is now a function of α and the position ξ. Choosing the position that produces the maximum tensile stress along the top of the wedge, the results have been plotted in figure 4. The reciprocal of the stress factor, $6P/(\sigma h^2 b_o)$, is shown as a function of α divided by the buckling load, $\alpha_B = 1.0352$. A lower limiting case of $Hh/P\ell =$ 0 is also plotted in figure 4.

Conclusion

The solution of a wedge on an elastic foundation has been presented. The buckling loads for an infinite wedge has been given for the free, simple, and rigid boundary condition at the truncated end. The bending of a wedge under both vertical and horizontal forces has been analyzed and the maximum stresses shown for a typical example.

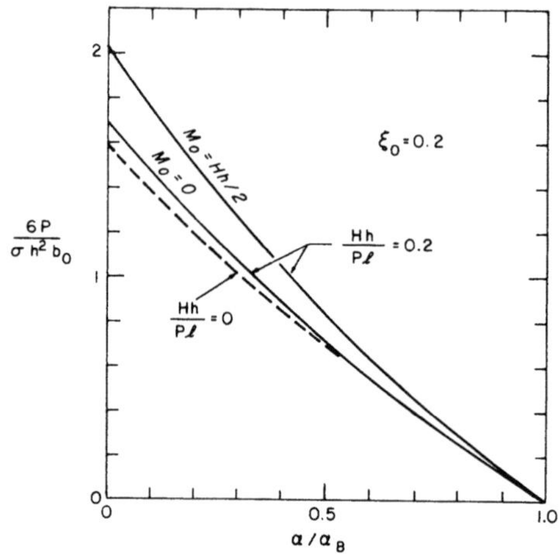

Figure 4. Typical example of maximum tensile stress in the wedge.

References

1. Nevel, D.E.: The narrow infinite wedge on an elastic foundation. SIPRE Technical Report 56, July 1958.

2. Nevel, D.E.: The narrow free infinite wedge on an elastic foundation. CRREL Research Report 79, July 1961.

3. Nevel, D.E.: The general solution of a wedge on an elastic foundation. CRREL Research Report 227, Nov. 1968.

4. Nevel, D.E.: The ultimate failure of a floating ice sheet. International Association of Hydraulic Research, Ice Symposium, Leningrad, 1972.

5. Sohdi, D.S.: Buckling analysis of wedge-shaped floating ice sheets. Port and Ocean Engineering under Arctic Conditions Conference, Trondheim, Aug. 1979.

Plastic Limit Analysis of Sheet Ice Loads on Conical Structures

T. D. RALSTON
Exxon Production Research Company
Houston, Texas

ABSTRACT

The techniques of plastic limit analysis are used to develop a description of the forces imposed on conical-shaped structures by moving ice sheets. A closed-form solution is presented in which the relative importance of the essential features of the failure process can be clearly viewed. The behavior of broken ice pieces as they move around the surface of the cone is an important design consideration for upward-breaking conical structures. The form of the yield criterion used to describe ice bending strength is not a critical parameter. The largely one-dimensional nature of the ice bending activity near a conical structure does not require an accurate two-dimensional description of ice bending strength. Downward-breaking conical structures offer some advantages in the form of reduced horizontal loads compared to upward-breaking cones. This reduction is largely a consequence of the reduced forces needed to submerge broken ice pieces rather than push them up over the surface of an upward-breaking cone. Additional factors, such as the contribution of the vertical force to the overturning moment and the potential jamming of ice ridges against a downward-breaking cone in shallow water, should also be considered when downward-breaking geometries are of interest.

INTRODUCTION

Plastic limit analysis is a well-known approach for addressing questions of ultimate strength and plastic collapse in a number of engineering disciplines. Many applications can be found in the theory of plates and shells, concrete structures, frame analysis, geotechnical engineering, and rock mechanics. The simplicity of this approach permits the construction of approximate solutions to engineering problems without allowing minor details to obscure the essential features of the problem. The general procedure usually consists of constructing rigorous upper and lower bounds for the plastic failure load. Upper bounds are constructed with the use of assumed velocity fields that describe the motion that occurs in the failure process. Observations of actual material failures in model tests and full-scale events are very useful for selecting the type of velocity field that might be most appropriate for a particular application. Lower bounds may be constructed with stress fields that are intended to approximate the actual stress distribution. Neither the stress field nor the velocity field need be exact. They must only satisfy certain restrictions imposed by the yield function and the displacement and stress boundary conditions for the particular application.

A plasticity procedure for estimating ice loads on conical structures has been previously discussed by Ralston [1], and found to be in good agreement with published model test data. This procedure used an elastic-perfectly plastic material model for both the ice sheet and the foundation. The present work generalizes this procedure and evaluates the need to develop more precise descriptions of ice bending strength. Detailed example calculations are presented to illustrate the relative importance of the essential features of ice action that are described by this approach for both upward- and downward-breaking conical structures.

PLASTIC ANALYSIS OF FLOATING ICE

This analysis idealizes a floating ice sheet as an elastic-perfectly plastic plate resting on an elastic-perfectly plastic foundation. Although the application to ice mechanics is recent, plastic analysis of plate bending is not a new subject. Modern discussions of this subject, e.g. Hodge [2], are usually formulated from the principles of plastic limit analysis; however, limit analysis of plates actually originated prior to the development of modern plasticity theory. Much of the early work is described in the literature under the heading of "yield-line theory". The basic concept was first developed independent of the limit analysis theorems in 1932 by Professor K. W. Johansen [3] of the Danish Institute of Technology in Copenhagen.

The bending strength of a plate is described by a bending moment yield criterion. This is a two-dimensional relationship between the principal bending moments. The present work is a pure bending analysis and thus the effect of in-plane forces on the moment capacity is neglected. One could extend this procedure to include in-plane forces if necessary; however, there would be a significant increase in both the computational complexity and the ice property data that would be required.

Moment yield criteria for an ice sheet can be determined either from bending tests on full thickness samples or from smaller scale tests (tension, compression, and small scale bending) used with appropriate scale and constitutive relationships. The bending strength of ice in model basins is usually determined from full-thickness bending tests. Similar work, Vaudrey [4], has also been conducted on full-thickness Arctic sea ice. These tests provide a good measure of the one-dimensional bending moment capacity. The bending moment capacity is not the parameter that is usually reported from such tests, however. Instead, investigators usually assume that the stress state through the ice thickness is elastic, that the variation of the elastic modulus through the thickness is negligible, and that the compression modulus is the same as the tension modulus. The "flexural strength", σ_f, is then reported as the value of the elastic tensile stress that occurs at the outer fibers of the ice sheet when the bending moment causes failure. The reported values for σ_f are usually followed by a disclaimer statement that points out the shortcomings of the assumptions that are used to transform the bending moment capacity into the flexural strength. The relevant input parameter for the present approach is the bending moment capacity, rather than the flexural strength. Thus the questions associated with the "flexural strength assumptions" are of little concern in this analysis. As a matter of convenience, the solutions developed in the present work can be expressed in terms of the flexural strength, σ_f, rather than the bending moment capacity using the expression

$$M_o = \sigma_f t^2/6,$$

where M_o is the bending moment capacity (per unit width) and t is the ice sheet thickness. This is done only to present the results in terms of the property, σ_f, that is usually reported in the literature.

The shape of the moment yield criterion in principal moment space is of interest when biaxial bending questions are addressed. If the ice bending strength is isotropic in the ice sheet plane, four points on the yield diagram can be determined by upward (M_o^u) and downward (M_o^d) bending test data from simple tests. There are numerous ways in which these points may be connected to complete the yield criterion and maintain the symmetry imposed by the isotropy assumption. If a detailed description of ice bending constitutive behavior were developed, the shape of the criterion could be determined analytically. It is not likely, however, that such a description will be developed in the near future. An alternative approach is to use completions that have proved to be useful for other materials and evaluate their appropriateness with model tests and analysis. Three potential completions, which differ in the second and fourth quadrants, are shown in Figure 1. Only the shape of the criterion in these quadrants is of interest in the present analysis.

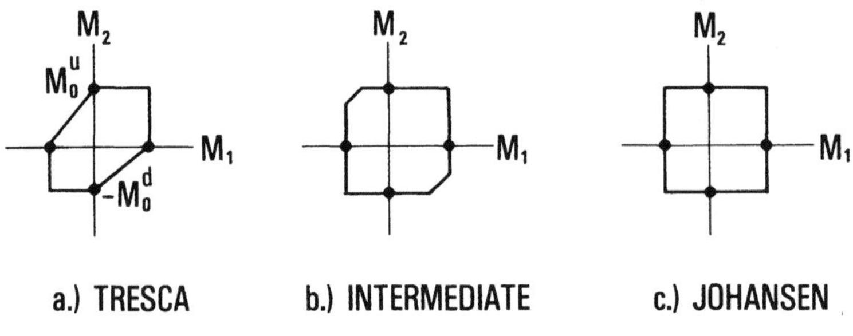

Figure 1. Moment Yield Criteria.

The Tresca completion is motivated by the Tresca yield criterion (Hodge [2]); however, other material assumptions could also lead to the same or very similar bending criteria. The stress states of interest for the description of ice bending strength (with the effect of in-plane forces neglected) are very simple. The aspects of ice behavior, e.g., anisotropy and pressure dependence, that deviate from the Tresca model are not of great importance for these states, but some difference between the actual bending behavior and the Tresca model would be expected. An important result discussed in the next section is that even a large departure from the Tresca model gives only small changes in calculated ice forces on a conical structure. The Johansen criterion in Fig. 1c has been used by Mansfield [5] to describe the strength of reinforced concrete plates. This criterion requires that the value of at least one of the principal moments must equal M_o^u or M_o^d before yield occurs. Fig. 1b is an example of an intermediate completion of the moment yield diagram.

Complex biaxial bending tests would be required to determine which of the criteria in Figure 1 gives the best representation of ice bending strength. A Tresca material is weaker than a Johansen material since its elastic range is a subset of the Johansen elastic range. The difference in ice loads on a conical structure that would correspond to these yield criteria can be determined analytically. This question is addressed in the following section in which the procedure developed by Ralston [1] for the Tresca model is extended to the Johansen criterion and example calculations are presented. Equal upward and downward bending strengths are assumed in this analysis; i.e., $M_o^u = M_o^d = M_o$. An analogous procedure can be applied if the upward and downward strengths are not equal.

The weight and buoyancy effects on a floating ice sheet act as a one-dimensional non-linear foundation. These effects are idealized by the elastic-perfectly plastic foundation response shown in

Fig. 2. The foundation reaction that corresponds to ice lifted out of the water is important for upward-breaking structures, whereas the "submerged" reaction is applicable to downward-breaking geometries. Strictly speaking, the actual foundation response is a non-linear elastic response since it is path independent. However, the present analysis is not intended to describe unloading conditions and the distinction between a one-dimensional non-linear elastic and plastic foundation is not significant.

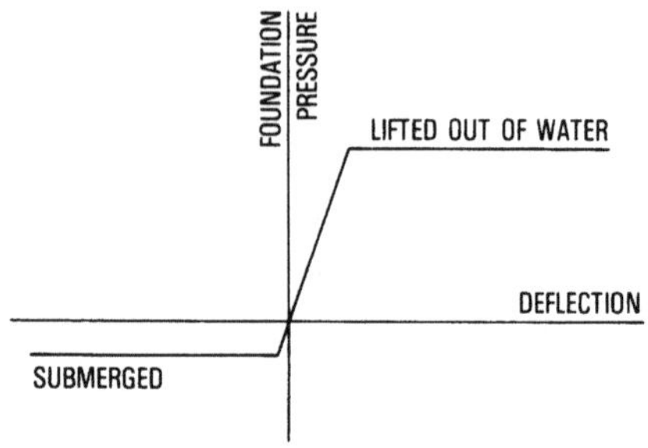

Figure 2. Elastic-Perfectly Plastic Foundation Response.

UPWARD-BREAKING CONE

The failure of an ice sheet that is advancing against a conical structure is depicted in Fig. 3. The motion of the ice is described by an assumed velocity field, Ralston [1]. The ice sheet is assumed to be translating with velocity $\underset{\sim}{V} = -V\underset{\sim}{e}_x$ defined for $r \geq R$, where the basis vector refers to the coordinate system shown in Fig. 3 and $R = D/2$ is the waterline radius of the cone. A deforming region given by

$$R \leq r \leq A, \quad -\pi/2 \leq \theta \leq \pi/2$$

exists in front of the cone in which a vertical velocity

$$\dot{\underset{\sim}{w}} = V \tan\alpha \, \frac{A-r}{A-R} \cos\theta \, \underset{\sim}{e}_z' \; ,$$

is superimposed on the horizontal translation. The value of A is later determined to optimize the upper bound. The broken ice riding on the surface of the cone is assumed to move in the tangent plane of the cone with velocity

$$\underset{\sim}{v} = -V\underset{\sim}{e}_x + V \tan \alpha \cos \theta \, \underset{\sim}{e}_z \ .$$

The possibility of interference to ice motion by the presence of the vertical section at the top of the cone is not addressed in this analysis.

The upper bound method of plastic limit analysis provides an estimate of the forces imposed on the cone. The rate of work done by external forces ($R_H V$) is equated to the rate of energy dissipation that occurs in the ice as a result of the assumed motion. The total rate of dissipated energy consists of contributions from several sources indicated in Fig. 3. This includes the formation of plastic hinges to represent the circumferential and side cracks, the work done against the foundation reaction, the continuous

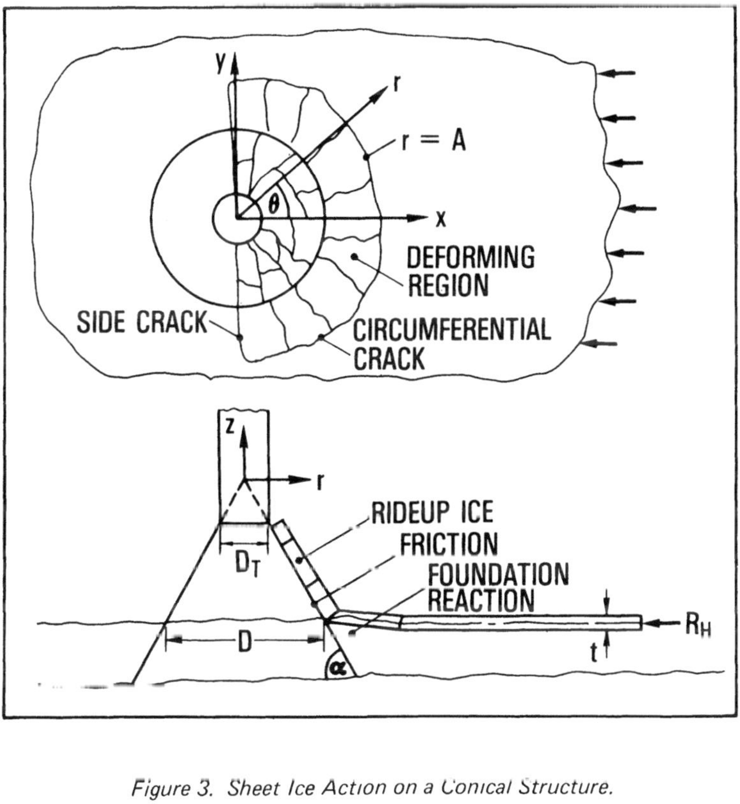

Figure 3. Sheet Ice Action on a Conical Structure.

deformation of ice within the deforming region, and frictional dissipation at the cone/ice interface.

I. Circumferential and side cracks.

The rate of dissipation of energy at the plastic hinge idealizations of the cracks is the integral of the product of the bending moment capacity and the jumps in the velocity gradient. This is expressed by

$$2\int_0^{\pi/2} M_o \left|\frac{\partial \dot{w}}{\partial r}\right|_{r=A} A d\theta + 2\int_R^A M_o \left|\frac{1}{r}\,\frac{\partial \dot{w}}{\partial \theta}\right|_{\theta=\pi/2} dr \quad .$$

The result is

$$2M_o\, V \tan\alpha \,\frac{\rho}{\rho-1} + 2M_o V \tan\alpha \left[\frac{1-\rho + \rho\ln\rho}{\rho-1}\right]$$

where the first term corresponds to the circumferential crack and the second represents the side cracks and ρ is defined by $\rho = A/R$.

II. Foundation reaction.

For the upward-breaking cone, the foundation energy dissipation rate is given by

$$2\int_0^{\pi/2}\int_R^A 0.9\,\rho_w g t\,\dot{w}\, r dr d\theta = \frac{0.9\,\rho_w g t\, V\tan\alpha\, R^2}{3}\,(\rho^2 + \rho - 2) \quad ,$$

where we have taken $0.9\,\rho_w g$ to be the weight density of the ice sheet.

III. Deforming region.

The principal curvature rates can be computed from the velocity field in the region $R \le r \le A$, and are given by

$$\dot{\kappa}_{1,2} = \frac{-V \tan \alpha \, A}{r^2 \, (A-R)} \left[\frac{\cos \theta}{2} \mp \sqrt{\frac{\cos^2 \theta}{4} + \sin^2 \theta} \right]$$

The corresponding energy dissipation rate is given by

$$2 \int_0^{\pi/2} \int_R^A [M_1 \dot{\kappa}_1 + M_2 \dot{\kappa}_2] \, r dr d\theta \quad ,$$

where (M_1, M_2) is the point on the principal moment yield criterion where the principal curvatures satisfy normality.

For the Tresca model, this reduces to

$$3.422 \, M_o \, V \tan \alpha \, \rho \ln \rho / (\rho - 1) \quad .$$

For the Johansen model, the equivalent expression is

$$(3.422 \, M_o + 1.422 \, M_o) \, V \tan \alpha \, \rho \ln \rho / (\rho - 1)$$

This is the only contribution to the total energy dissipation in which a difference between the Tresca and Johansen material assumptions appears.

If the upward and downward moment capacities are not equal, the first term in the Johansen expression would contain the downward strength, while the upward strength would appear in the second term. The moment capacity in the expression for circumferential and side cracks would also be the upward strength. The corresponding generalization for a Tresca material is somewhat more complex since the point (M_1, M_2) where (κ_1, κ_2) satisfies normality will depend on both θ and the ratio M_o^d / M_o^u.

IV. Rideup ice.

The rideup ice thickness is taken to be t_R, which may be different from the ice sheet thickness as a result of rafting and/or ice rubble accumulations. For a given structure geometry, the appropriate value of t_R may be determined by model tests. The rideup energy dissipation rate is given by

$$2 \int_{0}^{\pi/2} \int_{D_T/2}^{D/2} 0.9\, \rho_w g\, t_R V \tan\alpha \cos\theta \frac{r dr d\theta}{\cos\alpha} ,$$

which reduces to

$$\frac{0.9\, \rho_w g}{4}\, V\, t_R\, \frac{\tan\alpha}{\cos\alpha} (D^2 - D_T^{\,2})$$

where D_T is the top diameter of the cone.

V. Frictional dissipation.

This description of frictional forces acting at the ice/cone interface is approximate. The upper bound theorem of plastic analysis does not strictly apply to all situations where energy is dissipated by frictional sliding. However, as discussed by Chen [6], if we simply include frictional dissipation in the total rate of dissipation of energy, our upper bound calculation will provide useful information, if not the full answer. The rate of dissipation of energy resulting from friction at a sliding interface is the product of the relative tangential velocity, the coefficient of friction μ, and the normal force acting on the interface. The normal force for the ice pieces sliding (ride-up) on the surface of the cone is determined by the weight of the pieces. The normal force between the advancing ice sheet and the cone, however, is not known a priori. We therefore assume a cosine distribution of the normal force between the ice sheet and the cone around the circumference and relate the resultant force to the total horizontal force on the cone.

The total frictional dissipation rate can be shown to be equal to

$$R_H V\mu \, g(\alpha,\mu) - V\mu \frac{0.9 \, \rho_w g \, t_R}{4} [D^2 - D_T^2] \, f(\alpha,\mu) \, g(\alpha,\mu)$$

$$+ \frac{0.9 \, \rho_w g \, \mu V}{4 \cos \alpha} t_R \, [D^2 - D_T^2] \, E(\sin \alpha) \; ,$$

where $E(\cdot)$ is the complete elliptic integral of the second kind, i.e.,

$$E(\sin \alpha) = \int_0^{\pi/2} \sqrt{1 - \sin^2 \alpha \sin^2 \theta} \quad d\theta \; ,$$

and

$$g(\alpha,\mu) = \left(\frac{1}{2} + \frac{\alpha}{\sin 2\alpha}\right) \Big/ \left(\frac{\pi}{4} \sin \alpha + \frac{\mu\alpha \cos \alpha}{\sin \alpha}\right) \; ,$$

$$f(\alpha,\mu) = \sin \alpha + \mu \cos \alpha \, F(\sin \alpha) \; ,$$

$F(\cdot)$ = complete elliptical integral of the first kind.

The first term in the expression for the frictional energy dissipation represents the contribution from the advancing ice sheet in the absence of the rideup ice. The second and third terms arise from the presence of the rideup ice on the cone surface.

Horizontal and Vertical Forces

If the sum of the energy dissipation rates is set equal to the rate of work done by external forces, one can obtain the following expression for the horizontal force

$$R_H = \frac{\tan\alpha}{1-\mu g(\alpha,\mu)} \left\{ 2\,M_o \frac{\rho}{\rho-1} + 2\,M_o \left[\frac{1-\rho+\rho\ln\rho}{\rho-1}\right] \right.$$

$$+ (3.422\,M_o + 1.422\,M_o)\,\frac{\rho\ln\rho}{\rho-1} + 0.9\rho_w g t D^2(\rho^2+\rho-2)/12$$

$$\left. + \frac{0.9\rho_w g\, t_R\,(D^2 - D_T^2)}{4} \left[\frac{1}{\cos\alpha} + \frac{\mu E(\sin\alpha)}{\sin\alpha} - \mu\,\frac{f(\alpha,\mu)\,g(\alpha,\mu)}{\tan\alpha}\right] \right\}$$

This expression corresponds to the Johansen yield criterion. The analogous expression for the Tresca model lacks only the 1.422 M_o term in the second row.

After computing the horizontal force, the vertical force can be obtained from a static force balance on the cone. The result is

$$R_V = R_H \left[\frac{h(\alpha,\mu)}{\frac{\pi}{4}\sin\alpha + \frac{\mu\alpha}{\tan\alpha}}\right]$$

$$+ \frac{0.9\,\rho_w g\, t_R(D^2 - D_T^2)}{4} \left[\frac{\pi}{2}\cos\alpha - \mu\alpha - \frac{f(\alpha,\mu)\,h(\alpha,\mu)}{\frac{\pi}{4}\sin\alpha + \frac{\mu\alpha}{\tan\alpha}}\right] ,$$

where

$$h(\alpha,\mu) = \cos\alpha - \frac{\mu}{\sin\alpha}\left[E(\sin\alpha) - \cos^2\alpha\, F(\sin\alpha)\right] .$$

The expression for R_H contains the parameter $\rho = A/R$ which defines the size of the deforming ice region in front of the cone for this velocity field. This parameter can be selected to optimize (minimize) the bound for the ice failure force. The value of ρ affects the relative contributions to the cone force that arise from the weight and strength of the advancing ice sheet but it does not affect the contribution from the ice that is riding on the surface of the cone. The optimum value of ρ is given by the solution of

$$\rho - \ln\rho + 0.0830 \frac{\rho_w g\ t\ D^2}{\sigma_f t^2} (2\rho + 1)\ (\rho-1)^2 = 1.369$$

for the Tresca model and

$$\rho - \ln\rho + 0.0658 \frac{\rho_w g\ t\ D^2}{\sigma_f t^2} (2\rho + 1)\ (\rho-1)^2 = 1.292$$

for the Johansen model, where the ice flexural strength has been substituted for bending moment capacity in both equations. The selection of ρ depends only on the non-dimensional weight-strength ratio given by

$$\frac{\rho_w g\ t\ D^2}{\sigma_f t^2} \quad .$$

The numerator in this ratio is proportional to the weight of the floating ice sheet that affects the cone. The denominator is proportional to the bending moment capacity, which is the ice strength applicable to the cone. The value of this non-dimensional parameter is preserved in model tests when the usual scaling laws (Michel [7]) are observed.

The results presented by Ralston [1] are a simplified representation of the Tresca solution, written in the form

$$R_H = [A_1 \sigma_f t^2 + A_2\ \rho_w g\ t\ D^2 + A_3\ \rho_w g\ t_R (D^2 - D_T^2)]\ A_4 \quad ,$$

$$R_V = B_1\ R_H + B_2\ \rho_w g\ t_R (D^2 - D_T^2) \quad ,$$

in which the rideup ice thickness was taken equal to the ice sheet thickness and equal upward and downward bending strengths were also assumed. Graphical representations of the coefficients were provided to reduce the computational effort required to use these results.

Example Calculation

The following example calculation illustrates the relative importance of the different contributions to ice forces on a cone. The example cone geometry is used for illustrative purposes only; it does not necessarily correspond to a geometry being considered for any actual structure.

The diameter of the example cone is taken to be 18m at the waterline and 6m at the top. The ice sheet thickness is taken to be 1m; the flexural strength is assumed to be 700 kPa; and the ice/cone friction coefficient is fixed at 0.15. Cone angles of 20 to 70° from the horizontal are included for comparison. The rideup ice thickness is taken equal to the sheet thickness in this example. For a specific cone application the appropriate choice of t_R could be determined experimentally.

With these dimensions and properties, the value of the weight-strength ratio is 4.662. The corresponding values of ρ are 1.442 for the Tresca model and 1.431 for the Johansen yield criterion.

The calculated horizontal force is displayed as a function of cone angle in Fig. 4. The solid curve corresponds to the Tresca criterion. The dashed curve that is slightly above the solid curve represents the horizontal force calculated with the Johansen criterion. The difference between the two curves is only 6 to 8% and probably would not be detected in cone model test data.

The area beneath the Tresca curve in Fig. 4 is divided to indicate the proportion of the calculated force that is derived from the different features of the process. Because of the bounding nature of the analysis, this division need not be exact, but it should give a reasonable approximation to their relative importance. Each of the regions identified in Fig. 4 includes the effect of friction

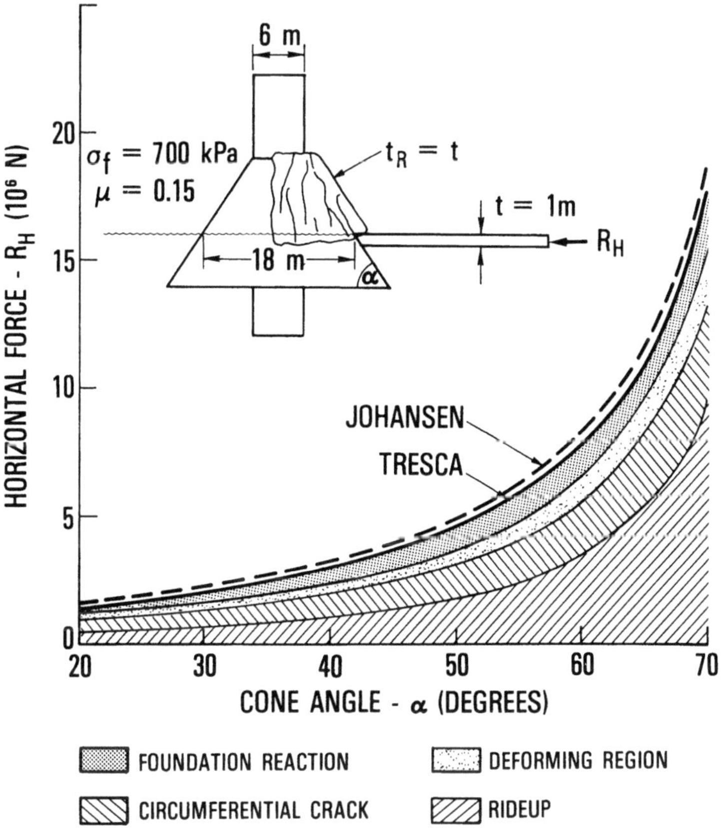

Figure 4. Ice Forces on a Upward-Breaking Cone.

and cone angle. For example, the circumferential crack portion of the force is given by

$$\frac{\tan\alpha}{1-\mu g(\alpha,\mu)}\left\{2M_0\,\frac{\rho}{\rho-1}\right\}.$$

The rideup ice on the surface of the cone accounts for 30 to 50% of the total force. The increase in rideup percentage with increasing cone angle is a consequence of the type of example used for the illustration. The top and waterline diameters of the cone were held constant while the cone angle was changed. Thus the surface area of the cone, and hence the amount of assumed rideup ice, increased with increasing cone angle.

The percentage contributions to the non-rideup portion of the force are: circumferential crack--46%, deforming region--29%, foundation reaction--22%, and side cracks--3%. These percentages are constant with respect to cone angle for this example, but would be somewhat different for other structure geometries and ice sheet characteristics. The circumferential crack formation is of major importance. The deformation within what is here called the deforming region frequently leads to the formation of radial cracks in model tests. If these fractures were to occur prior to the peak force and if there was no interference between the ice wedges that were formed, the present approach should overestimate the non-rideup portion of the force by about 29%.

The foundation reaction force is associated with lifting the floating ice sheet and creating the circumferential crack. The side cracks contribute a proportion of the total force that is too small to be identified in Fig. 4. It is unlikely that effects of this magnitude could be identified in model tests.

DOWNWARD-BREAKING CONE

The analysis presented for the upward-breaking cone can be extended to the downward-breaking geometry with only minor modifications. The velocity field and the energy dissipation rate calculations are all applicable to a downward-breaking geometry provided that we account for the difference in upward and downward foundation reactions. The upward-breaking weight density of the ice was taken to be 0.9 $\rho_w g$. The corresponding downward reaction force density is 0.1 $\rho_w g$. Thus all of the upward-breaking results apply to downward-breaking cones if we simply replace $\rho_w g$ by $\rho_w g/9$ in all of the expressions. The direction of the vertical force also changes. The vertical force on a downward-breaking cone tends to

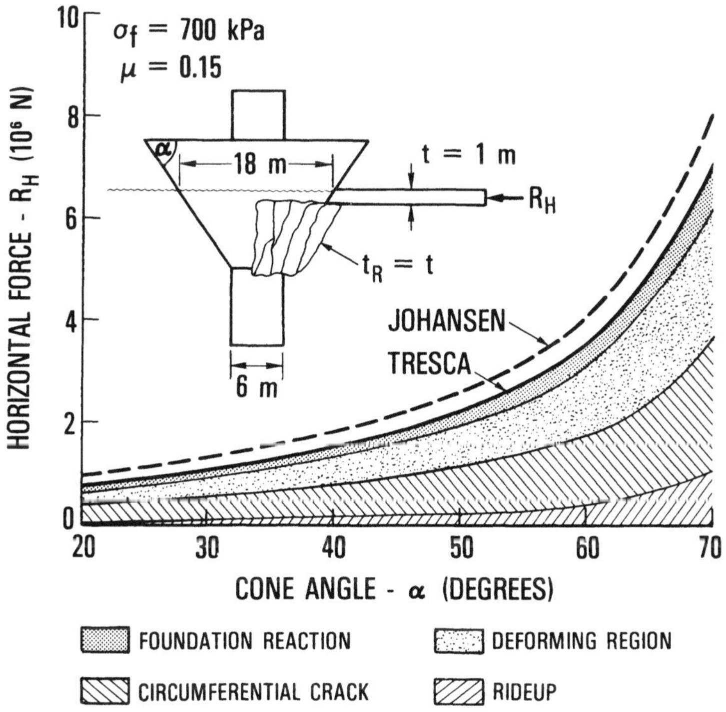

Figure 5. Ice Forces on a Downward – Breaking Cone.

lift the cone and adds to the overturning moment applied to the cone.

Example Calculation

Fig. 5 provides a breakdown for the horizontal force applied to a downward-breaking cone geometry. The inverted cone dimensions and ice properties are the same as those used in the upward-breaking example. Just as in Fig. 4 the solid curve corresponds to the Tresca criterion and the dashed curve gives the result for a Johansen material. The force for the Johansen criterion is 15 to 17% greater than the force for the Tresca model.

The shaded areas beneath the Tresca curve indicate that the rideup (perhaps ridedown?) ice force is a minor part of the total force for a downward-breaking cone. It varies from 6 to 15% of the total force. The percentage contributions to the non-rideup portion of the force are: circumferential crack--42%, deforming region--44%, foundation reaction--7% and side cracks--6%. This distribution is slightly different from the values for the upward-breaking cone, with the contribution from the deforming region increasing and that from the foundation reaction decreasing.

The comparison of the upward- and downward-breaking force curves in Figs. 4 and 5 shows an obvious advantage for the downward-breaking geometry. The calculated force for the inverted cone is about 1/2 of the force computed for the upward-breaking cone at all cone angles. This analysis, however, only addresses sheet ice loads. Ice pressure ridges, rubble piles, ice adhesion (adfreeze), and friction are also design considerations for most offshore structures. Some of the downward-breaking advantages may be lost if ice ridges and rubble jam between the conical surface and the sea floor in shallow water applications. Adfreeze and friction effects, on the other hand, may be easier to deal with on a downward-breaking geometry. The warmer temperatures on the submerged conical surface may increase the effectiveness of the special coatings and/or heating systems that have been suggested for adfreeze mitigation and friction reduction.

A significant difference between ice forces on upward- and downward-breaking conical surfaces has also been observed in model tests. A recent publication by Gerwick and Jahns [8] describes the conceptual design of a deep-water floating caisson vessel with a conical-shaped ice breaking surface. Their work included model tests from which they concluded that the downward-breaking mode effectively reduced the caisson's response by a factor between 1.5 and 3, compared to the upward-breaking mode.

CONCLUSIONS

A plastic limit analysis procedure has been used to develop a description of sheet ice failing against conical structures. This approach yields a closed form solution in which the relative importance of the essential features of the deformation can be analyzed. The solution depends on a non-dimensional weight-strength ratio $\rho_w g\ D^2 t/\sigma_f t^2$. The numerator is proportional to the weight of the floating ice sheet that affects the cone. The denominator is proportional to the ice sheet bending moment capacity, which is the ice strength that is applicable to a conical structure.

The precise form of the two-dimensional bending failure criterion is not critical in this description. The use of a significantly different form of this criterion made only small changes in the calculated load. Thus there is not much motivation to establish a more accurate two-dimensional bending failure criterion for sheet ice.

The motion of broken ice pieces on the surface of an upward-breaking cone is a major contribution to the ice load in the idealized situation of smooth single thickness sheet ice flow over the cone. Ice sheet rafting, rubble accumulations, and ice jamming against the vertical section of the structure could increase the rideup ice thickness and develop a greater proportion of the total force than indicated by our example calculations. Model tests can be used to evaluate the ice clearing effectiveness of specific cone geometries and determine if alternative descriptions of the rideup process are needed.

Downward-breaking geometries can be used to reduce horizontal ice forces developed by sheet ice. The present example calculations indicate that the reduction may be substantial for some geometries. This reduction is a consequence of the reduced forces required to submerge broken ice pieces rather than push them up over the surface of the cone. Additional factors, such as the contribution of the vertical force to the overturning moment and the potential jamming of ice ridges between the structure and the seafloor in shallow water, should also be considered when downward-breaking geometries are of interest.

REFERENCES

1. Ralston, T.D.: Ice force design considerations for conical offshore structures, Proceedings of the Fourth International Conference on Port and Ocean Engineering under Arctic Conditions, St. John's, Newfoundland (1977) 741-752.

2. Hodge, P.G.: Plastic Analysis of Structures, McGraw-Hill (1959).

3. Johansen, K.W.: Moments of rupture in cross-reinforced slabs, Proc. Intern. Assoc. Bridge and Structural Eng., 1 (1932) 277-296.

4. Vaudrey, K.D.: Determination of mechanical sea ice properties by large-scale field beam experiments, Proceedings of the Fourth International Conference on Port and Ocean Engineering under Arctic Conditions, St. John's, Newfoundland (1977) 529-543.

5. Mansfield, E.H.: Studies in collapse analysis of rigid-plastic plates with a square yield diagram, Proc. Roy. Soc. (London), A241, (1957) 311-338.

6. Chen, W.-F., Limit Analysis and Soil Plasticity, Elsevier Scientific Pub., New York, (1975).

7. Michel, B.: Ice pressure on engineering structures, U.S. Army Cold Regions Research and Engineering Laboratory, Hanover, New Hampshire (1970) 58-67.

8. Gerwick, B.C., and Jahns, H.O.: Conceptual design of floating drilling production and storage caisson for Arctic waters, (Preprint), The Fifth International Conference on Port and Ocean Engineering Under Arctic Conditions, Trondheim, Norway (1979).

Flow of Temperate Ice in a Vertical Shaft with Rough Wall Surfaces

NIELS REEH

Geophysical Isotope Laboratory,

University of Copenhagen, Denmark.

Summary

The glaciological aspects related to application of ice to support the rock walls in underground mining operations, are shortly discussed. Simple models for the states of stress and deformation in an ice body sinking into a vertical shaft are developed. Even if possible ice melt along the rock walls is neglected, it is concluded that the ice body is in general unstable, because cross contraction of the ice body will gradually reduce the intrusion depth of the wall obstacles into the ice. In order to avoid fall, the ice body should at intervals be supported by the shaft bottom, so that the necessary contact with the rock walls can be re-established. Estimates of the sink rate of the ice body are given.

Introduction

In sub-level mining of ore bodies two fundamentally different methods are applied: By one method, the side walls of the mine are allowed to cave in and fill up the mined-out space, resulting in dilution of the produced ore by rock waste, and leaving behind valuable ores in the caved wall material. By the other method - so called cut and fill mining - the mined ore is replaced by some other material, e.g. sand, which may even be stabilized with cement. Cut and fill mining permits complete recovery of ore with a minimum of dilution and may give a considerable reduction of costs in case of valuable ore, or when separation of ore and waste rock is relatively cost consuming.

It has been suggested to utilize ice as fill material in cut and fill mining, by creating an artifical sub-level glacier, sinking into the mine at a rate corresponding to that of ore extraction (Fangel [1]), see figure 1.

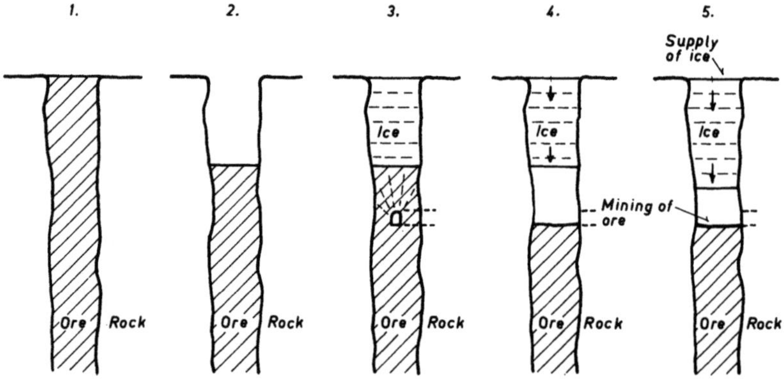

Figure 1. Sub-level mining with ice fill. Schematic. 1. Steep, narrow ore body in surrounding rock. 2. Ore extraction commenced by open pit methods. 3. Filling up of excavated room by ice, and preparation for blasting of ore immediately below the ice body. 4. Extraction of the blasted ore, while the ice body starts to sink due to lack of bottom support. 5. Mining process is continued underneath the continually sinking ice body, which is supplemented with ice from the surface.

This proposal raises several questions of glaciological nature, e.g. concerning the production and supply of the required ice volumes, the amount of ice melt, the sink rate and the risk of fall of the ice body, in parts or as a hole.

In this paper we shall deal with just two of these problems only which turn out to be closely related, viz. the rate of sinking and the risk of fall of the ice body.

General assumptions

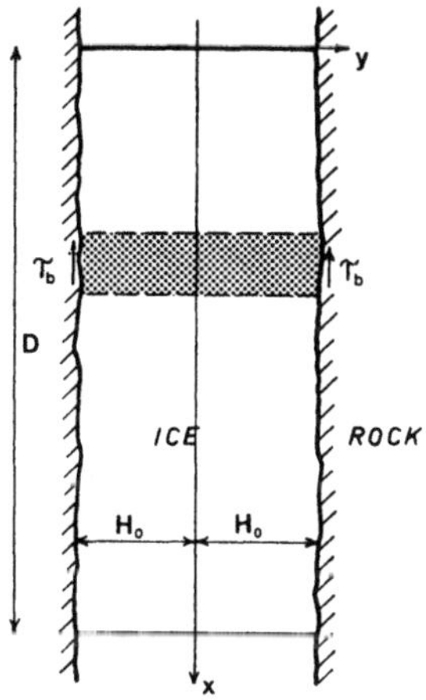

Figure 2. Ice body in a vertical shaft.

Consider an ice body of height D, flowing slowly down a vertical shaft of constant width $2H_0$, which is supposed to be small as compared to D, see figure 2. The extent of the ice body in the direction normal to the plane of the paper is assumed to be so large that flow may be considered two dimensional. Cartesian coordinates O-xy are introduced as shown in the figure. The upper and lower surfaces of the ice body are supposed to be free of load.

As the stress-strain rate relationship of ice we shall adopt Glen's law, which in the case of two-dimensional flow, reads

$$\frac{\partial v_x}{\partial x} = -\frac{\partial v_y}{\partial y} = \tfrac{1}{2} B\tau_e^{n-1} (\sigma_{xx} - \sigma_{yy}), \tag{1}$$

$$\tfrac{1}{2}(\frac{\partial v_x}{\partial y} + \frac{\partial v_y}{\partial x}) = B\tau_e^{n-1} \sigma_{xy}, \tag{2}$$

where the effective shear stress τ_e is given as

$$\tau_e^2 = \tfrac{1}{4}(\sigma_{xx} - \sigma_{yy})^2 + \sigma_{xy}^2, \tag{3}$$

and v_x, v_y = velocity components,
σ_{xx}, σ_{yy} = normal stresses,
σ_{xy} = shear stress
B, n = ice flow law constants.

Also it is assumed, that ice is an incompressible material, which imposes the following condition on the longitudinal strain rates

$$\frac{\partial v_x}{\partial x} + \frac{\partial v_y}{\partial y} = 0. \qquad (4)$$

Laminar glacier flow model

Disregarding terms due to inertia, which are negligible, an equilibrium consideration shows that the average shear stress at the ice rock wall contact has the magnitude

$$\bar{\tau}_b = \rho g H_0, \qquad (5)$$

ρ denoting density of ice and g acceleration due to gravity.

Now we assume the distribution of the wall shear stress τ_b to be uniform and hence equal to $\bar{\tau}_b$. Then, at any cross section, the average taken over the width of the ice body of the vertical normal stress $\bar{\sigma}_{xx}$ must be zero, the weight of each horizontal slice of ice being carried by the wall shear stresses, see figure 2. Apart from narrow edge zones adjacent to the free surfaces of the ice body, σ_{xx} will everywhere equal its average value and consequently be zero. As to the horizontal normal stress σ_{yy}, this is determined by the horizontal forces at the ice rock wall interfaces, since the ice is not subject to any other horizontal load. Therefore, if σ_{yy} is supposed to be zero at the rock wall, it will be so everywhere within the ice body. Since also σ_{xx} was found to be zero, eq. (1) shows, that in this case $\partial v_x/\partial x = \partial v_y/\partial y = 0$. Consequently the ice body will sink steadily into the shaft without any change of width. The vertical velocity v_x is independent of x and may be written

$$v_x(y) = v_s + v_d(y), \qquad (6)$$

where v_s is the sliding velocity of the ice body along the rock walls, and $v_d(y)$ is the contribution to the velocity from internal deformations. For a cold ice body frozen to the wall, the sliding velocity is equal to zero. For temperate ice (ice at the pressure melting point), v_s may be determined from the theory of basal sliding of a glacier.

Even though this theory is still a matter of debate, it is generally accepted, that in absence of pressure water at the glacier-bedrock contact the sliding velocity may be expressed as

$$v_s = k\ \tau_b{}^{\alpha}/R^{\beta}, \qquad (7)$$

where τ_b = basal shear stress

R = a measure of the relative roughness of the glacier bed, e.g. the ratio of amplitude to wavelength of the bedrock irregularities,

α and β = numbers (see table 1),

and k = a constant of dimension m year^{-1} bar$^{-\alpha}$ in which is incorporated the influence of the flow law parameter B, and if regelation sliding is important, also the influence of the latent heat of ice, the coefficient of melting point lowering of ice under pressure and the thermal conductivities of bedrock and ice, see Weertman [2].

In table 1 powers α and β in the sliding velocity expression (7) are given, as specified by different authors. With the ice flow law constant n = 3, it appears from the table that the question of how the sliding velocity depends on the relative roughness is a matter of opinion. Furthermore a similar difference of opinions is seen with respect to the factor k in eq. (7). This lack of clarification obviously makes even order of magnitude estimates of the sliding velocity a most doubtful undertaking.

Table 1. Powers α and β in sliding velocity expression, according to different authors.

Author	Weertman [2]	Lliboutry [3]	Kamb [4]
α	(n+1)/2	(n+1)/2	(n+1)/2
β	n+1	n	*(n+3)/(n+1)

* Actually in Kamb [4] a sliding velocity-bed roughness relationship of the form stated in eq. (7) is not given explicitly. However for R > 0.1, Kamb's eqs. (79) may be approximated by an expression of the form of eq. (7) with β as given in table 1. Also it should be emphasized, that the measure of relative roughness is differently defined by the different authors.

The velocity contribution from internal deformation $v_d(y)$ may be obtained from the equation valid for steady laminar glacier flow (see e.g. Paterson [5]):

$$v_d = \frac{2}{n+1} B(\rho g)^n H_o^{n+1} (1 - (y/H_o)^{n+1}), \tag{8}$$

where symbols are defined above.

The estimates of v_d that can be made from this equation are almost as doubtful as those made from eq. (7), due to large differences in quoted values of the ice flow law parameter B (0.05 - 1 bar^{-3} $year^{-1}$), reflecting the poorly known dependence on ice density, ice crystal-size and ice-texture.

Figure 3 shows plots of v_s and $v_d(0) = \frac{2}{n+1} B(\rho g)^n H_o^{n+1}$ as functions of the half shaft width H_o. Weertman's equation for sliding with cavitation has been used for v_s. Values of the physical parameters involved are given in the figure. It appears from the figure, that for shaft-widths up till 30 m's, the vertical motion of the ice body is dominated by sliding along the rock wall. However, with reference to the discussion above, the curves in figure 3 should obviously be taken as guidance only. As mentioned above a condition for applying the laminar glacier flow model is, that lateral forces do not occur. In the case of a temperate ice body a closer examination of the conditions at the ice rock wall contact will show, that this cannot be the case.

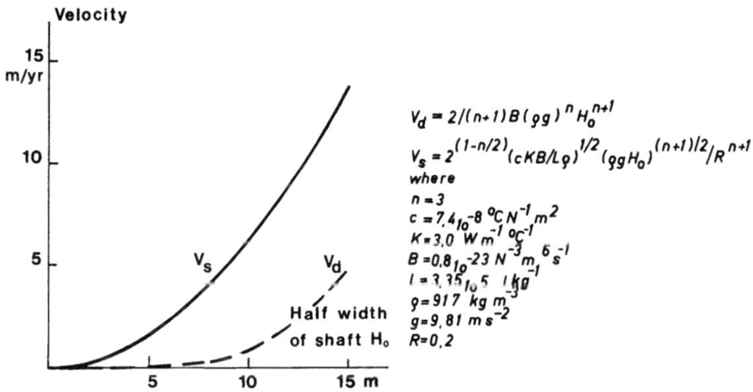

Figure 3. Sliding velocity v_s and velocity due to internal deformation v_d as functions of half width of shaft H_o.

Cross contraction of the ice body, (Nye-flow model).

Consider the ice rock wall contact shown schematically in figure 4. By analogy with the conditions at the base of a temperate glacier, a thin layer of water intervening between ice and rock will make the ice-rock wall interface itself shear-stress free. Accordingly, the ice body is prevented from falling only by the support of the rock wall irregularities, normal-stresses acting across the surfaces of contact between ice and rock. Since the pressure within the ice is small - in the laminar glacier flow model the pressure was found to be zero - cavities will form in the regions of potential tension, i.e. at the downwards turning faces of the irregularities, c.f. the analogy with the regions of potential cavity formation at the downstream faces of bedrock irregularities in glacier flow, Weertman [6].

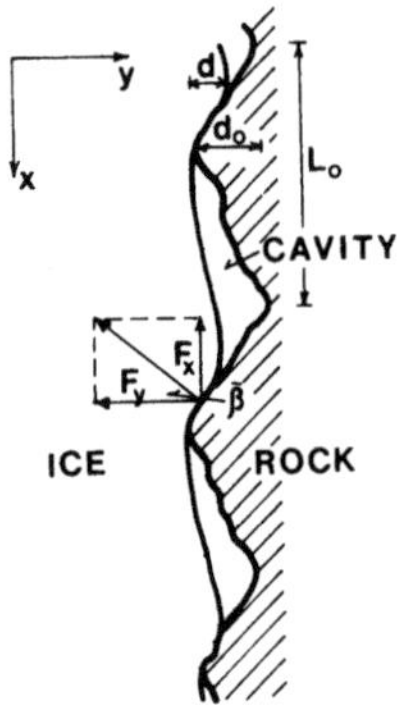

Figure 4. Contact between ice and rock wall, schematic.

Consequently the ice body is supported by upward-inward pointing normal stresses acting along the up-turning faces of the rock wall irregularities - or rather along parts of them only, depending on how far the process of cavitation has advanced, (see figure 4). On the average the vertical component of the resulting force F_x pr. unit height of the ice body equals the weight of half an ice slice, one unit high, i.e. $F_x = \tau_b$. Consequently, the horizontal component of the resultant force - also averaged over one unit height of the ice body - must have the magnitude

$$F_y = sF_x = s\tau_b,$$

where $s = \cot \bar{\beta}$, and $\bar{\beta}$ is an appropriate mean value of the angle β between the horizontal and the normal to the up-turning ice supporting face sections of the rock wall irregularities. The term "appropriate mean value" refers to the necessity of, in a strict treatment, to take into account the non-uniform distribution of the normal stress across the ice rock wall interface. We shall here neglect such refinements and put $s = \frac{1}{2} L_o/do, = 1/(2R_o)$, where $R_o = \frac{do}{L_o}$ is the relative roughness of the dominant rock wall irregularities, see figure 4.

Apart from narrow zones of approximate width L_o adjacent to the rock walls, where large stress variations will occur due to the irregular stress distribution on the ice body surface itself, the state of stress within the ice may be closely approximated by a linearly distributed shear stress field

$$\sigma_{xy} = - \tau_b \; y/H_o, \tag{9}$$

superposed by a uniform horizontal normal stress field

$$\sigma_{yy} = - s\tau_b . \tag{10}$$

This state of stress is very similar to the stress field considered by Nye [7] in his theory of compressional and extensional glacier flow, and in fact the flow problem of the ice body may be treated similarly. Accordingly, we shall assume that stresses are functions of y only and not of x, and that the normal strainrates $\partial v_x/\partial x$ and $\partial v_y/\partial y$ are independent of x and y. Hence we put $\partial v_x/\partial x = -\partial v_y/\partial y = r$ and assume stress components σ_{xy} and σ_{yy} to vary according to eqs. (9) and (10) respectively.

Using the stress strainrate relationship given by eqs. (1) - (3), the remaining stress component σ_{xx} and velocity components v_x and v_y are found to be

$$\sigma_{xx} = - s\tau_b + 2r/(B\tau_e^{\,n-1}), \tag{11}$$

$$v_x = v_o + rx - 2B\tau_b/H_o \int_{H_o}^{y} \tau_e^{\,n-1} \; y \; dy, \tag{12}$$

and $$v_y = - ry, \tag{13}$$

where the effective shear stress τ_e is related to y and r by the equation

$$\tau_e^{\,2n} - \tau_e^{\,2(n-1)} \; (\tau_b y/H_o)^2 - (r/B)^2 = 0. \tag{14}$$

For a given value of r, the stress- and velocity-fields are determined by eqs. (9) - (14). Still however the strainrate r seems to be undetermined. However, as already mentioned the shear stress distribution given by eq. (9) implies, that the average taken over the width of the ice body of the normal stress σ_{xx} be zero. This condition imposes a restriction on σ_{xx}, namely that

$$\int_0^{H_0} \sigma_{xx}\, dy = 0, \tag{15}$$

from which we obtain

$$- s\tau_b H_0 + 2r/B \int_0^{H_0} \tau_e^{\,1-n}\, dy = 0. \tag{16}$$

Combining this equation with eq. (14), r may be expressed in terms of s, i.e. in terms of the transverse normal stress σ_{yy}. Next the rate of thinning of the ice body is obtained from eq. (13) as $a = 2v_y\big]_{y=H_0} = -\,2r\,H_0$.

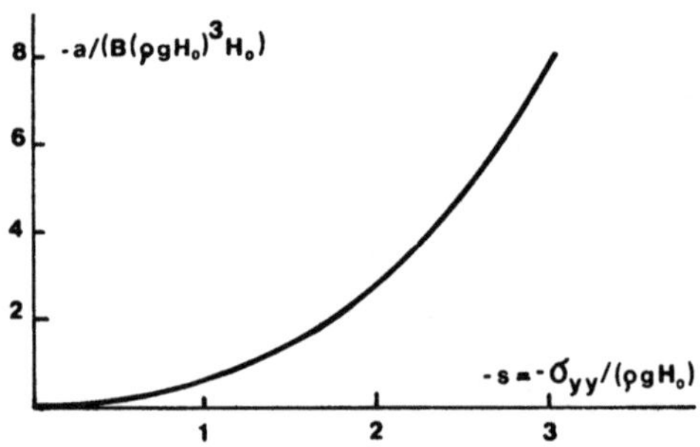

Figure 5. Dimensionless thinning rate as a function of dimensionless transverse stress.

The curve in figure 5 illustrates the relationship between dimensionless rate of thinning and transverse stress. For a rock wall surface of a driven shaft the amplitude to wavelength ratio of the irregularities can be reasonably estimated as 0.25, giving s = 2, see p. With this value of s, we obtain by means of figure 5, that $a \sim -1.0\ B(\rho g)^3 H_0^{\,4}$, which relation is plotted as the curve in figure 6.

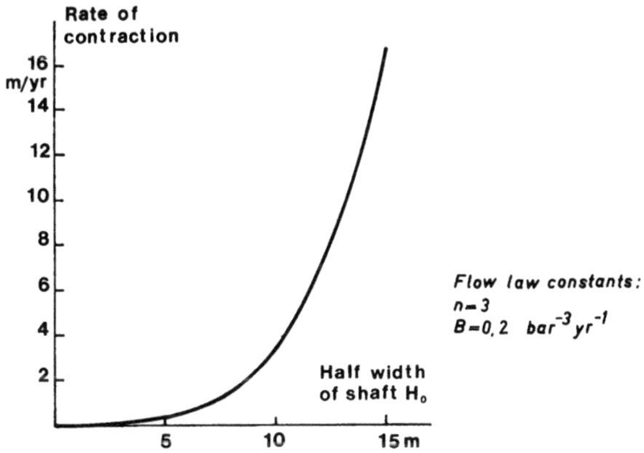

Figure 6. Rate of thinning of the ice body as a function of half width of shaft. cot $\bar{\beta}$ (for definition see figure 4) has been put equal to 2.

It appears from this figure, that even with shaft width's as small as 15 m's, the rate of thinning amounts to 1-2 m/year. If the surrounding rock is warm, the rate of thinning will even be increased by melting of ice at the rock wall contact.

The sliding velocity, predicted by the flow-model considered, is obtained from eq. (13) as

$$v_s = v_o + 1.0\ B\ (\rho g H_o)^3\ x.$$

Therefore according to the Nye flow model, the ice body will sink into the shaft at a rate that increases linearly from the top to the bottom, and is at the same time subject to a thinning at a constant rate.

Compared to the laminar glacier flow model, the Nye flow model is an improvement by accounting for the very important thinning of the ice body, which in the long run inevitably will lead to a fall due to the gradual reduction of the intrusion depth of the rock wall irregularities

into the ice. The fall of the ice plug can be prevented only if - at intervals - the sinking is brought to a stop by interrupting the mining procedure and letting the ice body be supported by the shaft bottom for a time sufficiently long, so the necessary ice/rock wall contact can be re-established.
At two points though, the Nye flow model does not give a satisfactory description: Firstly, the increase of the sliding velocity from top to bottom of the ice body, while the shear-stress at the rock wall is assumed to be constant, is in conflict with the sliding velocity law given by eq. (7). Secondly, the gradually decreasing intrusion depth of the rock wall obstacles into the ice caused by the continually thinning of the ice plug, will result in a decrease of the relative wall roughness and a contemporary increase of the sliding velocity, see eq. (7). In order to cope with the former of these short-comings, the concept of a uniform distribution of the shear-stress at the rock wall must be given up, attended by introducing a non-zero average over the width of the ice body of the vertical normal stress $\bar{\sigma}_{xx}$. To cope with the latter problem a step by step calculation in the time domain has to be performed.

Model of calculation with linear distribution of wall shear stress.

Giving up the assumption of a uniform distribution of the wall shear stress - or rather the shear stress at distance L_0 from the rock wall, see text above eq. (9), the condition for vertical equilibrium of the ice body may be written, see figure 7:

$$\tau_b = \rho g H_0 + H_0 \frac{\partial \bar{\sigma}_{xx}}{\partial x}, \qquad (17)$$

where $\bar{\sigma}_{xx}$, as above, denotes the average value over the width of the ice body of the vertical normal-stress, which is subject to the following boundary conditions:

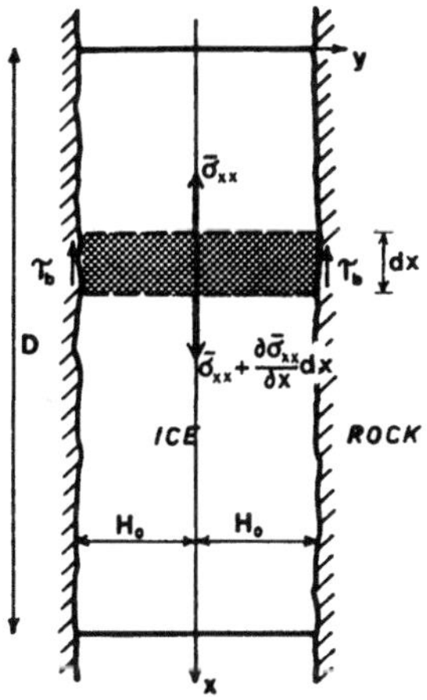

Figure 7. Vertical equilibrium of a slice of the ice body.

$$\bar{\sigma}_{xx} = 0 \text{ for } x = \begin{cases} 0 \\ D \end{cases}. \tag{18}$$

As an approximation it is now assumed, that $\bar{\sigma}_{xx}$ has a parabolic distribution with x, according to the law

$$\bar{\sigma}_{xx} = 4x(D-x)\sigma_M/D^2, \tag{19}$$

which satisfies the boundary conditions given in eq. (18). σ_M is the vertical mean normal stress halfway down the ice body, and is a function of time t. Combining eqs. (17) and (19), we get

$$\tau_b = \rho g H_o + (4D-8x)H_o\, \sigma_M/D^2, \tag{20}$$

i.e. a linear distribution of τ_b.
We shall now make the additional assumption that locally at each x, ice flow is approximate Nye-flow. This is a reasonable approximation, as long as stress- and strain-rate gradients in the x-direction are small, and so we shall assume this to be the case. Accordingly the stress- and velocity fields are determined by eqs. (9) - (14), with τ_b given by eq. (20). The restriction imposed on σ_{xx} by eq. (15) however should be changed in order to

agree with eq. (19). This is achieved by writing

$$\frac{1}{H_o}\int_o^{H_o} \sigma_{xx}dy = 4x(D-x)\sigma_M/D^2. \tag{21}$$

Except for the unknown time dependent stress σ_M, the states of stress and deformation within the ice body are determined by eqs. (9) - (14) and (20) - (21).

In order to determine σ_M and thereby the complete stress- and velocity solution, one further condition must be adopted. This is found by considering the motion of the ice body, the difference in sliding velocity Δv_s between either end of the ice body being at any time double determinable, since we have

$$\Delta v_s = v_s(D) - v_s(0) = \int_0^D r(x)dx. \tag{22}$$

In eq. (22), r(x) may be expressed in terms of σ_M as decribed above, while v_s can be calculated from eq. (7) as $v_s(x) = k\tau_b(x)^{\alpha}/R(x)^{\beta^s}$, where the relative roughness R(x) $= d/L_o$. d is the actual intrusion depth of the wall irregularities into the ice, see figure 4. Owing to the continuous thinning of the ice body, d will decrease with time in accordance with the following equation

$$d = d_o - \int_o^t H_o \; r(x)dt,$$

which shows that at any time, d and consequently R and $v_s(x)$ can also be expressed in terms of σ_M. This quantity may consequently be determined from eq. (22).

Due to the simplifying assumption of parabolic distribution of $\bar{\sigma}_{xx}$ (see eq. (19)), one unknown parameter only, - viz. σ_M - had to be introduced in order to describe the states of stress and velocity within the ice body. However, this simplification is not vital for applying the method of calculation outlined above, since in fact eq. (22) is valid with D replaced by any x, making possible a determination of the distribution of $\bar{\sigma}_{xx}$, - and thereby the complete stress- and velocity solution, -

to any desired accuracy.

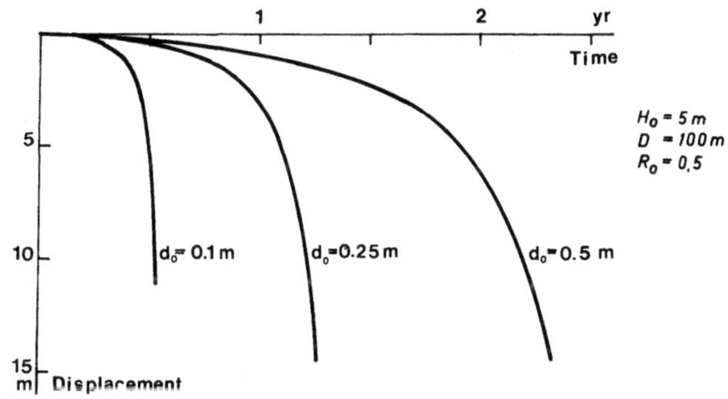

Figure 8. Displacement of lower surface of ice body as a function of time for different sizes of wall obstacles.

From the point of view of application, the most interesting quantity to be determined is the rate of sinking of the lower free surface of the ice body. To illustrate this, figure 8 shows the displacement as a function of time of the lower surface of a 10 m wide and 100 m high ice body for different sizes of the shaft wall irregularities. As appears from the figure, the rate of sinking of the ice body increases with time, mainly due to the decreasing depth of intrusion of the wall obstacles into the ice body, caused by its continuous thinning. As is the case with the other predictions made in this paper, also the curves in figure 8 should be taken as guidance only. With this reservation the figure suggests, that a sinking of several meters pr. year is attainable for shaft and ice body dimensions, which are relevant in a mining context. Once more however, the inevitable fall of the ice body, if not supported at intervals along the lower surface, should be stressed. This brings into question the necessity of planning the excavation of ore in such a way, that the ice body - or parts of it - now and then has the opportunity of resting on the bottom of the mine for a time sufficiently long, for good contact between the ice body and the latteral walls to be re-established by ice flowing into the cavities by the pressure of its own weight.

References

1. Fangel, H.: Utilization of dynamic and physical properties of ice in underground mining operations. Journal of Glaciology, vol. 19, No. 81, (1977) 672.

2. Weertman, J.: On the sliding of glaciers. Journal of Glaciology, vol. 3, No. 21, (1957) 33-38.

3. Lliboutry, L.: Traite de Glaciologie. Tom. 2. Paris: Masson et Cie (1965) 640-52.

4. Kamb, B.: Sliding motion of glaciers: Theory and observation. Rev. Geophysics and Space Physics, vol. 8, No. 4, (1970) 673-728.

5. Paterson, W.S.B.: The physics of glaciers. Oxford, Pergamon Press (1969) 89-93.

6. Weertman, J.: The theory of glacier sliding. Journal of Glaciology, vol. 5, No. 39, 287-303.

7. Nye, J.F.: The distribution of stress and velocity in glaciers and ice-sheets. Proc. Roy. Soc. Lond. A 239, (1957), 113-133.

Analytical Approach for the Determination of Ice Forces Using Plasticity Theory

K.M. Reinicke
BEB GEWERKSCHAFTEN BRIGITTA UND ELWERATH
Betriebsführungsgesellschaft mbH
Hannover, West-Germany

SUMMARY

A plasticity model is used to describe the strength and plastic failure of columnar grained ice. The model is applied within the framework of the plastic limit analysis method to predict analytically the ice crushing pressure in the initial indentation of a uniformly thick ice sheet by vertical flat, wedge shaped and circular indenters which are in perfect contact with the ice. For the ice-indenter interface both free-slip and no-slip conditions are considered to estimate the effect of ice adfreeze. The presented analyses assume plane strain and plane stress conditions to exist in the plane of the ice sheet. These special cases represent the limiting cases of very narrow indenters in thick ice sheets and very wide indenters in thin ice sheets respectively. The results of example calculations carried out, are in good agreement with laboratory observations. In particular the shape factors which may be derived from the results agree well with published data.

I. INTRODUCTION

The design of arctic offshore structures is controlled by the forces occurring in ice/structure interactions. For a safe and economical design the accurate prediction of these forces is therefore important. The most likely type of force in an ice-structure interaction is that of a horizontal thrust caused by a solid moving ice cover. To resist this force the structure must cause the impinging ice cover to fail. The corresponding failure forces are the subject of this work.

To predict ice failure forces accurately one must understand the mechanical behaviour of the ice. From laboratory experiments it is known that the deformation behaviour of ice exhibits elastic, ductile and brittle characteristics, depending on the nature and duration of loading, Gold and Krausz (1971). Under moderate strain rates, low levels of stress and small loading times ice behaves elastically. If any of these conditions is violated permanent deformations can be detected. These deformations may be either plastic,ductile,or brittle - the type depending primarily on the magnitude of the strain rate. Up to a strain rate of about $10^{-3}sec^{-1}$ ductile behaviour is usually observed. Beyond this range brittle failure is typical. Ice strength is also dependent on strain rate. Over the range of ductile behaviour, ice strength increases with increasing strain rate reaching a maximum at the ductile-to-brittle transition zone, Carter and Michel (1971), Gold and Krausz (1971). Beyond this zone the strength decreases approaching an asymptotic value at high strain rates. No conclusive information is available on whether ice work-hardens or work-softens during plastic deformations. For practical purposes then, when one is concerned about maximum ice forces in ice-structure interactions, ice may be modelled as an elastic-perfectly plastic material, using its maximum strength with respect to strain rate as its yield strength.

When an elastic-plastic material is loaded and the applied loads are gradually increased from zero, there is as a rule a three stage development in material response. These stages are: the initial elastic response, the intermediate contained plastic flow and finally the unrestricted plastic flow. When one is concerned about failure loads, only the latter stage is of interest. This stage can be assessed in a rather direct manner using the method of plastic limit analysis. This method enables definite statements to be made about failure loads without carrying out a step-by-step elastic-plastic analysis. The method is simple, since for its implementation stress and velocity fields may be employed, which need not represent the actual stress and strain states in the material at collapse. In this context, the term collapse is used to describe the conditions in an elastic-perfectly plastic material, where plastic flow would occur at constant load if accompanying geometrical changes were neglected. By using the upper and lower bound theorem of plastic limit analysis, Drucker et al. (1952), the load causing these conditions can be bracketed by upper and lower bounds:

Upper bound theorem

For a plastic situation, collapse must occur for any admissible velocity field if the rate at which the external forces do the work on the body equals or exceeds the rate of internal dissipation of energy.

A velocity field used in the upper bound analysis of a particular problem is called admissible if it matches the boundary condition of that problem and has the property that its associated strain rate field is normal to the yield surface $f(\sigma_{ij}) = 0$, i.e.,

$$\dot{\varepsilon}_{ij} = \lambda \frac{\partial f}{\partial \sigma_{ij}} \tag{1}$$

where f denotes the yield function and λ is a positive scalar function. (1) is known as the normality requirement or the associated flow rule.

Lower bound theorem

Collapse of a body will not occur if any state of stress can be found which satisfies the equations of equilibrium, the stress boundary conditions, and for which $f(\sigma_{ij}) \leq 0$ everywhere.

The scope of this work is limited to upper bound analyses for the initial indentation of a uniformly thick ice sheet by vertical flat, wedge shaped and circular indenters that are in perfect contact with the ice. For the ice-indenter interface, both free-slip and no-slip (adfreeze) conditions are considered. In this context the upper bounds are of special significance since they provide an upper limit for the maximum sheet ice design loads. Additional factors, such as the actual contact condition between the ice and the structure, the design condition strain rates, the ice floe size, etc. , all tend to reduce design ice loads to less than those computed by the present analyses. The present analyses assume plane strain and plane stress conditions to exist in the plane of the ice sheet. These special cases are particularly significant, since they represent the limiting cases corresponding to very narrow indenters in thick ice sheets ($D/H \rightarrow 0$) and very wide indenters in thin ice sheets ($D/H \rightarrow \infty$), respectively. In principle, the analyses presented here could be extended to three dimensions. A three-dimensional analysis would provide a curve of ice indentation pressure as a function of D/H, analogous to that presented Croasdale et al. (1977) and by Reinicke and Remer (1978).

II. YIELD FUNCTION AND RATE OF ENERGY DISSIPATION

The yield function used in this work to describe the strength behavior of ice is the anisotropic parabolic yield function discussed by Reinicke and Ralston (1977) restricted to planar isotropy. In terms of stress components relative to the Cartesian coordinate system of Fig. 1 it is given by

$$f(\sigma_{ij}) = a_1\left[(\sigma_y-\sigma_z)^2 + (\sigma_z-\sigma_x)^2\right] + a_3(\sigma_x-\sigma_y)^2 + a_4(\tau_{yz}^2+\tau_{zx}^2) + a_6\tau_{xy}^2 + a_7(\sigma_x+\sigma_y) + a_9\sigma_z - 1, \tag{2}$$

where

$$a_6 = 2(a_1+2a_3).$$

The function (2) describes a material which has different tensile and compressive strengths in the xy-plane and in the z-direction, i.e., the horizontal plane of the ice sheet and the thickness direction. The material is homogeneous in the sense that the strength values in the xy-plane and the z-direction are constant throughout. With increasing hydrostatic stress, the compressive strength increases non-linearly. Freshwater ice, as it appears in a laboratory-grown ice sheet, is well approximated by this material model due to its rather uniform crystal structure with the direction of crystal growth perpendicular to the plane of the ice sheet. Arctic sea ice has a somewhat different crystal structure, Weeks and Gow (1978), Wang (1979), and Vittoratos (1979), and the above assumption of planar isotropy may not be appropriate. The generalization of the present analyses to Arctic sea ice is straightforward; however, some additional ice strength data would be required.

The yield function given by equation (2) is applied in this work to predict extreme values of ice crushing pressure for the initial indentation of a laboratory-grown, freshwater ice sheet by indenters of various shapes. The present analyses are two-dimensional and assume plane strain and plane stress conditions to exist in the plane of the ice sheet.

Plane strain in the plane of the ice sheet requires that

$$\dot{\varepsilon}_{zz} = \dot{\varepsilon}_{xz} = \dot{\varepsilon}_{yz} = 0.$$

Imposing this restriction on the yield function (2) via the flow rule (1), results in the <u>plane strain yield function</u>

$$f(\sigma_{ij}) = a_6\left[\left(\frac{\sigma_x-\sigma_y}{2}\right)^2 + \tau_{xy}^2\right] + (2a_7+a_9)\left(\frac{\sigma_x+\sigma_y}{2}\right) - \left(1+\frac{a_9^2}{8a_1}\right). \tag{3}$$

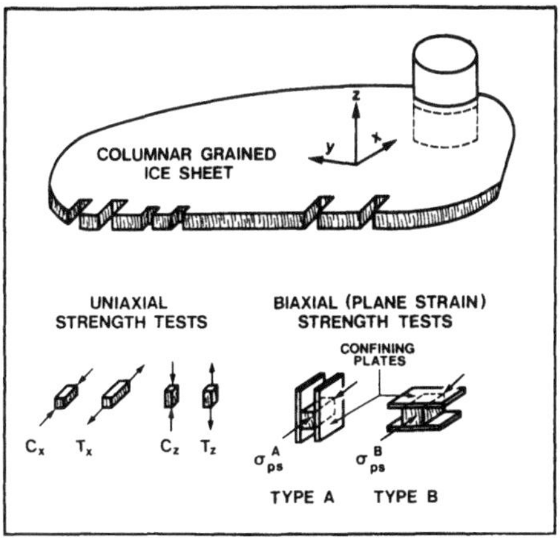

FIG. 1: COORDINATE SYSTEM AND STRENGTH TESTS FOR SHEET ICE.

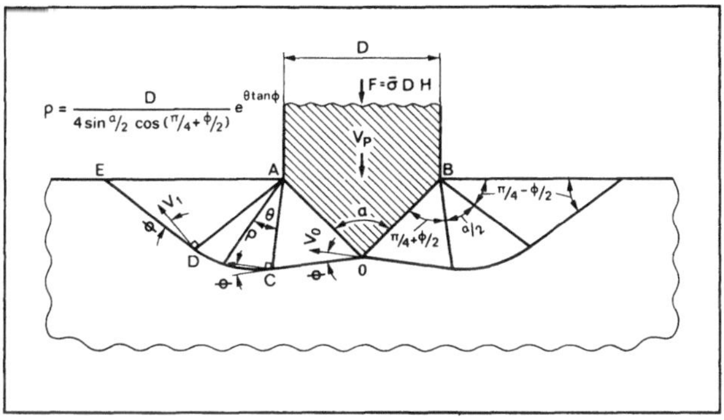

FIG. 2: FREE-SLIP UPPER BOUND VELOCITY FIELD FOR FLAT AND WEDGE SHAPED INDENTERS.

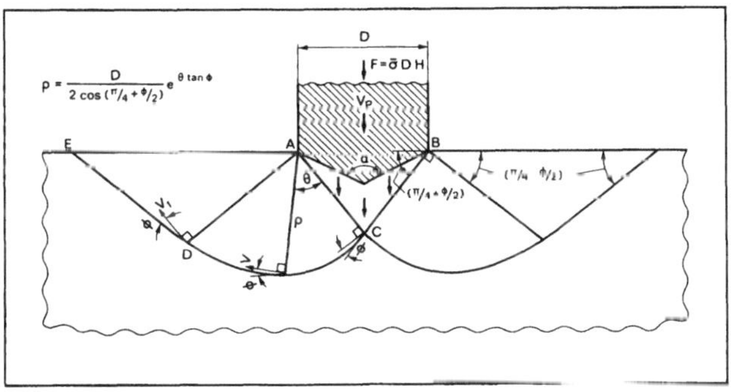

FIG. 3: NO-SLIP UPPER BOUND VELOCITY FIELD FOR FLAT AND WEDGE SHAPED INDENTERS.

Plane stress in the plane of the ice sheet requires that

$$\sigma_z = \tau_{xz} = \tau_{yz} = 0.$$

Imposing this restriction on (2) results in the plane stress yield function

$$f(\sigma_{ij}) = a_1(\sigma_x^2+\sigma_y^2) + a_3(\sigma_x-\sigma_y)^2 + a_6\tau_{xy}^2 + a_7(\sigma_x+\sigma_y) - 1. \qquad (4)$$

To determine the two-dimensional yield function (3) and (4), one needs four strength values. Let C_x, C_z, T_x and T_z denote the absolute values of unconfined compressive and tensile strength in the plane of the ice sheet and in the thickness direction, as in Fig. 1. Then these four values can be used to determine the yield function coefficients in (3) and (4) by

$$a_1 = \frac{1}{2C_zT_x}, \qquad a_3 = \frac{1}{T_xC_x} - \frac{1}{2C_zT_z},$$

$$a_7 = \frac{1}{T_x} - \frac{1}{C_x}, \qquad a_9 \ \frac{1}{T_z} - \frac{1}{C_z}. \qquad (5)$$

When biaxial, see Fig. 1, and/or triaxial strength data are also available, the yield function coefficients can be determined by the data that are most applicable to the situation that will be addressed in the subsequent analysis.

To carry out upper bound limit analyses one needs to determine the rate of dissipation of mechanical energy per unit volume of plastically deforming material. For the plane strain yield function (3) this dissipation rate is given by, Reinicke and Ralston (1977),

$$D_v = C_1(\dot{\varepsilon}_{xx}+\dot{\varepsilon}_{yy}) + C_2\left(\frac{\dot{\varepsilon}_{xy}^2-\dot{\varepsilon}_{xx}\dot{\varepsilon}_{yy}}{\dot{\varepsilon}_{xx}-\dot{\varepsilon}_{yy}}\right) \qquad (6)$$

where c_1 and c_2 are constants defined by

$$C_1 = \frac{2a_1a_6 + 2a_1a_7(a_7+a_9) + a_9^2(a_1+a_3)}{2a_1a_6(2a_7+a_9)}$$

$$C_2 = \frac{2a_7+a_9}{a_6}$$

In the special case of deformation concentrated at a surface of discontinuity, the rate of mechanical energy dissipation per unit area of the discontinuity surface is given by Reinicke and Ralston (1977),

$$D_A = C_1 \delta v + \frac{C_2}{4} \frac{\delta u^2}{\delta v}, \tag{7}$$

where δv and δu denote the changes in normal and tangential velocity across the discontinuity, respectively.

For the plane stress yield function (4) the rate of mechanical energy dissipation per unit volume is given by Prodanovic (1978),

$$D_V = C_3 \sqrt{(a_1 + a_3)(\dot{\varepsilon}_{xx}^2 + \dot{\varepsilon}_{yy}^2) + 2a_3 \dot{\varepsilon}_{xx}\dot{\varepsilon}_{yy} + 2a_1 \dot{\varepsilon}_{xy}^2} - \frac{a_7}{2a_1}(\varepsilon_{xx} + \varepsilon_{yy}), \tag{8}$$

where C_3 is the constant defined by

$$C_3 = \frac{1}{a_1}\sqrt{\frac{a_7^2 + 2a_1}{a_6}}.$$

The expression for the rate of energy dissipation per unit area of a velocity discontinuity is

$$D_A = \frac{C_3}{2}\sqrt{4(a_1 + a_3)\delta v^2 + 2a_1 \delta u^2} - \frac{a_7}{2a_1}\delta v. \tag{9}$$

III. UPPER BOUND ANALYSES

In the following, the material model is applied within the framework of plastic limit analysis to predict plane strain and plane stress upper bounds analytically for the ice crushing pressure developed by the initial indentation of a uniformly thick ice sheet by vertical flat, wedge shaped, and circular indenters that are in perfect contact with the ice. Both free-slip and no-slip conditions are considered at the ice-indenter interface.

The use of the upper bound method requires the assumption of a failure mechanism and uses virtual work concepts to determine the failure load. When the parabolic yield function (2) is used with the upper bound method, one can be rather flexible in choosing a failure mechanism, since the normality requirement is rather weak for the parabolic function. It requires only that plastic deformation is associated with some increase in volume, Reinicke and Ralston (1977), but it does not restrict the possible values of this increase as in the case of the linear Mohr-Coulomb material. To keep the analysis simple, the mechanisms presented in the following make use of this flexibility only to a certain extent. They could be easily extended to more complex mechanisms which would result in better (lower) upper bounds. In general the improvement achieved by such a generalization will be small and usually only in the order of a few percent.

Free-Slip Flat and Wedge Shaped Indenters

The deforming region of the failure mechanism used in the upper bound analysis of free-slip flat and wedge shaped intenters of wedge angle α is shown in Fig. 2. It consists of two rigidly moving triangular wedges similar to AOC, two log spiral shear zones similar to ACD, in which the velocity changes in magnitude, and two more rigidly moving wedges similar to ADE. If the change in magnitude of the velocity in the shear zone is taken to be $V = V_0 \exp(\Theta \tan\Phi)$ this mechanism is described by only one parameter, the angle Φ. This parameter is not determined by the normality requirement but may be chosen later to optimize the upper bound. Note that for $\alpha = \pi$, i.e., for flat indenters, this mechanism defaults into the Hill mechanism.

For the failure mechanism of Fig. 2, energy is dissipated on the velocity discontinuities similar to OC, CD, and DE as well as in the zones of radial shear similar to ACD. The dissipation rates are computed using the expressions (6) and (7) for a plane strain analysis and the expressions (8) and (9) for a plane stress analysis. Equating the total rate of dissipation of energy to the rate of work done by the boundary loads, $F\,V_p$, one obtains an expression for the average indenter pressure $\bar{\sigma} = F/DH$,

$$\bar{\sigma} = g(\Phi)\left\{\tan^2(\pi/4 + \Phi/2)\exp(\alpha\tan\Phi) - 1\right\}. \qquad (1o)$$

$g(\Phi)$ is a function of the parameter Φ defined by

$$g(\Phi)_{p.\,strain} = C_1 + \frac{C_2}{4}\cot^2\Phi \qquad (11)$$

for a plane strain analysis, and by

$$g(\Phi)_{p.\,stress} = \frac{C_3}{2}\sqrt{4(a_1 + a_3) + 2a_1\cot^2\Phi} - \frac{a_7}{2a_1} \qquad (12)$$

for a plane stress analysis. To arrive at (10)

$$V_P \sin(\alpha/2) = V_O \cos(\pi/4 + \Phi/2)$$

has been used, which is the free-slip condition i.e., the normal velocity of the ice relative to the indenter is zero. The ice, however, may freely slip along the indenter.

The expression (10) gives an upper bound for the collapse indentation pressure for any positive values of Φ less than $\pi/2$. The desired upper bound solution $\bar{\sigma}^{ub}$ is the minimum value of $\bar{\sigma}$ with respect to Φ.

No-Slip Flat and Wedge Shaped Indenters

To approximate the condition where the ice is frozen to the indenter, the failure mechanism shown in Fig. 3 is used. This mechanism is known in the literature as the Prandtl mechanism, Chen (1975). The material between the indenter and the surface ACB moves downward as a rigid body with the velocity of the indenter. Thus this particular velocity field will not describe any difference that may exist in the indentation forces for no-slip flat and wedge-shaped indenters. The rigidly moving region is followed by log spiral shear zones, in which the velocity is assumed to increase exponentially to the value $V_1 = V_O \exp(\pi/2 \tan\Phi)$ at AD, and by rigidly moving regions. In the mechanism here, the surfaces AC and BC are surfaces of velocity discontinuity. By requiring that the velocity difference vector across AC be inclined at the angle Φ to AC, implying

$$V_P = 2 V_O \cos(\pi/4 + \Phi/2),$$

this mechanism is again described by the angle Φ only.

The upper bound solution for this mechanism can be obtained in a manner analogous to the solution for the free-slip mechanism. For the average pressure $\bar{\sigma}$, one obtains

$$\bar{\sigma} = g(\Phi) \left\{\tan^2(\pi/4 + \Phi/2) \exp(\pi \tan\Phi) - 1\right\}, \tag{13}$$

where $g(\Phi)$ is the scalar function defined by (11) for a plane strain analysis and by (12) for a plane stress analysis. The expression (13) gives an upper bound for the collapse indentation pressure for any positive values of Φ such that $\pi/2 - \alpha < \Phi < \pi/2$.

Free-Slip Circular Indenters

The mechanism used in the upper bound analysis of free-slip circular indenters is shown in Fig. 4. The use of this mechanism for ice indentation analysis has been recently discussed by Prodanovic (1978). It is described by two parameters ψ_0 and Φ and consists of two rigidly moving regions, which are separated by log spiral discontinuity surfaces from the rest of the material which is at rest. In the moving regions the material is assumed to rotate with angular velocity ω about the centers 0 and 0', respectively.

This angular velocity is determined by

$$\omega = \frac{\tan\psi_0}{R} V_P ,$$

which is the algebraic form of the assumption that the ice may freely slip but not separate from the indenter.

By the upper bound method we get for the average indentation pressure

$$\bar{\sigma} = g(\Phi) \frac{1}{\sin(2\psi_0)} \left\{ \exp\left[2(\pi - \psi_0)\tan\Phi\right] - 1 \right\}, \tag{14}$$

where g(Φ) is defined as before. Expression (14) is a bound for the ice indentation pressure for any choice of $0<\psi_0<\pi/2$ and $0 \le \Phi_2 \le \psi_0$. These parameters can be selected to minimize $\bar{\sigma}$ resulting in the upper bound solution for this mechanism.

No-Slip Circular Indenters

A corresponding mechanism for no-slip circular indenters, also discussed by Prodanovic(1978),is shown in Fig.5. The material immediately adjacent to the indenter moves with the same velocity as the indenter. Separated from this region by surfaces of velocity discontinuity are two regions in which the material is assumed to be in a rotational motion with angular velocity ω about the centers 0 and 0', respectively. If one requires that the velocity difference vector across the surfaces of discontinuity be inclined at the constant angles $\Phi_1$and Φ_2 , the mechanism is determined by the three parameters ψ_0 , Φ_1 , and Φ_2 . The assumption that the angle between the velocity difference vector across the inner discontinuity and that surface is of constant value Φ_1, implies that

$$\omega = \frac{V_P \tan\psi_0}{R\exp(\pi/2 \tan\Phi_1)}$$

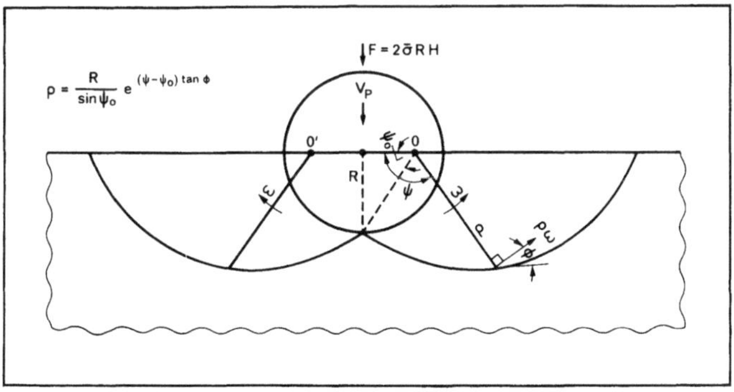

FIG. 4 : FREE-SLIP VELOCITY FIELD FOR CIRCULAR INDENTERS.

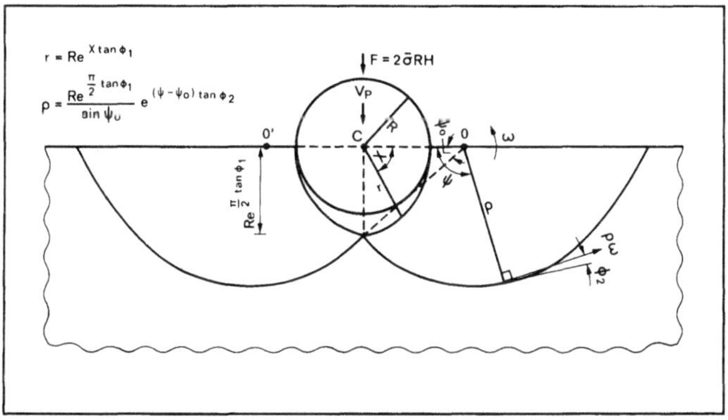

FIG. 5 : NO-SLIP VELOCITY FIELD FOR CIRCULAR INDENTERS.

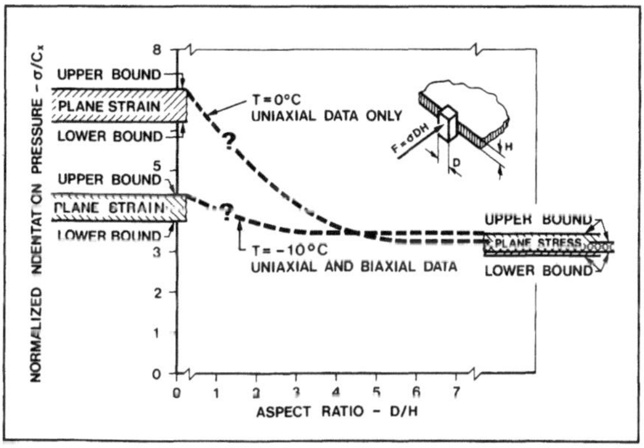

FIG. 6 : COMPARISON OF BOUNDS COMPUTED WITH DIFFERENT YIELD FUNCTION COEFFICIENTS FOR DIFFERENT ICE TEMPERATURES.

By the upper bound method we get for the average indentation pressure

$$\sigma = \frac{\tan\psi_0}{2\exp(\pi/2\tan\Phi_1)} \Big\{ g(\Phi_1)\,[\exp(\pi\tan\Phi_1)-1] + \frac{\exp(\pi\tan\Phi_1)}{\sin^2\psi_0}\, g(\Phi_2)\,[\exp(2(\pi-\psi_0)\tan\Phi_2)-1] \Big\}, \tag{15}$$

where $0<\psi_0<\pi/2$, $0<\Phi_1<\pi/2$, $0<\Phi_2<\psi_0$, and

$\psi_0 <$ are $\tan\,[\exp(\pi/2\tan\Phi_1)]$.

IV. NUMERICAL EXAMPLES

Example calculations are carried out for two different sets of yield function coefficients to determine plane strain and plane stress upper bounds for the ice pressure on vertical flat,9o^0 wedge shaped,and circular indenters. The analysis for the wedge shaped indenter is generally set up in terms of the wedge angle α and does not require this angle to be of a particular value. Other values could be used, which would result in other bounds.

For the first set of yield function coefficients, the ice strength data for C_X, C_Z, T_X, and T_Z at 0°C, reported by Carter and Michel (1971), are used. Values for these four strengths determine the yield function coefficients of the two-dimensional yield function (3) and (4) by (5). The coeffient values one obtains when using the average of the strength values reported by Carter and Michel C_X = 2.54 MPa, C_Z = 7.23 MPa, T_X = 1.02 MPa, T_Z = 1.10 MPa, are listed in Table 1.

Table 1: Yield Function Coefficients

Coefficients	Coefficients Values T = 0^{0}C	T = -10^{0}C Ralston (1978)
a_1	$0.41\ C_X^{-2}$	$1.77\ C_X^{-2}$
a_3	$2.08\ C_X^{-2}$	$5.27\ C_X^{-2}$
a_7	$1.49\ C_X^{-1}$	$6.04\ C_X^{-1}$
a_9	$1.96\ C_X^{-1}$	$-3.54\ C_X^{-1}$

This table also list a second set of yield function coefficients. This set was used by Ralston (1978) to analyse the laboratory ice indentation results by Michel and Toussaint (1977). Ralston used uniaxial and biaxial strength data at -10 °C as reported by Carter and Michel (1971) and Frederking (1977), respectively. The uniaxial data consisted of in-plane tests (C_X and T_X in Fig. 1). The biaxial data (σ_{ps}^{A} and σ_{ps}^{B} in Fig. 1) were also determined by active in-plane loading. These four measurements are sufficient to determine the yield function coefficients. Three of these measurements (C_X, T_X, σ_{ps}^{A}) represent points that appear directly on the plane stress yield surface. The fourth measurement (σ_{ps}^{B}) appears on the plane strain yield surface. These data give a better representation of the two-dimensional yield surface than one does obtain using uniaxial test data alone. The use of T_X, C_X, T_Z, C_Z provides good control on the yield surface only in the vicinity of the unconfined stress states. Only T_X and C_X appear on the plane stress yield surface, and no data points fall in the plane of the plane strain yield surface. Unfortunately, data comparable to σ_{ps}^{A} and σ_{ps}^{B} as used by Ralston (1978) at -10 °C are not presently available at 0 °C. Thus the best one can do at the present time is to use the unconfined strength data at the warmer ice temperature.

The upper bounds obtained for the two ice temperatures are listed in Table 2. For the flat indenter problem, the lower bounds derived by Reinicke and Ralston (1977), and Ralston (1978), are included for comparison.The bounds for the flat indenter problem are also shown in Fig. 6. A comparison of the upper and lower bounds for the flat indenter problems shows that the derived bounds are quite accurate. From the presented results a similar conclusion cannot be drawn for the bounds for wedge shaped and circular indenters. It is expected, however, that their accuracy is of about the same order as that for the flat indenter problem.

Table 2: Ice Pressure Bounds

Indenter Shape	Free-Slip Bounds				No-Slip Bounds			
	Plane Strain		Plane Stress		Plane Strain		Plane Stres	
	0 °C	-10 °C	0 °C	-10 °C	0 °C	-10 °C	0 °C	-10 °C
				Upper Bounds				
Wedge Shaped (α = 90°)	3.89	2.42	2.12	2.28	7.02	4.42	3.27	3.48
Circular	5.74	3.60	2.86	3. 9	9.21	5.81	3.95	4.23
Flat	7.02	4.42	3.27	3.48	7.02	4.42	3.27	3.48
				Lower Bounds				
Flat	6.15	3.77	2.94	2.98	6.15	3.77	2.94	2.98

The normalized plane strain bounds obtained for the two ice temperatures differ from each other by a factor of about 1.6 for all indenter shapes and ice-indenter boundary conditions. This difference may reflect the type of data used to determine the yield function coefficients, rather than a difference in ice behavior at -10 °C and 0 °C. Ralston (1977) has shown that if only uniaxial data are used to determine the yield function coefficients at -10 °C, the value of the type B plane strain strength (σ^B_{ps} in Fig. 1) will be overestimated by about 70 %. The measured value of σ^B_{ps} at this temperature is the same as the unconfined compressive strength C_x. If the mechanisms that lead to this equality at -10 °C are also operative at 0 °C, the coefficients given by equation (5) may overestimate plane strain indentation values. The value of σ^B_{ps} computed from equation (3) with the present 0 °C coefficients is about 65 % greater than C_x. Thus much of the difference in the normalized plane strain indentation pressures in Table 2 may reflect defficiencies in the available property data rather than differences in ice material bahavior at -10 °C and 0 °C.

The normalized plane stress bounds obtained for the two sets of coefficients agree within the range of accuracy of the limit analysis method. While there is almost no size effect for the -10 °C ice, an effect is suggested for the warmer ice data. As previously discussed, part of this difference in behavior may result from data deficiencies. One would also expect some size effect difference to develop with temperature changes, just as the various strength ratios (T_x/C_x, etc.) change with temperature. It is important to keep in mind that size effect curves developed by the present approach correspond to a fixed effective indentation strain rate. As previously discussed by Michel and Toussaint (1977), constant indentation strain rate requires that indentation tests be conducted at velocities that are determined by the size of the indenter, rather than constant velocities at all D/H values. Thus rate effects must be included if the present analyses are compared with experimental data.

The results listed in Table 2 show that the indenter shape and the slip condition at the ice indenter interface influence the ice pressure. This influence is as expected:

The no-slip (adfreeze) upper bounds are generally higher than the free-slip bounds. This difference is distinct for the wedge shaped and circular indenter. The slip condition at the ice-indenter interface has no or only little influence(within the range of accuracy of the limit analysis method)on the ice pressure on flat indenters.

Ice pressures on wedge shaped and circular indenters are lower than on flat indenters if a free-slip condition prevails at the ice indenter interface. In particular the pressure on free-slip, 90° wedge shaped indenters should be about 0.6 times the pressure on flat indenters, while the calculated pressure for circular indenters is about 0.85 times the flat indenter value. This is in good agreement with the experimental results by Korzhavin (1962) and Hirayama, Schwarz and Wu (1974), and the values for the shape coefficients used in many design guidelines. The load decreasing effect of the wedge and the semi-circle gets lost if the condition at the ice-indenter interface is no-slip. Then the pressure on the wedge shaped indenters should be the same as the pressure on flat indenters, the pressure on circular indenters may be even higher than that.

CONCLUSIONS

The analyses for the initial indentation of an ice sheet lead to the following conclusions:

- Plasticity theory provides a useful framework for describing ice strength. The anisotropic parabolic yield function used in this application can be used to describe the presently known ice strength behaviour.

- The plastic limit analysis solution procedure is an efficient means for estimating ice indentation pressures. This procedure can be readily used to transform basic ice strength data into prediction of ice indentation pressures. Since the ice strength data influence the pressure prediction it is important to recognize and use the most appropriate strength data.

- The presented results are confirmed by laboratory observations. In particular, the shape factors which may be derived from the results are in good agreement with previously published data. The results also imply that ice failure against wide indenters occurs at values larger than the uniaxial compressive strength C_X that has been used to determine or is implied by the yield function.

- The present theory is most applicable to laboratory tests conducted under carefully controlled conditions on carefully grown ice with homogeneous strength properties. The quantitative extension of this approach to full-scale applications requires that one is able to properly identify the (under full scale conditions) characteristic strength properties of the ice.

Acknowledgement - The author gratefully acknowledges BEB for granting permission to publish this work. The author also whishes to express his thanks to Dr. T.D. Ralston for this constructive criticism. The velocity fields for circular indenters were first suggested by Professor J.R. Rice of Brown University, Providence, Rhode Island.

REFERENCES

Carter, D., and B. Michel, 1971. "Lois et Mechanismes de l'Apparente Fracture Fragile de la Glace de Rivière et de Lac". Université Laval. Faculté des Sciences. Département de Génie Civil. Section Méchanique des Glaces. Rapport S-22.

Chen, W.F., 1975. "Limit Analysis and Soil Plasticity", Elsevier Press, Amsterdam.

Croasdale, K.R., Morgenstern, N.R., and Nuttall, J.B., 1977. "Indentation Tests to Investigate Ice Pressure on Vertical Piers", Journal of Glaciology, Vol. 19, No. 81, pp. 3o1-312.

Drucker, D.C., Prager, W., and Greenberg, H.J., 1952. "Extended Limit Design Theorems for Continuous Media", Q. Appl. Math., Vol. IX, No. 4, 1952.

Frederking, R., 1977. "Plane Strain Compressive Strength of Columnar-Grained and Granular Snow Ice", Journal of Glaciology, Vol. 18, No. 8o, pp. 5o5-516.

Gold, L.W., and Krausz, A.S., 1971. "Investigation of the Mechanical Properties of St. Lawrence River Ice" Can. Geotech. J., Vol. 8.

Hirayama, K., Schwarz, J., and Wu, H.-C., 1974. "An Investigation of Ice Forces on Vertical Structures", Iowa Institute of Hydraulic Research, Rep. No. 158.

Korzhavin, K.N., 1962. "Action of ice on engineering structures" U.S.S.R. Acad. Sci., Siberian Branch. Draft Transl. by Cold Regions Res. Eng. Lab. 1971.

Michel, B., and Toussaint, N., 1977. "Mechanisms and Theory of Indentation of Ice Plates", Journal of Glaciology, Vol. 19, No. 81, pp. 285-3oo.

Prodanovic, A., 1978. "Two-Dimensional Limit Analysis of Ice Failure Described by a Quadratic Yield Function" (Preprint), 15th SES Meeting, Dec. 1978, Univ. of Florida, Gainesville, Florida, U.S.A.

Ralston, T.D., 1977. "Yield and Plastic Deformation in Ice Crushing Failure" (Preprint), ICSI/AIDJEX Symposium on Sea Ice -- Processes and Models, Seattle, Washington.

Ralston, T.D., 1978. "An Analysis of Ice Sheet Indentation" Proc. IAHR Symposium on Ice Problems, pp. 13-31, Lulea, Sweden.

Reinicke, K.M., and Ralston, T.D., 1977. "Plastic Limit Analysis with an Anisotropic, Parabolic Yield Function", Int. J. Rock Mech. Min. Sci. & Geomech. Abstr., Vol. 14, pp. 147-154.

Reinicke, K.M., and Remer R., 1978. "A Procedure for the Determination of Ice Forces - Illustrated for Polycrystalline Ice -", Proc. IAHR Symposium on Ice Problems, pp. 217-234, Lulea, Sweden.

Vittoratos, E.S., 1979. "Existence of Oriented Sea Ice by the Mackenzie Delta" (Preprint), Fifth International Conference on Port and Ocean Engineering Under Arctic Conditions, Trondheim, Norway.

Wang, Y.S., 1979. "Crystallographic Studies and Strength Tests of Field Ice in the Alaskan Beaufort Sea" (Preprint), Fifth International Conference on Port and Ocean Engineering Under Arctic Conditions, Trondheim, Norway.

Weeks, W.F., and Gow, A.J., 1978. "Preferred Crystal Orientation in the Fast Ice along the Margins of the Arctic Ocean", Journal of Geophysical Research, Vol. 83, pp. 51o5-5121.

Ice Forces on Piles

H. SAEKI

A. OZAKI

Department of Civil Engineering

Hokkaido University, JAPAN

Abstract

Investigation was made on the ice forces on an isolated vertical pile and several vertical piles which stood in a line. In order to ascertain the various effects of factors which have influence on the ice forces on piles, many kinds of experiments were carried out. The factors considered in these experiments are the relative velocity between an ice floe and a pile, the sectional shape of pile, aspect ratio, and the interference among piles. Finally, it is found that the formula of the ice force on an isolated vertical pile proposed by the authors is widely applicable to the vertical pile structures.

Introduction

It is very important to investigate the ice forces on vertical piles for designing the off-shore structures in cold regions. The authors have experimented on the interaction between sea ice floes and piled structures. Factors which have influence on the sea ice force on an isolated vertical pile are the uniaxial strength and the thickness of the sea ice, the width and the cross-sectional shape of the pile structure, and the relative velocity between ice floe and the pile structure. For a row of piles the interaction between piles is a very important factor in addition to the factors mentioned above. It is the purpose of this paper to investigate experimentally the relations between ice forces on piles and the various factors mentioned above from the practical point of view. Experiments were carried out at Toppushi fishery harbour beside Lake Saroma which is connected with the Okhotsk Sea by a channel, 300 m wide. The ice used for these experiments was collected from Lake Saroma which had the salinity of 32 - 33 ‰. The mean diameter of grain was about 8 - 12 mm.

The effect of the relative velocity between a pile and an ice plate.

It is well known through the experimental results which were carried out by Schwarz (1) and Peyton (2) that the uniaxial compressive strength of sea ice is influenced by the strain rate and stress rate. Michel et al (3) carried out the indentation tests with indentation speeds varing from 6×10^{-5} cm/s to 3 cm/s and it was found that the ice force on rectangular indenter varied as a function of stress rate or strain rate.

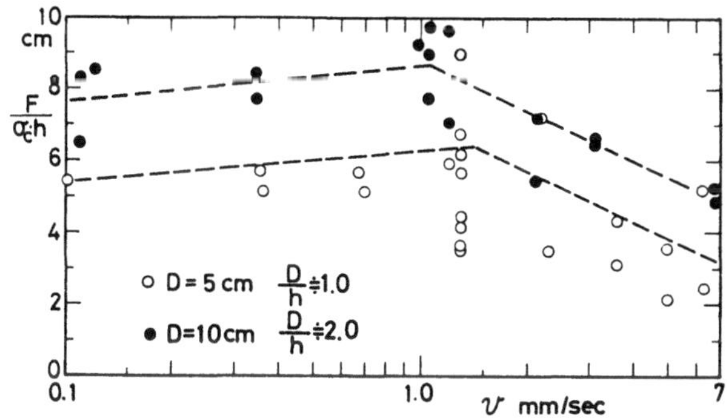

Fig. 1 Ice force on an isolated vertical pile as a function of relative velocity v.

The authors repeated the experiments on the relation between the ice force on an isolated vertical pile and the relative velocity v with a newly designed test apparatus (4). The result is shown in Fig. 1. The notation F, h, σ_c and v are the ice force on a pile, the thickness and the uniaxial compressive strength of sea ice plate, and the relative velocity between an ice plate and a pile. The uniaxial compressive strength σ_c in this experiment was obtained by a test with a cylindrical specimen of diameter of 10 cm and height of 20 cm at the strain rate $\dot{\varepsilon}$ = 0.001 - 0.003 sec^{-1} and at the st-stress rate $\dot{\sigma}_c$ = 1-3 $kg/cm^2 \cdot sec$ (5). The results shown in Fig. 1 suggest that the ice force $F/\sigma_c h$ has its maximum at the relative velocity v ≑ 1.00 mm/sec for both the aspect ratios.

Calculating the strain rate $\dot{\varepsilon}$ by using the equation, $\dot{\varepsilon} = v/4D$ which was obtained by Michel et al (3), the maximum ice forces on both piles, 5 cm and 10 cm in diameter, occur at the strain rate $\dot{\varepsilon} = 0.005\ sec^{-1}$ and $\dot{\varepsilon} = 0.0025\ sec^{-1}$ respectively and these strain rates are included in the transition region which was proposed by Michel et al.

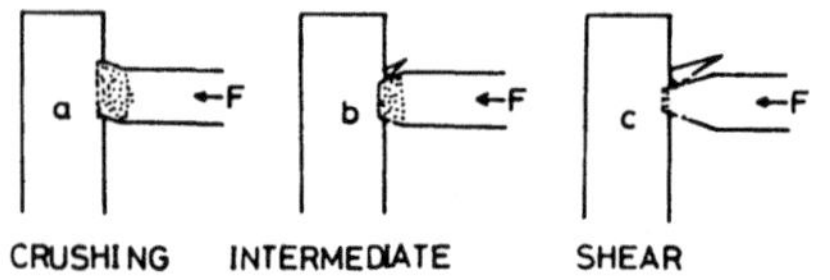

Fig. 2 Typical failure modes.

In the experiment, three modes of destruction of the ice plate colliding with the piles were: failure by crushing, shear or intermediate between them as shown in Fig. 2. The failure mode (a) occurs at low relative velocity and small strain rate, and the ice plate in front of a pile is destroyed by crushing. When the relative velocity is high and the uniaxial compressive strength or the elastic modulus of ice plate is large, the failure mode (c) occurs and at first sight the ice plate in front of a pile is destroyed by shear. The regions for three kinds of failure modes can be expressed by v and σ_c as shown in Fig. 3.

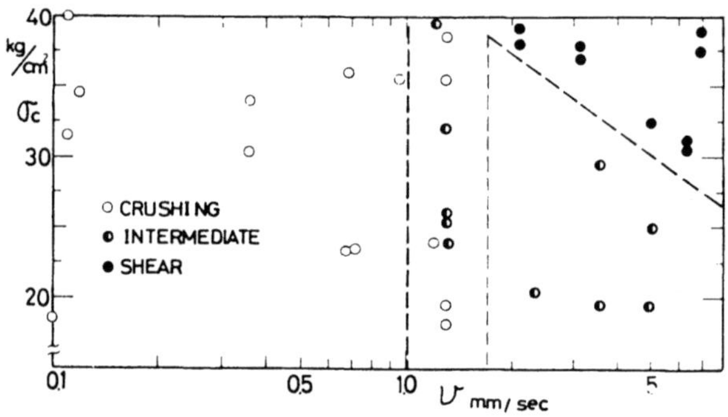

Fig. 3 Separation of failure modes by uniaxial compressive strength σ_c and relative velocity v.

The changes of ice forces with elapsed time for two failure modes, (a) and (c), are shown in Fig. 4.

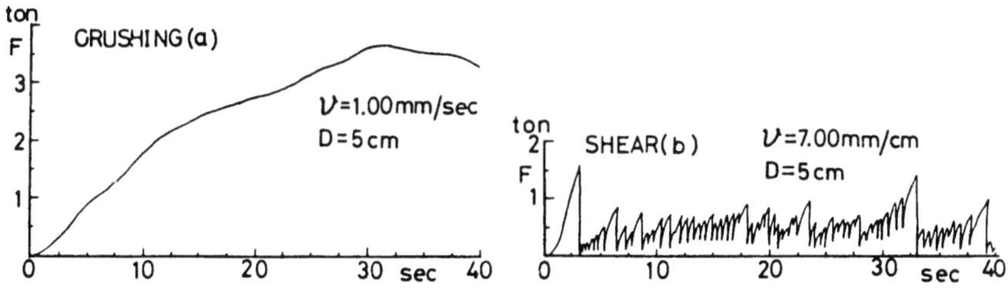

Fig. 4 Typical recorder traces for slow and fast relative velocities.

In the case of failure mode (c), the resonance between a pile structure and ice force is awaited with anxiety. Judging from the experimental results mentioned above, the maximum ice force on a pile occurs at the strain rate $\dot{\varepsilon} = 0.0025 - 0.005\ sec^{-1}$, and between the regions of two failure modes, crushing and intermediate. Authors carried out the same experiments on the floating infinite ice plate on Lake Saroma. The thickness of the sea ice was 20 cm and a rectangular pile 50 cm wide was used. In this experiment, the maximum ice force on the pile occurred at the relative velocity v = 2.5 - 3 mm/sec and this relative velocity corresponds to the strain rate $\dot{\varepsilon} = 0.00125 - 0.0015\ sec^{-1}$ which belongs to the transition region proposed by Michel et al (3). After this, experiments on ice forces are carried out at the relative velocity v = 1.0 mm/sec except the field test.

The effect of three different pile shapes.

Authors repeated the experiments for piles of three kinds of sectional shapes, circular, rectangular and wedged. The relative velocity in the experiments was 1.0 mm/sec. The experimental results for each pile are shown in Fig. 5, 6, and 7.
The notation B' is the diagonal length of a wedge-shaped pile with wedge angle of 90°. From those results, it can be concluded that the ice force on an isolated vertical pile is proportional to the ice thickness and the square root of the maximum width of the pile without referring to the shape of pile. Thus the ice force on a pile is expressed by Eq. 1.

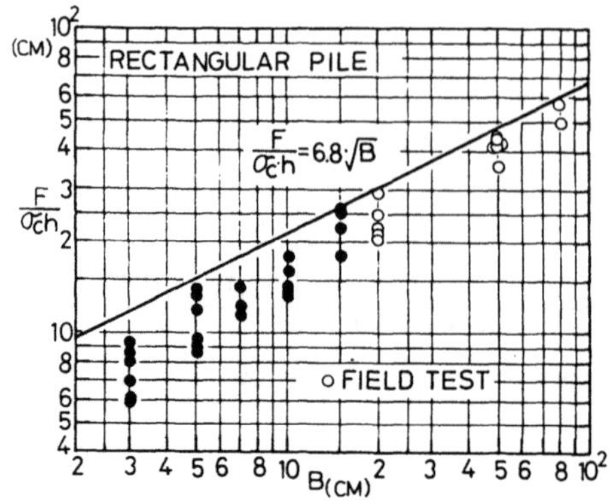

Fig. 5 Ice forces on a rectangular pile, (Saeki et al, 1977)

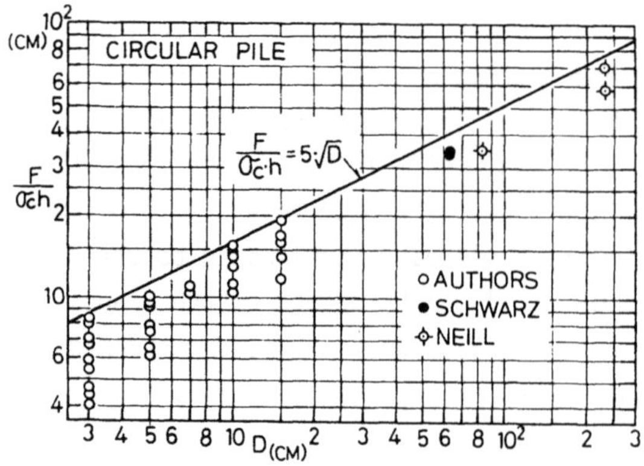

Fig. 6 Ice forces on a circular pile, (Saeki et al, 1977)

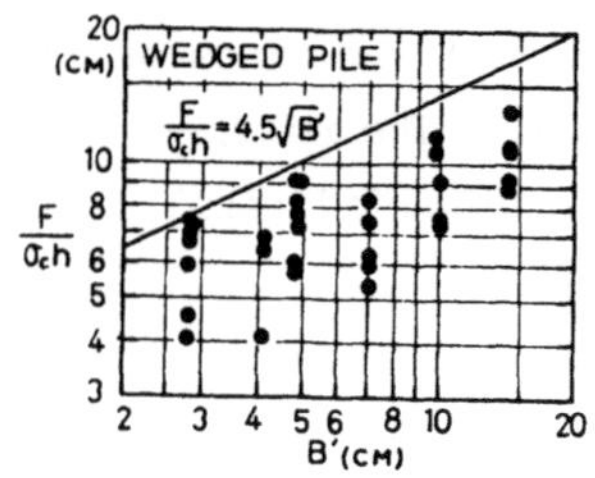

Fig. 7 Ice forces on a wedged pile with wedge angle of 90°.

$$F = C \cdot \sqrt{W} \cdot h \cdot \sigma_c \qquad (1)$$

W : width of pile (cm) h : thickness of ice floe (cm)

σ_c : uniaxial compressive strength measured under the established conditions (kg/cm^2).

C : shape factor of pile which is 5.0 for circular pile, 6.8 for rectangular pile and 4.5 for wedged pile with wedge angle of 90° ($cm^{1/2}$).

The scattering of experimental values is caused by the deviation of the uniaxial compressive strength σ_c. Korzhavin recommends the use of a shape factor for various pile shapes. Various investigators have indicated different sets of shape factors, but the mutual relations within each set coincides with what is found by the authors (6). Figure 5 shows the relation between $F/(\sigma_c \cdot h)$ and B. Both the experimental data in the small scale test and the data in the field test can be expressed by Eq. 1. This fact proves that the measuring apparatus for ice force which was developed by authors can represent the interaction between a pile and an ice floe well.

The effect of aspect ratio.

Recently, it is found by Ralston (7), Reinike et al (8) and Croasdale et al (9) by using the method of the plastic limit analysis that the effective ice pressure ($F/\sigma_c \cdot h \cdot B$) on a rectangular pile can be expressed as a function of the aspect ratio (B/h). Equation 1 proposed by authors may be written

$$\frac{F}{\sigma_c \cdot h \cdot W} = C \cdot \left(\frac{W}{h}\right)^{-\frac{1}{2}} \cdot \frac{1}{\sqrt{h}} \qquad (2)$$

Equation 2 suggests that the effective ice pressure can be expressed as a function of the aspect ratio and the thickness of ice.
In the experimental formula proposed by Hirayama et al (10), the effective ice pressure is expressed as a function of the aspect ratio and the ice thickness. In order to ascertain the effect of aspect ratio, the experiments on ice force on a rectangular pile were carried out for two aspect ratios, B/h = 1.0 and 2.0. The experimental results are shown in Fig. 8. Though the experimental data show scattering, the envelope of the data is in good agreement with Eq. 2. From the above complementary experimental results, it can be conclu-

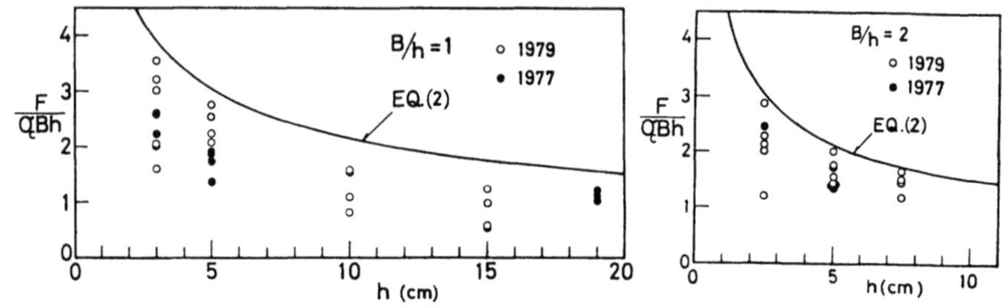

(a) (b)

Fig. 8 Effective ice pressure on a rectangular pile versus ice thickness in constant aspect ratio.

ded that Eq. 1 obtained by the authors is very useful for the design of coastal and off-shore structures.

Impact ice force on a pile

In testing the impact ice force on a pile, a special experimental apparatus was used (4). Experiments of the ice force on a circular pile were carried out in this test. The magnitude of the impact force which is exerted at the time of collisions is determined by the mass, velocity and strength of ice floe. The velocity of collision was kept constant (7.6 m/sec) but the mass was varied.

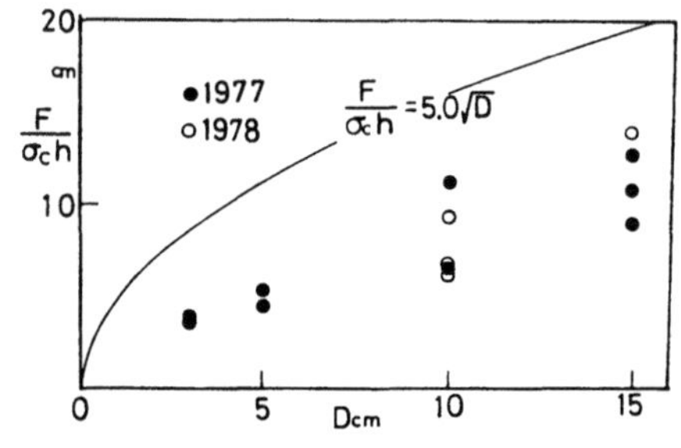

Fig. 9 Impact ice forces on a circular pile.

Figure 9 shows the relation between $F/(\sigma_c \cdot h)$ and D. The momentum of the ice plate fixed in the steel box is large enough to allow penetration into a pile instantaneously. The impact ice force does not exceed that found by use of Eq.1. This phenomenon can be understood by considering the facts that the compressive strength of sea ice has the maximum value at the stress rate of $\dot{\sigma}_c = 1 - 3$ kg/cm²·sec or at the strain rate of $\dot{\varepsilon} = 0.001 - 0.003$ sec^{-1}, the indentation pressure on a rectangular indenter has the maximum value at the strain rate of $\dot{\varepsilon} = 0.001$ sec^{-1} in the indentation test carried out by Michel et al, and the higher the relative velocity becomes the smaller the contact coefficient becomes in Korzhavin's formula (6).

Ice forces on piles which stand in a line.

It is expected that the ice forces on piles which stand in a line are influenced by the intervals between piles. If the intervals are large enough in comparison with pile diameter, the interference between piles would not occur. Saeki et al carried out the experiments on the ice forces on three vertical piles which stood in a line at intervals of K · D by using a newly designed experimental apparatus (11).

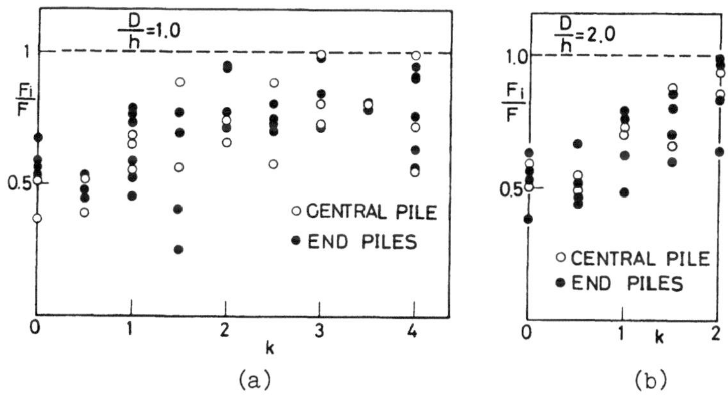

Fig. 10 Ice forces on three piles which stand in a line.

Figure 10 shows the relation between F_i/F and K. F can be calculated by Eq. 1 and F_i is the ice force on each pile. As the values of F_i/F are smaller than 1.0, the ice force on each pile is smaller than F. Judging from the experimental results mentioned above, within a range of K $\geq$ 1.0, the interference between piles does not occur and Eq. 1 is applicable to the estimation of ice forces on piles which stand in a line.

Acknowledgement

We wish to express our profound gratitude for the extensive cooperation given by Mr. Toshiyuki Ono, Mr. Ken-ichiro Hamanaka and Mr. Kazunori Nishiya of Hokkaido University in the experiments. This work was partly supported by the Scientific Research Funds from the Ministry of Education.

References

1 Schwarz, J.: The pressure of floating ice-fields on pile. Proc. IAHR Ice Symposium, Reykjavik, (1970).

2 Peyton, H. R.: Sea ice strength. Geophysical Institute, University of Alaska, Report No. NR-307-247/7-6-55, (1966).

3 Michel, B.; Toussaint, N.: Mechanics and theory of indentation of ice plates. Jour. of Glaciology, Vol. 19, No. 81, (1977) 285-300.

4 Saeki, H.; Hamanaka, K.; Ozaki, A.: Experimental study on ice force on a pile. Proc. POAC 77, Vol. 2, (1977), 195-706.

5 Saeki, H.; Nomura, T; Ozaki, A.: Experimental study on the testing methods of strength and mechanical properties for sea ice. Proc. IAHR Symposium on Ice Problems, Luleå, Sweden, (1978), 135-150.

6 Michel, B.: Ice pressure on engineering structures, U. S. Cold Regions Research and Engineering Laboratory. Cold Regions Science and Engineering. Hanover, N.H., Pt III, Sect. Bib. (1970).

7 Ralston, T. D.: An analysis of ice sheet indentation. Proc. IAHR Symposium on Ice Problems, Lulea, Sweden, (1978), 13-32.

8 Reinicke, K. M.; Remer, R.: A procedure for the determination of ice forces- illustrated for polycrystalline ice-. Proc. IAHR Symposium on Ice Problems, (1978), 217-238.

9 Croasdale, K. R.; Morgenstern, U. R.; Nuttall, J. B.: Indentation tests to investigate ice pressures on vertical piers. Jour. of Glaciology, Vol. 19, No. 81, (1977), 301-311.

10 Hirayama, K.; Schwarz, J.; Wu, H. C.: Ice forces on vertical piles indentation and penetration. Proc. IAHR Symposium on Ice Problems, Hanaver, U. S. A., (1975), 429-446.

11 Saeki, H.; Ono, T.; Ozaki, A.; Abe, S.: Estimation of sea ice forces on pile structures. Proc. IAHR Symposium on Ice Problems, Luleå, Sweden, (1978), 465-478.

Failure of Sea Ice by Repeated Compression

T. TABATA, Y. NOHGUCHI
The Institute of Low Temperature Science
Hokkaido University, Sapporo, Japan

Abstract

Sea ice samples of Lake Saroma, Hokkaido, Japan, and Barrow Alaska were subjected to two uniaxial compression tests: the normal compression test and the fracture test in which a cyclically changing load was applied to a sample so that compression and stress relaxation were repeated between two fixed levels of stress until the sample fractured, the average of them being about 30% to 60% of the compressive strength. The latter test seems to be equivalent to the creep test in that the average stress is kept constant throughout the test. The duration of time for a sample to fracture decreased with an increase in average stress or temperature or both. In the latter test, although the stress in each compression process is small, the total amount of mechanical work for a fracture to take place is much larger than that by the former test.

Introduction

A behavior of a sea ice failure due to a cyclically changing load was first observed from a change in force applied to a marine tower (built by Mitsui Shipbuilding & Engineering Co., Ltd., in 1975 about 600 m off Monbetsu located in the Okhotsk Sea coast of Hokkaido, the northernmost island of Japan) by a pack ice movement. Figure 1 gives an example of a recorded curve of the force, which oscillates in a range between 90 tons and 30 tons, for a 3.5-minute long, that is, ice force increased to 90 tons by a pack ice movement and then, relaxed to 30 tons, this period is followed by a sudden drop in force to about 10 tons; that is, the failure of ice takes place. The frequency of oscillation being about 20 sec, and the resonance frequency of the tower being about 1/15 sec. In light of little knowledge of

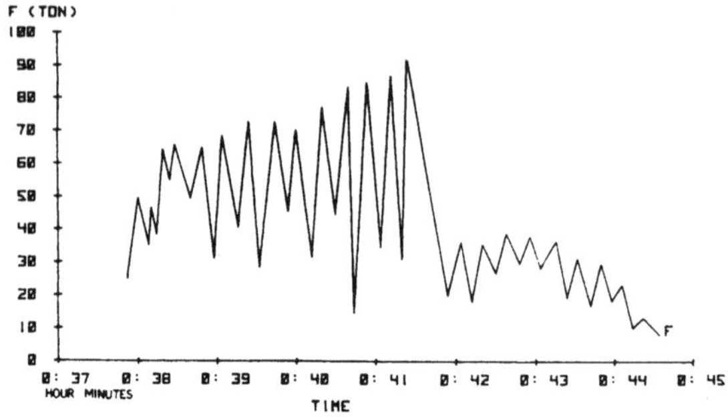

Figure 1. A recorded curve of force applied to a marine tower by a pack ice movement.

this behavior, an uniaxial compression test was undertaken by repeating compression and relaxation of stress between two levels of stress by applying a cyclically changing load to a sea ice sample.

A test was carried out in February, 1977 and 1978, in salt Lake Saroma located in the Okhotsk Sea coast of Hokkaido (Nohguchi and Tabata 1977,1978). The sea water of the lake has a salinity of 30% and the maximum thickness of sea ice is about 40 cm. Another test was carried out at Naval Arctic Research Laboratory at Barrow, Alaska in April, 1978 (Nohguchi and Tabata 1978). In both tests observations were made not only of the fracture of sea ice by a repeated application of a load but also of the normal uniaxial compression of sea ice.

Test method

A cylindrical ice sample was obtained from the ice cover so that the long axis of the sample was vertical or parallel to the surface of the ice cover, the former being referred to as a vertical sample and the latter as a horizontal sample. The length of the sample ranged from 75 mm to 200 mm; the diameter was 75 mm or 100 mm.

A portable compression machine, 10 tons in capacity, was used for both the uniaxial compression test and the fracure test by use of repeated compression. In the compression machine a platen was moved by an electric motor so that the speed of its movement could be varied from 7 mm/min to 50 mm/min; a slower speed was obtained by moving the platen manually. The load applied to a sample was measured by an electric load cell and the displacement of the platen was measured by an electric dial gauge. Both were recorded.

Because the friction coefficient between sea ice and stainless steel is very small (Tusima and Tabata, 1979) a stainless steel plate was inserted between the sample and the platen.

The conventional uniaxial compression test was made with the constant strain rate. Each test had a different strain rate but it was ranged between 10^{-2} to 10^{-3}/sec.

In the fracture test in which a load was cyclically changed two stress levels were determined. When the stress reached the high level while a sample was compressed at a constant strain rate, the movement of the platen was stopped so that the applied stress was relaxed to allow lowering to the scheduled low level. When it reached this level, the sample was compressed again at the constant strain rate until the stress reached the scheduled high level. By repetition of this sequence the stress applied to the sample was kept nearly constant until the sample fractured. Each test had a different high level of stress, the largest one being nearly the uniaxial compressive strength of the sample and the smallest one being about a half of it. The amplitude between the high and the low stress was changed from several percent to several tenth percent of the average

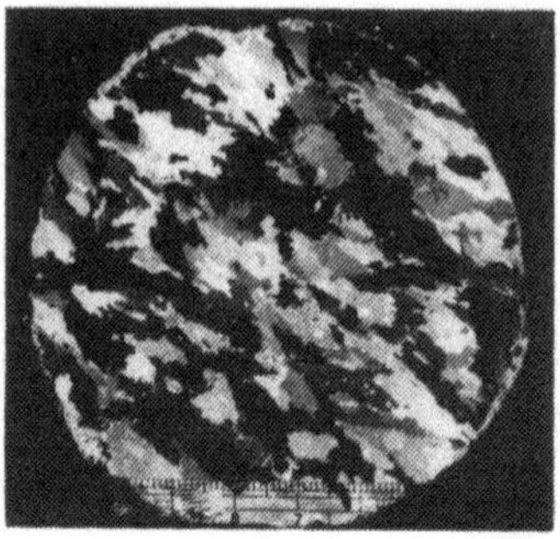

Figure 2. A cross section of an ice sample.

value of the applied stress. The strain rate of compression was 5 x 10^{-4}/sec. After the fracture, some samples were melted for making a salinity measurement. Meantime, a thin section of ice (Figure 2) was prepared to have the crystallographic structure of the sample examined by cross-polarization.

Test results

1. Type of fracture

Three main types of fracture, which are seen in Figure 3, were observed. When the strain rate was larger than 10^{-4} /sec, all the samples bursted, whereby most of the samples showed that they broke into many small vertical columnar segments (Figure 3-a) and others showed a shear fracture (Figure 3-b and -c). Cases like Figure 3-c were seen only in horizontal samples. When the strain rate decreased and the temperature was relatively high, some ice samples did not break up at the maximum stress, but they broke up at the moment the stress was relaxed to 2-3 kg/cm^2 from the maximum stress. The percentage of such a fracture increased with decreasing strain rate and increasing temperature. In the case in which the strain rate was low, the stress decreased slowly after having reached the maximum stress, whereupon the sample did not break up although many small vertical cracks were seen in the ice.

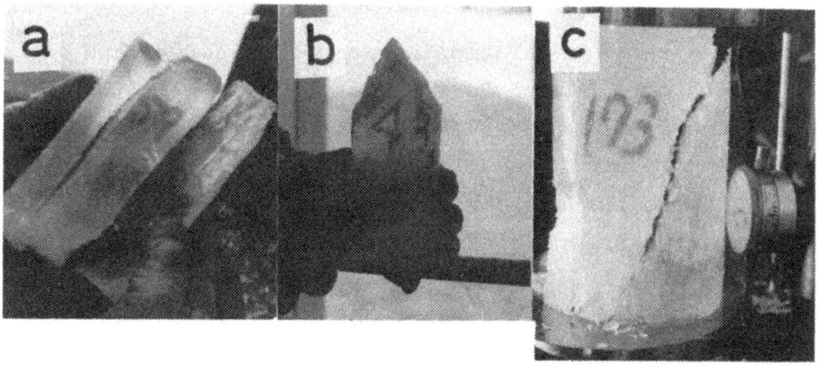

Figure 3. Three types of fracture of samples.

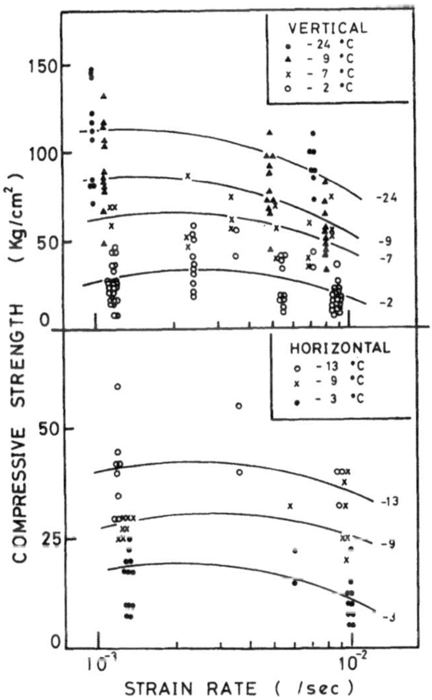

Figure 4. Strain rate dependency of compressive strength.

2. Strength

Figure 4 shows a relation between strain rate and strength obtained at Lake Saroma in 1977. It is clear from the figure that the strength increases with decreasing temperature. The strength of vertical samples is about twice that of horizontal samples. Despite a large number of discrepancies among each group, as was stated by Dr. Schwarz, the strength appears to have a maximum value in the strain rate between 10^{-3}/sec and 10^{-2}/sec. The strength decreases with increasing salinity of ice, as is seen in Figure 5.

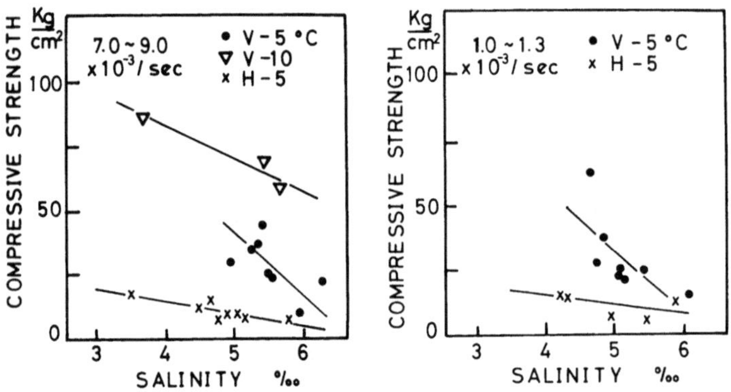

Figure 5. Salinity dependency of compressive strength.

Repeated compression test

1). Stress-time curve

An example each of stress-time and strain-time curves is presented in Figure 6. Time t_1 during which the stress is increasing and time t_2 during which the stress is getting relaxed are obtained for each sequence of the stress. Time t_1 is proportional to the strain and almost constant during each sequence; time t_2 increases gradually in the initial period of the test and then decreases gradually, as is seen in the figure, when an amplitude

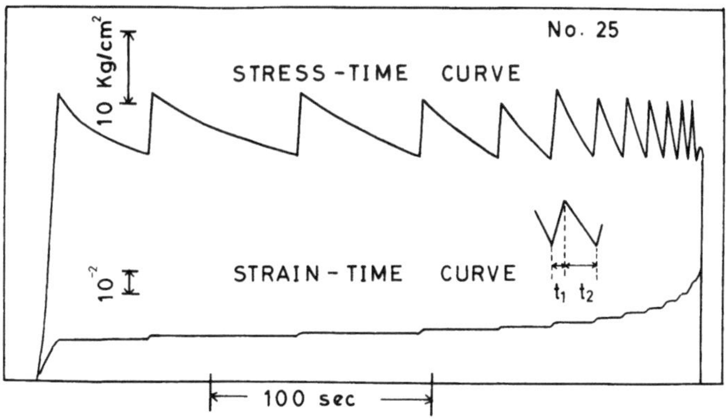

Figure 6. An example each of stress-time and strain-time curves of the repeated compression test.

of stress between the high and the low level is small, compared with the average stress applied. It is also seen in the figure that the initial period of the test during which time t_2 increases in this strain-time curve is equivalent to a range from the primary creep to the secondary creep in the strain-time curve of the uniaxial compressive creep test; that the following during which time t_2 decreases with the repeated sequences is equivalent to the tertiary creep. Since the average stress is kept constant throughout the cyclic test, the repeated compression test seems equivalent to the uniaxial compressive creep test.

In the repeated test with decreasing average stress and also with decreasing amplitude between the high and the low level of stress, the time required for a sample to fracture increased. With lowering temperature, such time and also the number of repeated sequences increased.

2) Time for fracture

The obtained relation between average stress and duration of the time required for a sample to fracture is pres-

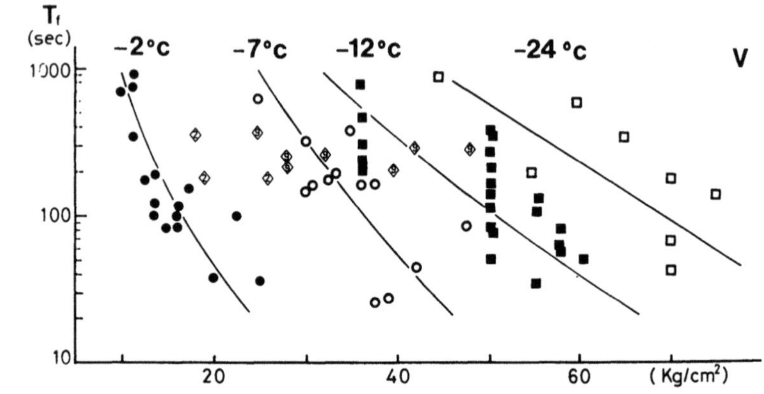

Figure 7. Relation between average stress applied and time for fracture. (a) : Vertical samples.

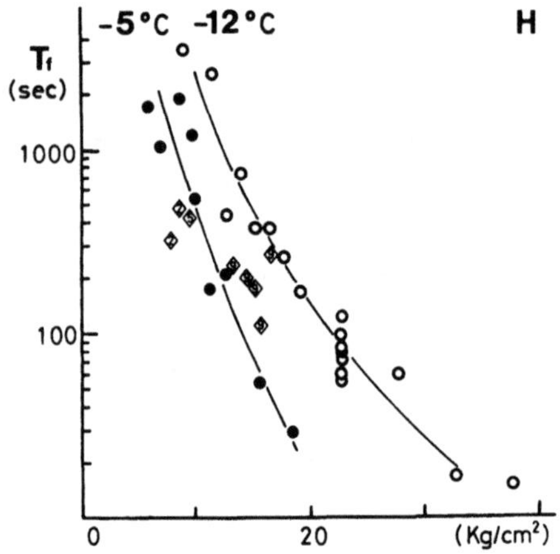

(b) : Horizontal samples.

ented in Figure 7 (a) (vertical samples) and 7 (b) (horizontal samples). The dimension of samples are : 79 mm in diameter; 84 mm and 100 mm in length. Rhombic plots in the figure show data obtained by the uniaxial compression test with the strain rate of 10^{-4}/sec to 10^{-5}/sec; numerals in them represent temperatures short of the minus sign at which the test was carried out. It follows from this fiqure that the duration of time for a sample to fracture increases with decreasing stress and lowering temperature.

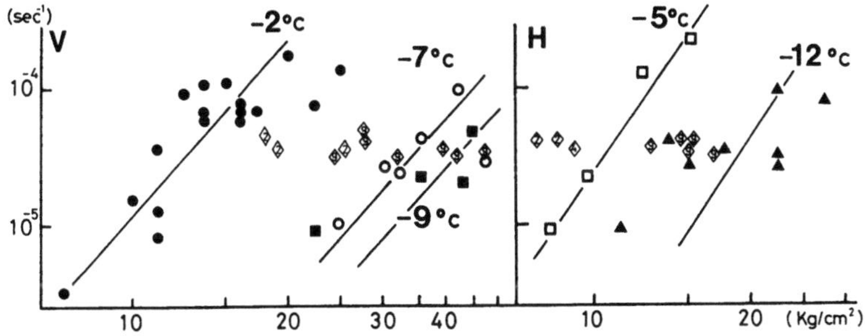

Figure 8. Relation between average stress and observed minimum strain rate.

3) Stress and strain

Figure 8 shows a relation between average stress and observed minimum strain rate. In the figure, V and H mean the data obtained from vertical and horizontal samples respectively. Also the same data of the uniaxial compression test showing in Figure 7 are again seen in the figure. As is the case with the uniaxial compression test at a constant strain rate and at the same temperature, the minimum strain rate increases with increasing average stress. The minimum strain rate of horizontal samples for the same average stress is much larger than that of vertical samples. Since the strain rate in the strain-time curve is inversely proportional to the time required for stress relaxation, the minimum strain rate may be equivalent to the minimum stress relaxation velocity. A relation of both quantities is presented in Figure 9. As is seen in the figure, similar information can be obtained from the stress record of the repeated compression test even when data of strain are lacking.

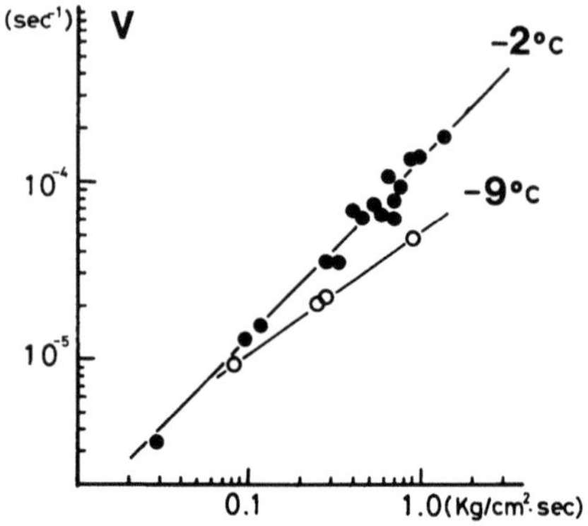

Figure 9. Relation between minimum strain rate and minimum stress relaxation velocity in the repeated compression test.

4) Number of repetition

The number of repeated sequences of stress may depend on the nature of a sample, average stress, amplitude of the two stress levels and stress rate during compression. If an increase in stress takes a long time until it attains the high level of stress, the number of repeated sequences becomes small. In the present test, the total time during which relaxation takes place is much longer than the total time during which compression takes place; hence, the effect of the speed of an increase in stress on the number of repetition may be neglected. The number counted ranged from 2 to 24.

5) Equivalent strength

Considering the total amount of energy required for a sample to fracture by the repeated compression test, the authors obtained the following equation, which converts the number of repetitions, n, the amplitude between two stress levels a and average stress b into the compressive strength c.

$$C = \sqrt{2abn + b^2} \qquad (1)$$

The relation between strength thus defined and temperature is presented in Figure 10. The strength increases with lowering temperature, and it is greater than that obtained from the uniaxial compression test.

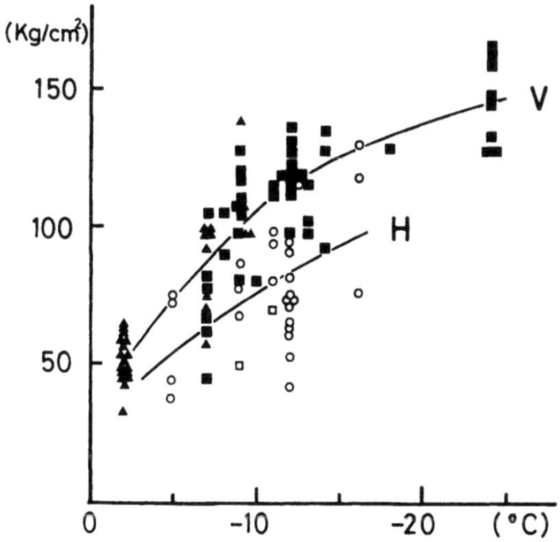

Figure 10. Temperature dependency of calculated strength for vertical and horizontal samples.

It is concluded that in the repeated compression test, although the stress in each compression is small, the total amount of energy required for a sample to fracture is much larger than that in the normal uniaxial compression test.

Conclusion

Sea ice was subjected to two uniaxial compression tests: the normal compression test at a constant strain rate and the fracture test under a cyclically changing load, with the results obtained as follows :

(1) The uniaxial compressive strength obtained from a vertical sample was about twice that obtained from a horizontal sample.

(2) The uniaxial compressive strength increased with a decrease both in temperature and salinity.

(3) A repeated compression test seems to be equivalent to a compressive creep test since the average stress throughout the cyclic test is kept constant. In fact, the primary, secondary and tertiary creeps were obtained from changes in relaxation time during each sequence of

repetition.

(4) With a smaller stress than that of the compressive strength of sea ice, even less than a half of it, sea ice broke up by repeated compression.

(5) The time required for a sample to fracture decreased with an increase both in average stress and temperature.

(6) Although the stress was small while a sample was compressed in the repeated compression test, the total amount of energy required for a sample to fracture was much larger than that by the normal compression test.

Acknowledgement

The authors would like to express their heartfelt thanks to: Drs. N.Ono and M.Aota, T.Kawamura and T.Takizawa of Sea Ice Research Laboratory, and also to students T. Saito, K. Ogawa, and Drs. K.Nishimura for their kind assistance. They are also much indebted to Drs. J.Kelley, W.Weeks, T.Gow, L.Shapiro and Mr. L.Metzner for their kind help in the field work at Barrow, and to the Office of Naval Research for the permission to use the facility of NARL , Barrow, Alaska.

References

1) Nohguchi,Y.;Tabata, T. : Failure of sea ice by compression, Low Temperature Science, Ser.A.35.(in Japanese with an English summary)(1977).

2) Nohguchi, Y.; Tabata, T.: Failure of sea ice by repeated compression, Low Temperature Science, Ser. A.37. (in Japanese with an English summary)(1978) 63-68.

3) Nohguchi, Y.; Aota, M.; Tabata, T.: Arctic Sea Ice Research IV. Measurement of uniaxial compressive strength, Low Temperature Science, Sea.A.37.(1978) 131-139.

4) Tusima, K.; Tabata, T.: Friction measurements of sea ice on flat plates of metals, plastics and coatings, POAC, 13-18 Aug. 1979, Trondheim, Norway, Proceedings.

5) Schwarz, J. : The pressure of floating ice field on piles, In "IAHR Symposium, ice and its action on hydraulic structures" (Reykjavik, Iceland), IAHR Committee on Ice Problems, Paper 6.3.12pp.(1970).

Discussion of Papers

1. Introduction
2. Written questions and answers
3. Resumé of discussion
4. Movies shown at symposium

1. Introduction

Only a limited number of written questions were submitted. They are quoted together with the answers in the following. I had intended to have a more detailed part of the discussion presented in the proceedings, but the problem is that only some of the questions were short and simple. Most questions were rather broad elaborations on the problems - i.e. a dialog between lecturer and questioner- which it is very difficult to put in writing from a recording. A short resumé of some of the topics, which were raised and discussed during the symposium is also presented in the following.

2. Written questions and answers

Question to A. Assur from K. Hutter.

Do you know how a limit analysis like yours would compare with a calculation using a flow law. My suspicion is that with a creep analysis you might obtain smaller loads.

Answer:

An analysis using a creep (flow) law or a rate dependent theory in general should result in smaller indentation pressures. Such an analysis is desirable but to my knowledge has not been performed yet. I would like to emphasize that the flow law used in such an analysis must be capable of describing the (initial) anisotropic and pressure sensitive strength behaviour of the ice.

——— o ———

Question to A. Assur from P.R. Kry.

The concept of a strain failure criterion for ice is potentially a very useful device, especially in long term, or fatique-related bearing capacity problems where failure of the ice could be predicted by monitoring strain. However, for the problem of ice forces on structures, the use of a strain failure criterion does not necessarily simplify the prediction of forces at failure. It would appear that it shifts the problem from determining the ice failure as a function of applied force, to determining the applied force of the strain corresponding to failure. Do you have a comment as to whether this is a simpler problem considering uncertainties in constitutive equations.

Answer:

The differential equations should be solved in terms of the location and time of allowable strain rather than allowable stress. This has not been done as yet even for simple cases.

——— o ———

Questions to R. Frederking from A. Assur.

1. Eqn. 2 if expressed in terms of the friction angle ϕ = arc tan μ becomes tan $[\alpha + \phi]$ which is mathematically more tractable [1].

[1] Assur, A., 1971. POAC, Trondheim.

2. The shape of the structure, discussed by Frederking is similar to barges with a straight front bow, which are remarkably efficient in breaking ice on the Mississippi River. The reason: The path in the ice broken at the front of the barge is slightly wider than the barge avoiding excessive side friction with crushing failures of ice pieces pressed against the existing ice sheet. In comparison the shape of the MANHATTAN bow virtually produced such undesirable effects. It is a promising path to explore.

Answer:

Assur is quite correct in stating that Equation 2 can be expressed in a mathematically more convenient form. In doing so, however, one must be careful not to lose contact with the physics of the problem.

The basis of selecting the structure shape for this study was more for analytical convenience than for optimizing ice breaking resistance. A simple inclined plane structure, as used here, causes the ice to fail mainly in bending. The resulting broken ice translates along the plane with relatively low contact pressures and hence low frictional forces. Using a similar shape for the bow of an ice breaking ship would likely result in improved breaking performance due to the absence of ice crushing and the corresponding high frictional loads as well as reduced side friction loads. Other factors, however, would have to be considered; such as sea-keeping characteristics and clearing ice around propellers.

——— o ———

Question to L.W. Gold from D. Goodman.

While I agree that your model provides a good description of transient creep, should you not include an additional term to represent diffusional creep by vacancy transport. This mechanism will operate independently of the dislocation mechanism which is probably responsible for your empirical equation.

The constitutive equation for diffusional creep is:

$$\dot{\gamma} = \frac{a_0 \Omega \, \text{Deff} \, \sigma}{KT \, d^2}$$

where $\text{Deff} = Dv \left| 1 + \frac{\zeta\delta}{d} \frac{D_\beta}{D_v} \right|$ and

$$D_\beta = D_{\beta 0} \exp\left(-\frac{\Delta F}{RT}\beta\right) \qquad D_v = D_{v0} \exp\left(-\frac{\Delta F}{RT}v\right)$$

Where $a_0 = 42/\pi$, $\Omega = 3.27 \times 10^{-29} M^3$, K is Boltzmann's constant, T the absolute temperature, d the grain size, δ the grain boundary width, ΔF_β activation energy for boundary diffusion, ΔF_v activation energy for volume diffusion, $D_{\beta 0}$ and D_{v0} preexparential constants and σ is the shear stress.

Answer:

The model presented in the paper is based on observed behaviour and not on an assumed contribution of individual deformation mechanisms. It is very probable that diffusion creep contributes to the permanent strain, but for the stress levels at which we have been working, and for observations available in the literature, it would appear that the term involving stress raised to the power 3 still dominates the viscous contribution. These observations indicate that diffusion controlled creep only becomes significant relative to creep described by a power law dependence for stress appreciably less than $0.1 \, MN/M^2$.

——— o ———

Question to D.J. Goodman from A. Assur.

Using crack propagation theory it is possible to derive scale or size effects for the failure of beams. Did you find experimental evidence of this nature?

Answer:

No. The specimen size chosen ensures that the centre of the crack remains in plane strain, and the apparent plastic zone size is considerably smaller than the distance from the end of the specimen to the back surface of the specimen.

——— o ———

Question to K. Hutter from A. Assur.

It is gratifying to see from Dr. Hutter's work that the concepts of adjusted flexural properties, used by us since the middle fifties and published in 1966 (see also improved version for design purposes [1]) withstood the test of times and the most rigorous analysis. The method, of cource, strictly speaking does not work for Poisson's ratio varying with depth but Dr. Hutter showed that errors introduced by a constant Poisson's ratio are negligible.

A varying Poisson's ratio can be considered exactly in its effect on the plate stiffness but only approximately in summing up movements. Nevertheless I suggest to use an average, weighted by the effect of each layer on the bending resistance at the place of maximum flexural stress. This, in principle, is not exactly correct but it is a reasonable compromise; it assumes, of course, that we know more about the variation of Poisson's ratio with salinity, temperature etc., than we do know now.

[1] Weeks W. and Assur A. (1967): The mechanical properties of sea ice USA CRREL Meteograph II-c-3.

Answer:

I appreciate your comment and would like to respond to it in the following way: In my paper, Ref. [9], the Poisson ratio may arbitrarily vary with depth. Yet the demonstration that errors induced by neglecting this variation in sea ice are negligibly small is based on a formula given in the paper of yours and Weeks you mentioned. This proof simply involved numerical integrations of expressions involving Young's modulus and Poisson's ratio errors the thickness of the plate. Of course, I agree that once Poisson's ratio may without essential loss of accuracy be set constant there still remains the problem which value this constant should have. What you propose seems to be a valid approach I can agree to. Yet, since I do not expect the dependency of Poisson's ratio to have a greater effect than to lead to variations between 0.3 and 0.4, roughly, I do not think that it would matter much whether one takes a certain average of .7, or a mean value 0.35. My position regarding a better understanding of the dependency of Poisson's ratio on brine is clear. Since it will not matter much anyhow, it need not be known.

——— o ———

Question to K. Hutter from A. Assur.

The method shown may lead to overdesign. In nature tensile cracks form in front of the indenter before the sliplines develop. This limits the extent of slipline formation. Also the behaviour in an infinite field should be explored in addition to the indentation of an edge.

Answer:

The method presented has been used to describe laboratory observation. Extension to design would require an assessment of full scale ice behaviour. The presence of the cracks that you describe would indeed reduce failure loads for both half plane and infinite field ice sheets.

——— o ———

Question to A.D. Kerr from (not identified)

Are there other implications of the analytical findings of the simple model study for actual ice cover problems?

Answer:

The parameter k* represents the ratio of the stiffness of the base to the stiffness of the floating ice cover. For floating covers the base stiffness is constant whereas the plate stiffness reduces with reducing thickness of the ice cover. Thus, the k* value that corresponds to a thin floating plate will be much larger than the k* value for a thick plate, which is made of the same ice.

The post-buckling branch depends on the magnitude of the k*-value. Therefore, the buckling loads obtained from tests with very thin ice plates should be used with caution for the interpretation or prediction of the buckling response of thick ice covers. The results shown in Fig. 5 also suggest that when using thin plates in tests special attention should be paid to the possible occurrance of lift-off and its effect on the obtained results.

——— o ———

Question to A.D. Kerr from (not identified)

What value of Young's modulus should be used when evaluating a buckling load, especially noting that the deformational behavior of ice is strongly time dependent.

Answer:

An analytical model should correspond as closely as possible to the physical problem under consideration. Since an analysis usually has many simplifying assumptions the coefficients which enter a given formulation should be determined in such a manner that the analytical results agree, as closely as possible, with the results of the actual structure.

The formulations presented in the paper assume an elastic response. Hence, the obtained results are valid when buckling takes place very shortly after the plate is subjected to the axial loads. For this case Young's modulus is the averaged modulus (along the cover thickness) as shown in "Forces an Ice Cover Exerts on Rows or Clusters of Piles due to a Change of the Water Level," by A.D. Kerr, Proc. IAHR Symposium on Ice Problem, Luleå, Sweden 1978, page 515. If the axial loads are increased very slowly, and time dependent in-plane deformations do take place before buckling failure, an elastic Young's modulus may still yield reasonable results when it is justified to assume that the buckling deformations occurred instantaneously.

As another example of how Young's modulus E or flexural rigidity D should be chosen, consider the case when the buckling region is subjected prior to buckling to a closely spaced fine network of cracks, as is sometimes observed in tests. Then, although buckling may take place in the elastic range, an effective (reduced) D value should be used, which reflects this situation.

——— o ———

Question to B. Michel from K. Hutter

Your theory and tests seem to have been developed and performed for increased loads. Can you foresee what will go on for unloading?

Answer:

In this theory the micro-cracking activity ceases as soon as there is a decrease in load and the strain recovery curve can be well predicted with the basic model $\alpha\nu$ creep without cracking activity.

——— o ———

Question to B. Michel from R. Frederking

Could you comment on the physical meaning of the parameters α, κ and m and if it is yet possible to predict these values for a particular set of test conditions?

Answer:

The α and m parameters relate to the increased number of mobile dislocations in function of the deformation. The m factor will eventually be known for a given type of ice once a larger number of creep curves have been analyzed with this model. As for the α factor, it is essentially a function of the initial number of mobile dislocations and the number of dislocation loops controlling their multiplication. This depends essentially on the way the ice has crystallized and the amount of impurities. Ranges of these values will eventually be found. As for the κ parameter which characterizes the micro - cracking activity it will depend very much on the types of ice and the grain sizes. Its physical meaning is an increase in the time scale when micro - cracking occurs but not much can be said about it, because very few curves have been analyzed.

——— o ———

Question to B. Michel from D.J. Goodman

Taylor showed that a polycrystalline aggregate can only deform in a ductile way if there are five independant slip systems (so that all grains remain in contact). The basal system can only provide three such systems. When a polycrystalline ice sample deforms, crystals which are oriented favourably for basal slip do so easily but there will also be some grains which are forced to slip on either the prismatic or pyramidal system to maintain contact with all grains. The stress acting on these systems will be high because of the easy slip on the basal plane, and it is the constitutive equation of the non-basal system which will control the creep rate of the whole sample. Cracks may form before the non-basal planes slip if the hydrostatic pressure is sufficiently low.

Answer:

Taylor's analysis is based on a continuity of shear stresses at crystal boundaries. The model we propose emphasises the role of the grain boundary as an area of shear stress concentration produced by change in direction of the grain boundary. The multiple contacts of one crystal with the others produce different stress levels on the crystal boundaries which produces an average stress different from one crystal to the next on a slip band. This process is not accounted for in classical continuum mechanics. There is thus no need for slip in the prismatic directions to account for creep of ice. The easy glide in the basal plane with pile-ups and climb of dislocations is enough to explain well the creep curves as shown in the paper.

——— o ———

Question to D. Nevel from D.J. Goodman

I would like to make a comment on the general analysis of ice forces acting on conical structures. In general the assumption is that the ice should not ride over the structure. To design for this case it is necessary to not only break the ice in flexure but allow it to be moved off to one side by the sloping face of the cone. Would it not be interesting to study what happens when ice rides over a structure. In this case the ice is broken by its weight as it rides over the structure. The cone angle can be kept low to reduce the loads. If submerged structures are then placed around the lighthouse or other structure, they break the ice before it reaches the structure. This will reduce the loads acting on the lighthouse. Clearly changes of height due to the tide must be accounted for.

Answer:

Unfortunately Nevels answer is not available in written form.

P. Tryde wishes to make the following answer:

1. It would be interesting to study what happens when ice rides over the structure.

2. If the cone angle is kept low the structure would be wider i.e. a possible increase of the ice forces.

3. To place structures all around a lighthouse is not a very good idea, as the cost of such measures would be extremely high.

——— o ———

3. Resumé of discussion.

Some topics were raised during the afternoons for a more detailed discussion. A few of the important and vital subjects are related in the following.

A. Stress/strain.

Several speakers discussed whether stress or strain should be measured. It became quite clear that no one can measure stress. What it is possible to measure is the strain. It appeared to me that the general opinion was that strain should be used. It all becomes even more complicated, when we consider the three-dimensional strain field including shear strain. Looking from the point of view of the material all it knows is that it has been "shuffled" around. (Miller) When it comes to theories, which attemp to explain the behavior of materials, all the existing classical theories fail to work. (Miller). So it becomes very clear that future basic research must be oriented towards this problem.

B. Fracture mechanics.

The understanding of fracture mechanics is very important in a world of technology. Think about micro cracks in metal and the research directed towards the understanding of 'when' and 'why'. The research within this field was presented by K.J. Miller and D.J. Goodman and it was generally accepted that the experience gained by studies of fracture in metals could with some modifications also be applied to ice.

C. Testing methods.

V. Schwarz presented the recommendatations of the IAHR committee on ice testing. Uniaxial tests are very complicated to perform. The use of compliant type pattern adds to the complexity of the system (B. Michel).

The influence of the stiffness of the testing machine was discussed, this is a very important problem because you do not want the stress to be a function of your machine.

D. Ice forces acting on cones or inclined planes.

Several of the papers dealt with the dynamic aspects of ice forces acting on cones and inclined planes. It was generally agreed that from a theoretical point of view many of the physical and mechanical problems had been solved, but the prediction of the movement of the ice fragments under influence of friction and gravity still remains to be solved. C. Sørensen and R.M. Frederking reviewed the problems based on experimental work performed to verify theories.

E. Icebergs.

The importance of being able to estimate the size of an iceberg based on the visual appearance above water was discussed, and also the problems envolved in predicting the movements caused by currents and wind.

F. Icebreakers.

New trends in icebreaking was discussed and comments on the design of a new Danish icebreaker was offered to K. Kure.

G. Miscellaneous subjects.

In many of the papers various important physical and mechanical parameters have been discussed in detail:
a) creep, b) strain rate, c) uniaxial compression strength, d) grain size effects, e) porosity, f) linear elasticity, g) non-linear viscosity, h) Glen's law, just to mention a few of the subjects.

4. Movies shown at symposium:

1) 'Iceberg Dynamics Project'.
   Amoco, Imperial oil, Mobile &
   Canadian Superieur.

2) Movies showing ice around a conical
   structure. (Japan)

MIX
Papier aus verantwortungsvollen Quellen
Paper from responsible sources
FSC® C105338

If you have any concerns about our products,
you can contact us on
ProductSafety@springernature.com

In case Publisher is established outside the EU,
the EU authorized representative is:
Springer Nature Customer Service Center GmbH
Europaplatz 3, 69115 Heidelberg, Germany

Printed by Libri Plureos GmbH
in Hamburg, Germany